Reservoir Management Strategies for Control and Degradation of Algal Toxins

Awwa Research Foundation

About the Awwa Research Foundation

The Awwa Research Foundation (AwwaRF) is a member-supported, international, nonprofit organization that sponsors research to enable water utilities, public health agencies, and other professionals to provide safe and affordable drinking water to consumers.

The Foundation's mission is to advance the science of water to improve the quality of life. To achieve this mission, the Foundation sponsors studies on all aspects of drinking water, including supply and resources, treatment, monitoring and analysis, distribution, management, and health effects. Funding for research is provided primarily by subscription payments from approximately 1,000 utilities, consulting firms, and manufacturers in North America and abroad. Additional funding comes from collaborative partnerships with other national and international organizations, allowing for resources to be leveraged, expertise to be shared, and broad-based knowledge to be developed and disseminated. Government funding serves as a third source of research dollars.

From its headquarters in Denver, Colorado, the Foundation's staff directs and supports the efforts of more than 800 volunteers who serve on the board of trustees and various committees. These volunteers represent many facets of the water industry, and contribute their expertise to select and monitor research studies that benefit the entire drinking water community.

The results of research are disseminated through a number of channels, including reports, the Web site, conferences, and periodicals.

For subscribers, the Foundation serves as a cooperative program in which water suppliers unite to pool their resources. By applying Foundation research findings, these water suppliers can save substantial costs and stay on the leading edge of drinking water science and technology. Since its inception, AwwaRF has supplied the water community with more than $300 million in applied research.

More information about the Foundation and how to become a subscriber is available on the Web at **www.awwarf.org.**

Reservoir Management Strategies for Control and Degradation of Algal Toxins

Prepared by:
Justin D. Brookes, **Robert Daly**, **Rudi H. Regel**, **Michael Burch**, **Lionel Ho**, **Gayle Newcombe**, **Daniel Hoefel**, **Chris Saint**, and **Thomas Meyne**
Cooperative Research Centre for Water Quality and Treatment
Private Mail Bag 3, Salisbury, South Australia, 5108, Australia

Michele Burford
Griffith University, 170 Kessels Rd, Nathan, Queensland, 4111, Australia

Maree Smith and **Glen Shaw**
University of Queensland, Brisbane, Queensland, 4072, Australia

Peyi Peyi Guo and **David Lewis**
The University of Adelaide, Adelaide, South Australia, 5005, Australia

and

Matthew Hipsey
Centre for Water Research, 35 Stirling Highway, Crawley, Western Australia, 6009, Australia

Jointly sponsored by:
Awwa Research Foundation
6666 West Quincy Avenue, Denver, CO 80235-3098

and

Cooperative Research Centre for Water Quality and Treatment.
Private Mail Bag 3, Salisbury, South Australia, 5108, Australia

Published by:

Awwa Research Foundation

Distributed by:

American Water Works Association

IWA Publishing

DISCLAIMER

This study was jointly funded by the Awwa Research Foundation (AwwaRF) and the Cooperative Research Centre for Water Quality and Treatment (CRCWQT). AwwaRF and CRCWQT assume no responsibility for the content of the research study reported in this publication or for the opinions or statements of fact expressed in the report. The mention of trade names for commercial products does not represent or imply the approval or endorsement of AwwaRF or CRCWQT. This report is presented solely for informational purposes.

ISBN: 978-1-84339-208-8

CONTENTS

TABLES

FIGURES

FOREWORD

The Awwa Research Foundation is a nonprofit corporation that is dedicated to the implementation of a research effort to help utilities respond to regulatory requirements and traditional high-priority concerns to industry. The research agenda is developed through a process of consultation with subscribers and drinking water professionals. Under the umbrella of a Strategic Research Plan, the Research Advisory Council prioritizes the suggested projects based upon current and future needs, applicability, and past work; the recommendations are forwarded to the Board of Trustees for final selection. The foundation also sponsors research projects through the unsolicited proposal process; the Collaborative Research, Research Applications, and Tailored Collaboration programs; and various joint research efforts with organizations such as the U.S. Environmental Protection Agency, the U.S. Bureau of Reclamation, and the Association f California Water Agencies.

This publication is a result of one of these sponsored studies, and is hoped that its findings will be applied in communities throughout the world. The following report serves not only as a means of communicating the results of the water industry's centralized research program but also as a tool to enlist the further support of the non-member utilities and individuals.

Projects are managed closely from their inception to the final report by the foundation's staff and large cadre of volunteers who willingly contribute their time and expertise. The foundation serves a planning and management function and awards contracts to other institutions such as water utilities, universities, and engineering firms. The funding for this research effort comes primarily from the Subscription Program, through which water utilities subscribe to the research program and make an annual payment proportionate to the volume of water they deliver and consultants and manufacturers subscribe based on their annual billings. The program offers a cost-effective and fair method for funding research in the public interest.

A broad spectrum of water supply issues is addressed by the foundation's research agenda: resources, treatment and operations, distribution and storage, water quality and analysis, toxicology, economics, and management. The ultimate purpose of the coordinated effort is to assist water suppliers to provide the highest possible quality of water economically and reliably. The true benefits are realized when the results are implemented at the utility level. The foundation's trustees are pleased to offer this publication as a contribution toward that end.

David E. Rager
Chair, Board of Trustees
Awwa Research Foundation

Robert C. Renner, P.E.
Executive Director
Awwa Research Foundation

ACKNOWLEDGMENTS

The authors of this report wish to acknowledge the following organizations for additional cash and in-kind support for this project:

Awwa Research Foundation
CRC for Water Quality and Treatment
SEQ Water
Veolia Water

The advice of the Project Advisory Committee (PAC) and Misha Hasan, The AwwaRF Project Manager, is greatly appreciated.

The project team thanks Mark O'Donohue, Fiona McKenzie Smith, Dennis Steffensen, Geoff Eaglesham, Najawa Slyman, Lani Pascual, Paul Rasmussen and Thibault Mas for input into various aspects of the project.

Field assistance by SEQ Water staff was greatly appreciated.

Dr Dennis Steffensen has provided on-going support and encouragement for cyanobacteria studies, including this project, within the Source Water Management Program of the CRC for Water Quality and Treatment.

EXECUTIVE SUMMARY

The cyanobacteria are an extremely well adapted group of organisms which dominate the freshwater phytoplankton community during periods of calm, stratified conditions. Cyanobacteria are notorious as indicators of degraded aquatic systems and compromise water quality for both potable supply and recreational use. There is a general opinion that "cyanobacterial blooms" are increasing in frequency due to anthropogenic eutrophication (Hallegraeff 1993) and the regulation of waterways with weirs and dams.

The toxins produced by the cyanobacteria have been responsible for livestock deaths (Falconer 1993) and recently human deaths in a dialysis clinic (Jochimsen et al. 1998), which has stimulated considerable public interest and concern. However, recent research has shown that with appropriate treatment most toxins can be effectively removed from drinking water (Drikas et al. 2001) significantly reducing the risk to consumers.

Cyanobacteria also produce taste and odour compounds that are not readily removed during conventional water treatment and compromise the aesthetic quality of potable water. Even low numbers of cyanobacteria can produce concentrations of these compounds, which exceed the odour threshold and attract customer complaints. The water industry has been active in seeking methods to control cyanobacterial populations, however, controlling to a level that ensures no taste and odour problems is difficult. Nonetheless, the most sustainable long-term solution for reducing toxins, taste and odours is to control cyanobacterial growth through mixing and a reduction in nutrient levels.

This project was initiated to examine cyanobacterial toxin production and degradation by microorganisms in sand filters and the water column. The production and biodegradation of cyanobacterial toxins was incorporated into an ecological model to facilitate knowledge transfer from this project to any water supply reservoir or lake.

The principal aim of this project was to investigate cyanobacterial toxin degradation in reservoirs and to develop reservoir management approaches for the control of toxin production. To satisfy this aim a series of experimental and modelling objectives were developed:

1. Examine factors that lead to the growth of *Cylindrospermopsis* in water supply reservoirs.
2. Determine mechanisms for toxin degradation and evaluate opportunities for optimising degradation prior to entry to treatment plants.
3. Determine the rate of microcystin and cylindrospermopsin degradation in natural waters.
4. Determine whether pre-treatment oxidation affects the cell integrity of cyanobacteria.
5. Develop integrated hydrodynamic and kinetic models to describe cyanobacterial growth in response to light, nutrients and mixing.
6. Incorporate toxin, taste and odour production and degradation into the ecological models to facilitate knowledge exchange from this project.
7. Predict the response of cyanobacteria, toxins and taste and odour concentrations to catchment and reservoir management.

The project included a review of the available literature on the ecology of cyanobacteria, cyanobacterial toxin production and degradation and a series of experiments and modelling studies to address the knowledge gaps.

Controlling cyanobacteria in reservoirs relies upon an understanding of the conditions that favour their growth. Historical data and short-term field experiments assessed the conditions favouring Cylindrospermopsis in North Pine Dam (NPD), a subtropical reservoir in Queensland. This enabled attributes of Cylindrospermopsis ecology to be modelled with an ecological model

(CAEDYM), from which predictions of risk magnitude and possible management interventions could be modelled.

Biodgredation studies were undertaken on two cyanobacterial toxins, cylindrospermopsin and microcystin. These consisted of spiking natural waters with known concentrations of toxin and measuring degradation kinetics through time. Sand column biofilters were also challenged with microcystin to examine biofilm establishment and the evolution of a microbial community able to degrade the toxins. The microbial community was screened, using molecular techniques, for organisms capable of degrading toxins.

The toxin degradation kinetics and rates describing degradation were incorporated into the ecological model CAEDYM. The ecological model can be coupled to the hydrodynamic models DYRESM (1D) or ELCOM (3D) to enable prediction oncyanobacterial growth, toxin production and degradation. The model was tested on two reservoirs, North Pine Dam in Queensland, Australia and Eagle Creek Reservoir in Indiana, USA.

The dominance of *C. raciborskii* in the three water reservoirs in this study is consistent with many other tropical and subtropical reservoirs (Bouvy et al. 2000, McGregor and Fabbro 2000). In addition, there is evidence that *C. raciborskii* is becoming established in an increasing number of waterbodies, including in temperate regions (Padisák 1997; Saker et al. 2003; Ryan, Hamilton and Barnes 2003). A number of factors have been suggested as possible causes of *C. raciborskii* dominance in reservoirs, including nitrogen fixing capacity, a high phosphate affinity and intracellular storage, the presence of a thermally stratified water column, water temperatures > 25 C, shade tolerance, long hydraulic residence times, and resistance to grazing (Padisák 1997, McGregor and Fabbro 2000).

Microcystin degradation within two water natural water sources that experience *Microcystis* blooms was reasonably rapid, with microcystin concentrations reaching undetectable levels after 6-8 days from an initial concentration of 20 $\mu g\ L^{-1}$. However, these rates were slower than those within sand-filters, which degrade microcystin toxin from 20 $\mu g\ L^{-1}$ to undetectable concentrations with 4 minutes (Ho et al. 2006).

Interestingly, although microcystin degradation occurred within a sample collected from the Torrens Lake, a gene that has been identified to be responsible for microcystin degradation, *mlrA*, was absent from these samples suggesting that there are other mechanisms or other genes responsible for microcystin degradation. In addition, *mlrA* was absent from a number of water samples that were collected from a number of Australian water sources that experience *Microcystis* blooms.

Some natural aquatic organisms or their extracellular enzymes have the capacity to degrade soluble Cylindrospermopsin (CYN). At North Pine Dam, QLD, Australia, CYN degradation was observed at all sites investigated from initial concentrations ranging from 3 to 76 $\mu g\ L^{-1}$. However, complete degradation did not occur at 3 and 7.6 $\mu g\ L^{-1}$ and no degradation was evident at lower CYN concentrations. The average maximum degradation rate indicated that the observed rate of reaction was proportional to the concentration of CYN in the concentration range 3 to 76 $\mu g\ L^{-1}$.

The lack of complete degradation at the lower CYN levels indicates that, perhaps within NPD the existence of a threshold CYN concentration that sustains growth of the CYN degrading organism but which the degraders cannot proliferate exists, resulting in little or no loss of toxin is present. At low CYN concentrations the organisms may utilize a more energetically favorable carbon source.

When degradation did occur it was usually delayed for a period of time, i.e. lag period. Repeated CYN exposure reduced the lag period suggesting that biodegradation of CYN may commence more rapidly in water bodies that have had frequent exposure to toxic blooms of organisms that produce CYN. The water temperature also affects the average maximum rate of degradation of CYN and the time taken for total loss of toxin. For CYN degrading organisms located in NPD 25°C appears to be the optimum temperature for degradation when compared to 20°C and 35°C.

Dyresm-CAEDYM successfully simulated the reservoir hydrodynamics (Figures 7.1-7.3) and growth of the phytoplankton species. The timing and magnitude of blooms was similar for the field and the simulated data sets. The good fit between predicted *Cylindrospermopsis* growth and the observed concentrations meant that the Cadeym model could be extended to include toxin production and degradation. Toxin concentration was measured in North Pine dam between Spetember 2001 and February 2002. The peak Cylindrospermopsin toxin concentration did not correspond with the peak *Cylindrospermopsis* concentration, which highlights that this feature may be difficult to simulate. The failure to predict the toxin concentration is predominately due to the fact that in the reservoir the cyanobacterial biomass and cyanobacterial toxins did not correlate and timing of the peak concentration of the two parameters did not correspond. In the toxin production and degradation model toxin concentration is tied to cyanobacterial growth and so relies on a relationship between biomass and concentration. If there is only a weak relationship and high variability in real situations it becomes very difficult to predict with a numerical ecological model. The disparity between biomass and toxin concentrations highlights the complexity of toxin dynamics and the difficulty water quality managers' face in predicting, controlling and treating cyanobacteria. A model for the production and degradation of secondary metabolites was also configured for the study of geosmin production by *Anabaena*. The concentration of geosmin in the water supply to the Eagle Creek treatment plant was simulated enabling prediction of a range of management scenarios.

The rate of total toxin degradation during chlorine pre-treatment was assessed under different conditions. The slowest degradation rate was at pH 10, because all of the chlorine was present as the less reactive hypochlorite species. The fastest degradation rate occurred in untreated water that was free from cyanobacterial cells. While chlorine pre-treatment may increase the number of cells removed during coagulation, any resulting extracellular toxin would be poorly removed by coagulation/filtration. If insufficient pre-oxidant is added for complete degradation of released toxin, it would need to be removed using other treatment processes such as activated carbon, ozonation or by additional chlorine before distribution.

Enhanced algal cell removal is not the only reason chlorine pre-treatment is used. For situations where pre-treatment is necessary or where chlorine disinfection is the only treatment, toxin degradation appears to be the rate limiting step and can effectively be used to describe total toxin destruction. In the presence of NOM and algal cells, the observed degradation rates were at least as fast as those predicted using the pH relationship described by Acero, Rodrigez and Meriluoto (2005) and were often faster. The recommendation of Nicholson, Rositano and Burch (1994) that complete toxin destruction can be achieved by maintaining a chlorine residual >0.5 mgL-1 for 30 min at pH less than 8 was also tested. Under these conditions 90-100% of total toxin was removed. Coagulation without pre-treatment is effective at removing toxin inside intact cells and is likely to reduce the total amount of toxin post filtration compared to the previous scenario. In both scenarios chlorine disinfection post filtration could be used to degrade any remaining toxin

The Framework for Management of Drinking Water Quality contained in Chapter 2 of the Australian Drinking Water Guidelines (ADWG), (http://www7.health.gov.au/nhmrc/publications/synopses/eh19syn.htm) outlines the methodology for providing safe drinking water by managing the complete catchment to tap water supply system. This document is achieving global recognition as the best way to manage our drinking waters as we move into the 21st Century and is being incorporated into National and State Health Guidelines.

It is important to understand the level of risk that the different cyanobacteria and toxins pose to drinking water. This allows managers of catchments and urban water utilities to focus their efforts on policies, works and operational practices to not only lower risks to public health but also improve the environmental health of these waters.

A series of factsheets are included that present the findings of a major research program carried out by the Australian Cooperative Research Centre (CRC) for Water Quality and Treatment into areas such as understanding cyanobacterial growth, detection methods for cyanobacterial toxins and water treatment options for cyanobacterial cells and toxins.

The production of the metabolite sub-routine describing the production and degradation of toxin or taste and odour compounds is an important addition to the reservoir and lake model but requires further validation. By using the model to predict the risk from cyanobacteria in any given reservoir the operators are better equipped to manage the hazards in the both the reservoir and the water treatment plant. There is a data requirement for model use but the expense associated with this is considered modest compared with the improved operation, health and aesthetic outcomes that are realised with better system understanding and management.

The application of the model in a predictive capacity allows operators the ability to test management scenarios before implementation. Candidate scenarios could include, but are not limited to:

- investigating what reduction in catchment derived nutrient load is required to reduce the cyanobacterial population to a concentration that will ensure the toxin or taste and odour thresholds are not reached
- predicting the impact of artificial destratification on cyanobacterial growth and toxin production or geosmin production
- determining the likelihood and likely concentration of cyanobacterial toxins in the source water and what treatment options should be incorporated into plant design to remove these compounds and produce aesthetically acceptable water?
- assessing the probability that recreational waters will experience cyanobacterial blooms

This study has led to the development of the following recommendations:

The best way to control cyanotoxins is to control the growth of cyanobacteria. This can be achieved by reducing nutrients and light. Light limitation can be induced by mixing with artificial destratifiers. Nutrient reduction is more difficult to achieve but must be part of a long term strategy for water quality improvements.

If algicides are to be used it is recommended that they be used to arrest growth when cell numbers are low (<5,000 cells/mL) to minimize the concentrations of toxins and geosmin released into the water. The use of algicides when cell numbers are high, and the subsequent cell lysis, can lead to high concentrations of toxins and taste and odour compounds which are difficult to remove. Algicides can also destroy bacteria responsible for the degradation of toxins and so released toxins will persist for longer in water following algicide treatment than they would naturally.

Pre-treatment with oxidants such as chlorine, ozone or potassium permanganate can improve the removal of contaminants but can also lead to cell lysis. If cyanobacteria are in high numbers it is recommended that pre-treatment not be used so that the majority of toxins, taste and odour compounds can be removed during coagulation/flocculation. If pre-treatment must be used, a chlorine residual of 0.5 mg/L must be maintained for 30 minutes at pH <8 to ensure toxin degradation. This will vary depending upon the dissolved organic carbon concentration and should not be used as a rule of thumb without testing on specific water.

Conventional treatment processes such as coagulation, flocculation, sedimentation and filtration will remove up to 90% of the total toxin present if it is contained within healthy cyanobacterial cells (less for cylindrospermopsin) and should be used in supplies that have cyanobacterial issues. Dissolved toxin (ie toxin that has been released from the cells) must be removed using additional treatment such as powdered activated carbon (PAC), granular activated carbon (GAC), ozone or chlorine. The doses of oxidant and PAC, and the type of activated carbon required for treatment is dependent on the type of toxin and the water quality. More information is provided in factsheet #6.

We cannot rely on biodegradation to control the cyanobacterial toxins, microcystin or cylindrospermopsin in drinking water reservoirs. Toxins can be present in the water column without degrading bacteria being present. Extracellular cylindrospermopsin can persist for months suggesting that biodegradation does not always occur. Additionally, screening with PCR for microcystin degrading organisms revealed that these organisms were not always present in toxic blooms of *Microcystis aeruginosa*.

When a toxic *C. raciborskii* bloom occurs in a water body that has had previous exposure to toxic blooms of CYN producing cyanobacteria it is highly probable that biodegradation of the soluble fraction of toxin will be biodegraded. If no previous history of toxic CYN producing organisms has been recorded it is unlikely, in the first instance, that biodegradation will occur. Biofilters have been very successfula at degrading microcystin and this simple technology warrants further investigation as a cost-effective method for removing toxins in water treatment plants.

CHAPTER 1: BACKGROUND, MOTIVATION &SCOPE

The cyanobacteria are notorious in lakes and water supply reservoirs were they compromise water quality by producing toxins, and taste and odour compounds. The potential health risks, the aesthetic problems and the costs associated with removing the toxins has prompted considerable research over the past four decades on cyanobacterial toxins, their control and appropriate treatment processes for their removal in potable water supply.

The best method of controlling the risk from toxins, and taste and odours compounds is to manage the growth of cyanobacteia. This relies on managing nutrient inputs into the reservoir and promoting mixed conditions that favour other phytoplankton over the cyanobacteria. However, it is not always possible to manage to acceptable levels and water treatment barriers need to adequately remove these compounds prior to distribution. An important aspect of managing hazrds such as cyanobacteria is predicting what risk they present to the treatment plant and to human health. International guidelines are available through the World Health Organization (Chorus and Bartram, 1999), but tools are still required to manage the cyanobacterial risk. This project has incorporated new findings on toxin production and degradation into a reservoir model to enable prediction on toxin risk and the expected outcome from management interventions.

The principal aim of this project was to investigate cyanobacterial toxin degradation in reservoirs and to develop reservoir management approaches for the control of toxin production. To satisfy this aim a series of experimental and modelling objectives were developed:

8. Examine factors that lead to the growth of *Cylindrospermopsis* in water supply reservoirs.
9. Determine mechanisms for toxin degradation and evaluate opportunities for optimising degradation prior to entry to treatment plants.
10. Determine the rate of microcystin and cylindrospermopsin degradation in natural waters.
11. Determine whether pre-treatment oxidation affects the cell integrity of cyanobacteria.
12. Develop integrated hydrodynamic and kinetic models to describe cyanobacterial growth in response to light, nutrients and mixing.
13. Incorporate toxin, taste and odour production and degradation into the ecological models to facilitate knowledge exchange from this project.
14. Predict the response of cyanobacteria, toxins and taste and odour concentrations to catchment and reservoir management.

The report is structured into chapters describing discrete aspects of the work. The literature review summarises our knowledge on the structure, toxicity and current management of cyanobacterial toxins. Subsequent chapters describe cyanobacterial ecology; Cylindrospermopsin and Microcystin degradation; modelling cyanobacterial growth, toxin production and toxin degradation; the effect of preteatment with oxidants on cell lysis and toxin release. A series of factsheets have also been developed to provide a quick reference on cyanobacteria, their toxins and available management strategies.

CHAPTER 2: LITERATURE REVIEW

THE CYANOBACTERIA

The cyanobacteria are an extremely well adapted group of photo-autotrophic organisms which dominate the freshwater phytoplankton community during periods of calm, stratified conditions. Cyanobacteria are notorious as indicators of degraded aquatic systems. Whilst the cyanobacteria have existed on earth for more than 2.5 billion years (Lau, Sapienza and Doolitte 1980), there is a general opinion that "cyanobacterial blooms" are increasing in frequency due to anthropogenic eutrophication (Hallegraeff 1993) and the regulation of waterways with weirs and dams.

The toxins produced by the cyanobacteria have been responsible for livestock deaths (Falconer 1993) and human deaths in a dialysis clinic (Jochimsen et al. 1998) which has stimulated considerable public interest. However, recent research has shown that with appropriate treatment most toxins can be effectively removed from drinking water (Drikas et al. 2001) significantly reducing the risk to consumers. Cyanobacteria also produce taste and odour compounds that are not readily removed during conventional water treatment and compromise the quality of potable water. Even low numbers of cyanobacteria can produce concentrations of these compounds, which exceed the taste threshold and attract customer complaints. The water industry has been active in seeking methods to control cyanobacteria populations, however, controlling to a level that ensures no taste and odour problems is difficult. Nonetheless, the most sustainable long-term solution for reducing toxins, taste and odours is to control cyanobacterial growth through mixing and a reduction in nutrient levels.

THE CYANOTOXINS

Cyanotoxins can be classified according to their chemical structures as cyclic peptides (microcystins and nodularin) or alkaloids (anatoxin-a, anatoxin-a(S), saxitoxins, and cylindrospermopsin) but are more commonly discussed in terms of their toxicity to animals.

Hepatotoxins

Hepatotoxins are low molecular weight toxins (800-1100 daltons) that generally act more slowly than the alkaloid toxins because the main target is the liver. They can cause extensive necrosis of this organ, resulting in death by haemorrhage shock and/or liver failure within a few hours to several days (Gorham and Carmichael 1988, Sivonen and Jones 1999). They contain either five (nodularins) or seven (microcystins) amino acids, with the two terminal amino acids being condensed to form cyclic compounds (Sivonen and Jones 1999).

Microcystins

Microcystins are monocyclic heptapeptides (Gorham and Carmichael 1988, Dawson 1998) and were first identified from the cyanobacterium *Microcystis aeruginosa* by Bishop, Anet and Gorham (1959). These toxins have subsequently been found to occur in the cyanobacterial genera *Anabaena*, *Aphanocapsa*, *Hapalosiphon*, *Microcystis*, *Nostoc*, *Anabaenopsis* and *Oscillatoria* (*Planktothrix*) (Sivonen and Jones, 1999).

Structure. The microcystins are medium molecular weight cyclic peptide toxins (Hunter 1998) with microcystin-LR (L, L-leucine; R, L-arginine) being the most common having a molecular weight of 994 daltons (Chong et al. 2002). These toxins consists of a seven amino acid ring structure (Figure 2.1), of which two, N-methyl-dehydroalanine (Mdha) and 3-anmino-9-methoxy-2,6,8-trimethyl-10-phenyldeca-4,6-dienoic acid (ADDA) are almost unique to this toxin (Hunter 1998, Dawson 1998).

Toxicity. Although human illnesses attributed to the microcystins has included such symptoms as gastroenteritis and allergic or irritation reactions, the primary target is the liver. The mechanism of toxicity for microcystin involves the inhibition of protein phosphatase 1 and 2A (Dawson 1998). Within the liver this biochemical effect leads to the disruption of the hepatocyte skeletal structure and cell integrity. In the case of acute toxicity resulting from exposure to high doses of the toxin, this can lead to more widespread disintegration of liver integrity and hemorrhage (NHMRC and NRMMC 2001). Pallor, prostration and sometimes convulsions and twitching can develop. Respiration can become laboured and irregular. Death can occur in one to three hours. On necropsy of experimental mice mild jaundice, dark, slowly clotting blood, haemorrhages and excess yellow clotted liquid in the body cavities was observed (Gorham and Carmichael 1988).

Microcystins have also been reported as being potent tumour promoters in experimental animals (Falconer 1991, Humpage 1999). The significance of this for humans, who may be subject to chronic exposure via drinking water, is still unclear.

The i.p LD_{50} for microcystin-LR in mice, is in the range of 25 to 150 μg kg $^{-1}$ with the commonly accepted value being reported to be 50 μg kg $^{-1}$. The oral LD_{50} is reported to be 5000 and 10900 μg kg $^{-1}$ for two different strains of mice. Even higher values have been demonstrated in rats (Kuiper-Goodman, Falconer and Fitzgerald 1999).

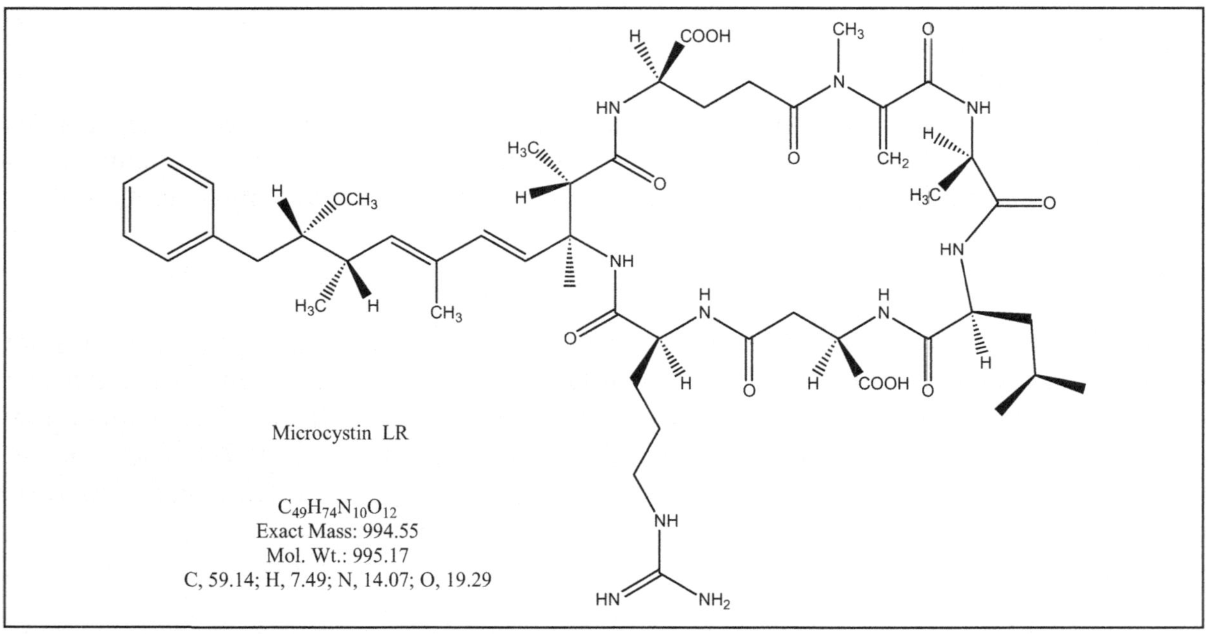

Figure 2.1 The chemical structure of Microcystin-LR.

Health effects. Significant human illness has been associated with exposure to microcystins in drinking and recreational waters (Turner et al. 1990), as well as in dialysis waters (Hunter 1998). In one of the earliest cases of a reported illness from exposure to microcystins, a male physician developed abdominal pains and nausea several hours after falling into a lake with a bloom of *Microcystis* sp. and *Anabaena circinalis.* He developed painful diarrhoea and vomiting five hours after his initial exposure. The next morning he was reported to have had severe headache, myalgia, arthralgia and malaise (Dillenberg and Dehnel 1960).

During September 1989, a team of Army recruits had taken part a canoe exercise lasting several hours in the Rudyard reservoir, during which some of the recruits swallowed water containing toxic *Microcystis aeruginosa* cells (Turner et al. 1990). Subsequent to these exercises a number of the recruits had symptoms that might have been associated with cyanobacterial poisoning. Symptoms included sore throats, headaches, abdominal pains, dry coughs, diarrhoea, vomiting and blistered mouths (Turner et al. 1990)

Human deaths have also been attributed to microcystin poisoning. In 1996 in Caruaru, Brazil, over seventy deaths were attributed to accidental intravenous exposure of kidney patients to microcystins from dialysis water. After routine hemodialysis treatment, it was reported that 116 out of 131 patients treated experienced visual disturbances, nausea, and vomiting. Subsequently, one hundred patients developed acute liver failure and as of 2001, 76 have died. Examination of phytoplankton from the dialysis clinic's water source, treatment system and serum and liver tissue of clinic patients led to the identification of the cyanotoxins microcystin and cylindrospermopsin (Carmichael et al. 2001).

Nodularins

Nodularin is a cyclic pentapeptide (Figure 2.2) produced only by the cyanobacterium *Nodularia spumigena*. It is structurally similar to the microcystins and exerts similar toxicity to microcystin-LR (NHMRC and NRMMC 2001).

Structure. Nodularin is a structurally distinct five amino acid ring hepatotoxin (Hunter 1998). There are 6 known variants of nodularin (Sivonen and Jones 1999), however, these are either rare or non-toxic.

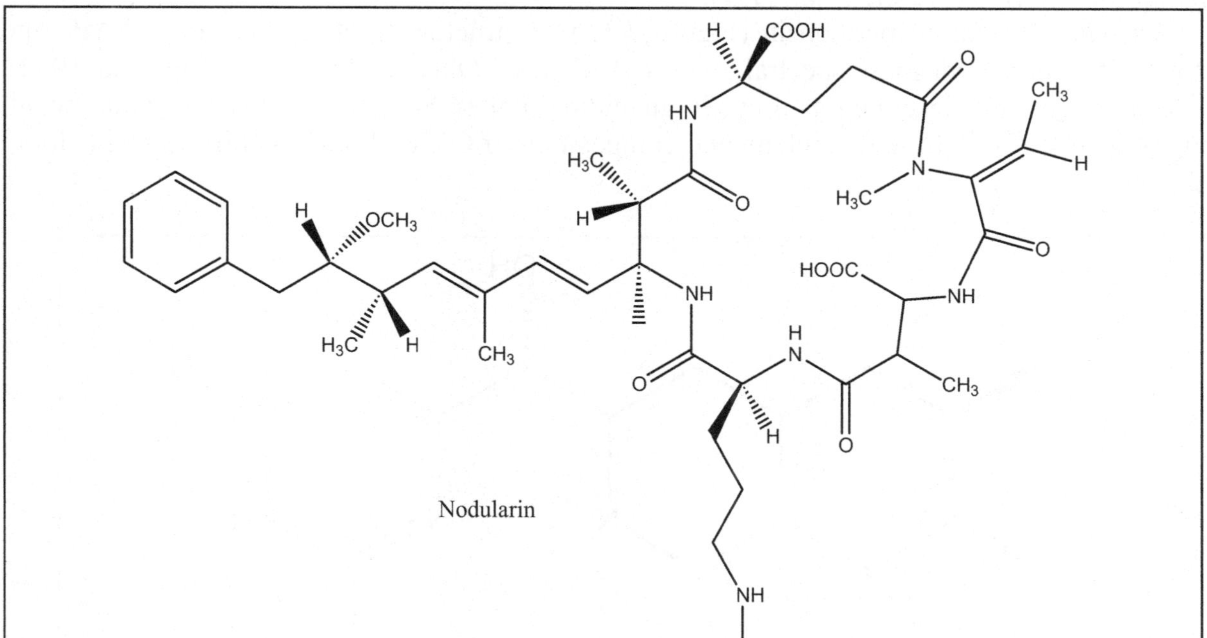

Figure 2.2 The chemical structure of Nodularin.

Toxicity. The toxicology of nodularin is similar to microcystin-LR, with the liver being the primary organ of toxicity. Nodularin inhibits protein phosphatases 1 and 2A and results in hemorrhaging of the liver when present there. The liver becomes swollen and engorged with blood, which can cause a doubling of liver weight to about 10% of total body weight in mice (Runnegar, Jackson and Falconer 1988). Nodularins have also been reported to hasten tumour development, but not initiate a cell progression towards becoming cancerous (Carmichael 1994).

The i.p. LD_{50} for mice using purified toxin has been reported as 70 μg kg^{-1} (Runnegar, Jackson and Falconer 1988), where death occurred one hour after injection.

Health effects. There have been no reports of human health effects from consumption of water containing nodularin. However, nodularin has been reported to be at least as hepatotoxic as microcystin for i.p. exposure and given its identical mode of action, should therefore be regarded as presenting at least the same risk to human health as microcystin if ingested (Falconer, Choice and Hosja 1992).

Nodularin is known to accumulate in mussels in estuaries and therefore the consumption of contaminated mussels represents a potential route of exposure to the toxin (Falconer, Choice and Hosja 1992). During the summer of 2001 a bloom *of Nodularia spumigena* occurred in the Gippsland Lakes area of Southern Victoria. During the bloom, seafood samples were collected and nodularin concentrations were estimated. Nodularin concentrations reached levels of concern in mussels and in prawn viscera at cell count as low as 30,000 cells mL^{-1} (Van Buynder et al. 2001).

Cylindrospermopsin

Cylindrospermopsin is a cyclic alkaloid that was first isolated from the cyanobacterium *C. raciborskii* and has subsequently been found to occur in the cyanobacteria *Aphanizomenon ovalisporum* (Banker et al. 1997, Shaw et al. 1999), *Umezakia natans* (Harada et al. 1994) and *Raphidiopsis curvata* (Wayne et al. 2001).

Structure. Cylindrospermopsin (Figure 2.3) was structurally elucidated using NMR and fast atom bombardment mass spectral studies (FABMS) (Ohtani, Moore and Runnegar 1992). This cyanotoxin was described as a cyclic guanidine moiety bridged to a hydroxymethyluracil group with a MW of 415 and an elemental composition of C15H21N5O7S (Figure 2.3) (Moore et al. 1993).

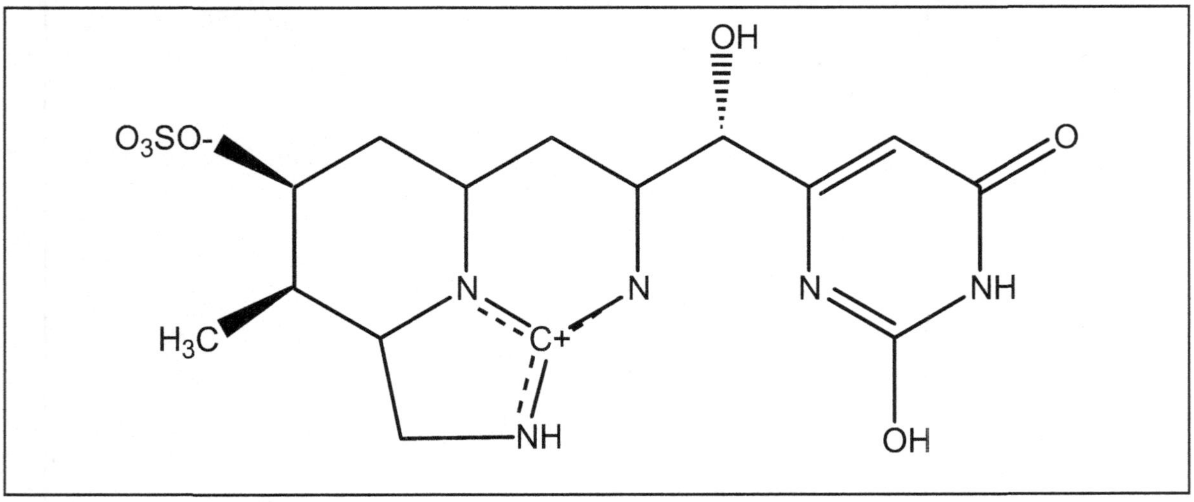

Figure 2.3. The chemical structure of cylindrospermopsin.

Toxicity. Terao et al. (1994) reported the i.p LD_{50} of purified cylindrospermopsin as 0.2 mg kg^{-1} after five days. Necropsy revealed that the livers and kidneys were swollen and pale yellow in colour. There was also extensive hepatocellular fatty infiltration. Seawright et al. (1999) reported the oral LD_{50} as 6 mg kg^{-1} after five days. The toxicological syndrome for oral ingestion of purified clylindrospermopsin was similar to that of i.p. injection. The mechanism of action of clylindrospermopsin appears to be the inhibition of protein synthesis and inhibition of glutathione (GSH) synthesis (Terao et al. 1994, Runnegar et al. 1994), which may be initiated by binding of activated metabolites to DNA or RNA (Shaw et al. 2000).

Toxicity studies have also been conducted by Hawkins et al. (1997) via a single i.p. injection of a freeze-dried extract of a *C. raciborskii* culture (AWT 205). Post mortem examination showed haemorrhage within the liver, kidneys and small intestine and congestion of the lungs at the higher doses administered. The 24-hour LD_{50} was calculated to be 52 mg kg^{-1}.

Seawright et al. (1999) also conducted studies into the toxicity of cylindrospermopsin freeze-dried extract peritoneally. These studies showed that death usually occurred from two to six days after dosing. Pathological changes included fatty liver, periacinar coagulative necrosis, acute renal tubular necrosis, atrophy of the thymic cortex and the lymphoid follicles in the spleen, subepicardial and myocardial haemorrhages plus multiple ulceration of the esophageal part of the gastric mucosa. The kidneys of lethally affected mice were also recorded as often being pale and swollen.

Health effects. Cylindrospermopsin is believed to have been the causative agent in the Palm Island "mystery disease" poisoning incident in Queensland, Australia, in 1979, where 138 children and ten adults of Aboriginal/Islander descent were afflicted with an illness (Bourke et al. 1983, Hawkins et al. 1985). The illness occurred a few days after Solomon Dam was treated with copper sulphate to control a dense cyanobacterial bloom. The copper was reported to cause lysis of cyanobacterial cells, with the subsequent release of cyanotoxins from the cells into the surrounding waters (Hawkins et al. 1985). A study into the incident indicated that only people connected to the reticulated water supply from the dam were affected (Bourke et al. 1983), suggesting that the causative agent was most likely associated with this water body.

The illness presented in succession over a three-week period beginning with hepatitis like symptoms followed by malaise, anorexia, vomiting and tender hepatomegaly. In the 24-48 hours after the onset of symptoms, 39% of the children developed profuse diarrhoea, which commonly lasted five-days. In 94% of these cases, diarrhoea contained frank (Byth 1980, Bourke et al. 1983).

Although no reports of human poisoning attributable to cylindrospermopsin have been documented since the Palm Island incident, cattle deaths in Queensland have been ascribed to this toxin (Saker, Neilan and Griffiths 1999; Thomas et al. 1998).

During 1997, a herd of 300 cattle were drinking from a dam showing heavy growth of toxic *C. raciborskii* in Queensland. One cow and three calves were found dead around the dam. Necropsy results from the cattle included severe abdominal and thoracic haemorrhagic effusion, hyperaemic mesentery, pale and swollen liver, and distended gallbladder. During the three weeks following the onset of the bloom a further two cows and six calves died (Thomas et al. 1998; Saker, Neilan and Griffiths 1999).

The majority of the *C. raciborskii* strains tested so far from Queensland waters appear to produce cylindrospermopsin. Toxin concentrations of 1 μg L^{-1} have been associated with 20,000 cells mL^{-1} (McGregor and Fabbro 2000). It is therefore likely that most blooms of *C. raciborskii* encountered will have some degree of toxicity (Chiswell 1999).

Cylindrospermopsin and *C .raciborskii* cells have been observed in the gut contents of crayfish harvested during a bloom, indicating consumption of shellfish containing relatively high concentrations of toxin may pose a threat to public safety (Saker, Neilan and Griffiths 1999).

Neurotoxins

There are three types of known cyanobacterial neurotoxins, anatoxin-a, anatoxin-a(S) and the saxitoxins. The saxitoxins include saxitoxin, neosaxitoxin, C-toxins and gonyautoxins (Sivonen and Jones 1999). These alkaloid toxins are diverse in both their chemical structures and in their toxicity, but are generally heterocyclic and of low molecular weight (< 1000 daltons).

Saxitoxins

Saxitoxin and neosaxitoxin are carbamate alkaloid neurotoxins that were originally isolated from shellfish that had concentrated the toxin from marine dinoflagellates (Sivonen and Jones 1999). These toxins have since been confirmed in the cyanobacteria *Aphanizomenon flos-aquae, Lyngbya wollei* (Lagos et al. 1999) *Anabaena circinalis* (Humpage et al. 1994) and *Cylindrospermopsis raciborskii* (Lagos et al. 1999).

Structure. The saxitoxins are a group of carbamoyl and decarbamoyl alkaloids (Figure 2.4) that are either non-sulfated (saxitoxins), singly-sulfated (gonyautoxins), or doubly-sulfated (C-toxins). The various types of toxins vary in potency. The least potent neurotoxin is the C toxins, the more potent is the gonyautoxins and the most potent is saxitoxin (Jones and Negri 1997).

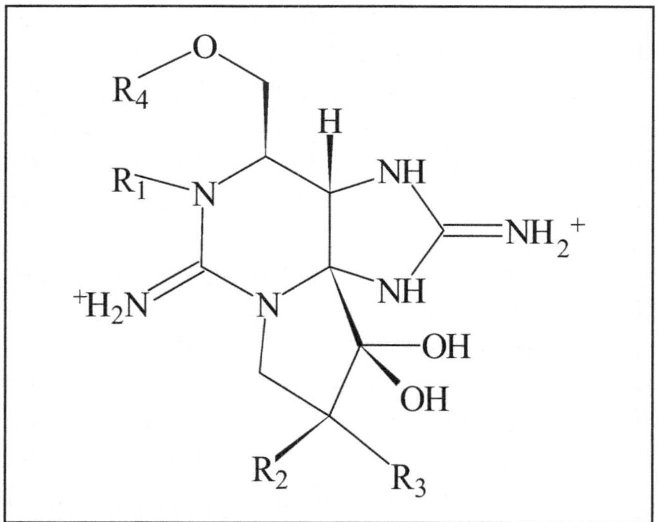

Figure 2.4 The general chemical structure of saxitoxins.

Toxicity. Saxitoxin and neosaxitoxin disrupt communications between neurons and muscle cells. They do so by preventing acetylcholine from being released from the neurons. They block the inward flow of sodium ions across the membrane channels without affecting permeability to potassium. This results in the snuffing out of any impulse and hence the prevention of the secretion of acetylcholine (Carmichael 1994, Gorham and Carmichael 1988).

This can lead to paraesthesia of the mouth and throat progressing to the neck and extremities, dizziness, weakness, ataxia and muscular paralysis. Nausea, vomiting, thirst and tachycardia can also occur. Symptoms can occur within five minutes of intoxication and in fatal cases death can occur within two to twelve hours. In non-fatal cases intoxication generally resolves within one to six days. The toxin is rapidly cleared by urinary excretion. There are no known chronic effects but long-term animal studies are lacking (NHMRC and NRMMC 2001).

The i.p LD_{50} for mice using PSP toxin extract has been reported as 10 μg kg^{-1} and 263 μg kg^{-1} for oral administration (Kuiper-Goodman, Falconer and Fitzgerald 1999).

Health effects. There is no evidence of human health effects caused directly by consuming water-containing saxitoxins. However, El Saadi and Cameron (1993) identified 20 patients who had drunk water from the Murray Darling River during a bloom of *A. circinalis* in 1991-92 and developed symptoms including sore throat, respiratory difficulties, diarrhoea, vomiting, nausea, muscle weakness, or headaches (El Saadi and Cameron 1993).

Numerous reports of human toxicity associated with consumption of shellfish containing relatively high concentrations of saxitoxins have been made (Kao 1993). Negri and Jones (1995) reported that the Australian freshwater mussel *A. condola* accumulated high levels of PSP toxins when fed *A. circinalis*. Their study indicated that greater than 80 mg of PSP toxins (C-toxins and gonyautoxins) were detected in 100 g of mussel flesh following two to three days of exposure to water containing 2×10^5 cells mL^{-1} of toxic *A. circinalis* (Negri and Jones 1995).

Anatoxin-a

Anatoxin-a is a secondary amine that was first isolated from *Anabaena* spp. and has subsequently been found to occur in the cyanobacterial genera *Aphanizomenon*, *Cylindrospermum*, *Microcystis*, *Oscillatoria* and *Planktothrix*.

Structure. Anatoxin-a is a low molecular weight (166 daltons) secondary amine with a molecular formula of 2-acetyl-9-azabicyclo (4-2-1) non-2-ene (Figure 2.5; Huber 1972, Gorham and Carmichael 1988).

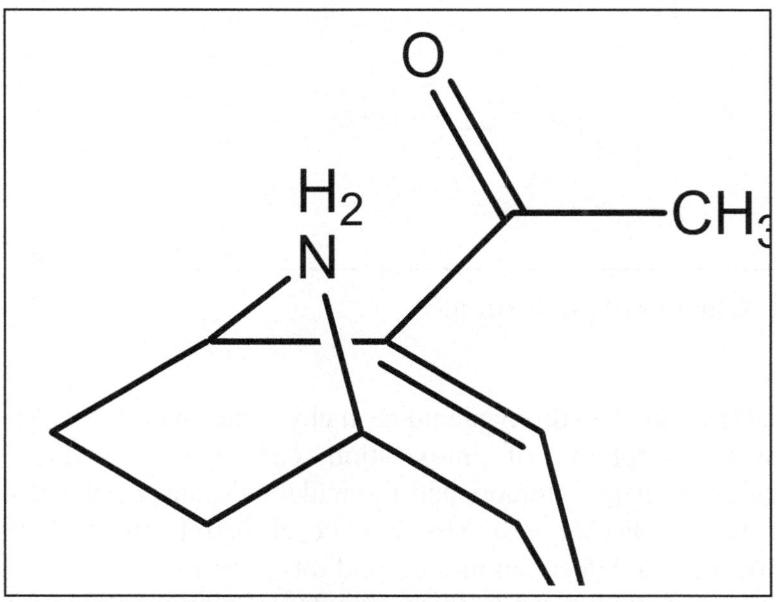

Figure 2.5 The chemical structure of Anatoxin-a.

Toxicity. Anatoxin-a is a potent post-synaptic cholinergic nicotinic agonist, which causes a depolarising neuromuscular blockage (Carmichael, Briggs and Gorham 1975). Symptoms of intoxication can include staggering, muscle fasciculation's, gasping, convulsions and opisthotonus. Death is due to respiratory failure and can occur within minutes of administration. The neuromuscular blockage produced by anatoxin-a is very long lasting, making artificial respiration impractical (Gorham and Carmichael 1988, Hunter 1998).

The i.p LD_{50} for mice has been reported as 375 μg kg $^{-1}$ and > 5000 μg kg $^{-1}$ for oral administration (Kuiper-Goodman, Falconer and Fitzgerald 1999).

Health effects. There have been no reports of human health effects from consumption of water containing anatoxin-a. However, this toxin has been reported to have been involved in many animal and wildfowl poisonings (Ressom et al. 1994).

Anatoxin-a(S)

Anatoxin-a(S) is an organophosphate produced by the cyanobacteria *Anabaena flos-aquae* and *A. lemmermannii* (Kuiper-Goodman, Falconer and Fitzgerald 1999).

Structure. Anatoxin-a(S) is a phosphate ester of a cyclic N-hydroxyguanine (Figure 2.6; Sivonen and Jones 1999) with a molecular weight of 252 daltons (Gorham and Carmichael 1988).

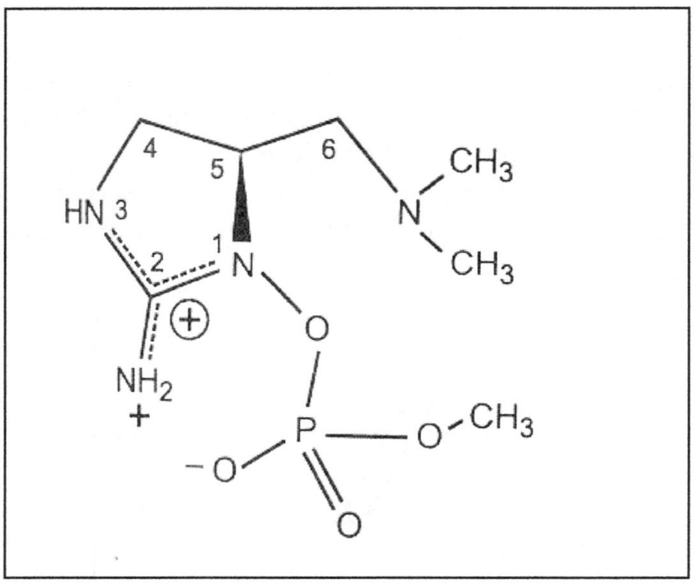

Figure 2.6 The chemical structure of anatoxin-a(S).

Toxicity. Anatoxin-a(s) is an N-hydroxyguanidine mehyl phosphate ester and is a potent inhibitor of cholinesterase. Symptoms of intoxication can include marked salivation, hypersalivation, muciod nasal discharge, tremors and fasciculation's, ataxia, diarrhoea, bruxism, dyspnoea, cyanosis, lachrymation (discharge of tears) in mice, chromodacryorrhea (bloody tears) in rats and urinary incontinence and defecation in mice and rats prior to respiratory arrest. Death from repiratory failure may occur within ten to thirty minutes of administration (Hunter 1998, Mahmood and Carmichael 1986, Gorham and Carmichael 1988). The i.p LD_{50} has been reported as 20 μg kg $^{-1}$ (Kuiper-Goodman, Falconer and Fitzgerald 1999).

Health effects. There have been no reports of human health effects from consumption of water containing anatoxin-a(S). However, this toxin has been reported to have been involved in dog deaths in the USA (Kuiper-Goodman, Falconer and Fitzgerald 1999).

ENVIRONMENTAL FATE OF CYANOTOXINS

Once cyanotoxins are released from the cyanobacterial cells into the surrounding water several mechanisms may determine their fate in the environment;

- dilution with non-contaminated water,
- adsorption to suspended solids and sediments,
- incorporation into aquatic plants and animals,
- degradation via photolysis and
- microbial degradation by natural aquatic bacteria (Welker, Steinberg and Jones 2001).

Many of the above mechanisms have been investigated or discussed in relation to the microcystins and it has been suggested that microbial degradation (biodegradation) was responsible for the greater part of microcystin removal in aquatic systems (Welker et al., 2001).

Biodegradation

Biodegradation can be divided into two processes. The first process is where the compound being metabolised can be used as a source of carbon (C), some other nutrient element (e.g. nitrogen) or energy. When the compound is introduced into the natural environment, the heterotrophic bacteria present are able to grow and proliferate, and the small initial bacterial population present increases in size over a period of time. Parallel to increases in cell number is an increasing rate of degradation of the compound. The second process involves those compounds subject to co-metabolism. Co-metabolism is defined as the metabolism of a compound from which microorganism cannot use the product as a nutrient or energy source. Such metabolic activities require the utilization of existing enzymatic and other physiological defense systems (Matsumura 1974). Compounds that serve as a nutrient source can provide the active bacterial populations with a selective advantage, as they become more abundant over time. However, a compound that does not serve as a nutrient source does not give the organism metabolizing it a selective advantage; hence its cell density does not rise over time. An absence of an increase in population size would be reflected in the lack of a rapid rise in the rate of biodegradation, as is typical of those substrates that support microbial growth (Alexander 1994).

Several conditions must be satisfied for biodegradation to take place in a particular environment. Firstly, the degrading microorganism/s must exist in the environment and have the necessary enzymes to bring about degradation of the compound in question (Daubaras and Chakrabarty 1992, Alexander 1994). Secondly, the organism must be present in the environment containing the compound and thirdly, the compound must be accessible to the organism. A compound not readily accessible to such microorganisms or their extracellular enzymes will not be degraded (Alexander 1973).

Often before biodegradation occurs, a period of time is noted in which no destruction of the compound is evident. This time interval is commonly known as the acclimation or lag period. It is the length of time between the addition or entry of the compound into the environment and evidence of its loss (Alexander 1994). The length of the acclimation phase can vary enormously.

It may be a matter of hours to many months and the duration can vary among compounds and environments.

A number of explanations have been proposed for the length of the acclimation phase. Firstly, the length can be attributed to the time needed for enzymes participating in the initial phases of degradation of the compound to be induced (Alexander 1973; Torstensson, Stark and Goransson 1975; Stephenson, Lester and Perry 1984). The time required for enzyme induction is usually minutes to hours (Richard 1968) although the period for enzymes to become fully induced may be longer, especially at low substrate concentrations (Button and Robertson 1985). Enzyme induction is the process whereby the synthesis of new enzyme proteins occurs in response to a stimulus. The stimulus is normally a signal in the environment such as metabolic intermediate or other compounds (e.g. toxin). In *E. coli* certain enzymes are produced only when the cells are grown on a certain substrate. For example, when cells are grown in the absence of lactose, the cells contain very few molecules (5 per cell) of the enzyme β-galactosidase, which cleaves lactose into glucose and lactose. There is no need for this enzyme in the absence of lactose. If lactose is added to *E. coli*, in a very short amount of time there will be approximately 5000 molecules of β-galactosidase per cell. If lactose is removed from the media, synthesis of β-galactosidase stops.

Should such a compound not induce the appropriate enzyme/s needed for the degradation and should there be no other inducers in the immediate vicinity, the compound will not be degraded. A lack of induction may come about because the substrate or other inducers does not penetrate the potentially active cell or because the concentration in aqueous solution is too low to promote induction (Alexander 1973).

Mutation or the appearance of new genotypes by genetic exchange may also account for the length of the acclimation phase (Walker and Newman 1956; Smidt, Hellwig and Knackmuss 1983). It has been suggested that acclimation phases that have reproducible duration among replicates are not the result of mutations (Fournier, Codaccioni and Soulas 1981; Spain et al. 1984).

The acclimation phase can often be a reflection of the time it takes for small populations of microorganisms capable of degrading compounds of interest to reach cell densities sufficiently high to give detectable loss of the compound (Spain, Prichard and Bourquin 1980; Ventullo and Larson 1986). Because each microbial cell consumes small amounts of substrate, high cell densities would be required to give readily detectable loss of a substrate (Wiggins, Jones and Alexander 1987). However, some constraints may be present that prevents the low density of these active microorganisms from proliferating readily. The identity of the constraint may vary from one environment to another. The presence of toxins, the lack of adequate supply of essential nutrients, temperature or predation by protozoa may delay the increase in cell numbers or activity of the degrading organism/s (Wiggins, Jones and Alexander 1987). For compounds that exhibit acclimation periods longer than one month, it is unlikely that the growth of cells initially present in the sample can account for acclimation (Wiggins, Jones and Alexander 1987).

Many organic compounds such as cyanotoxins are quite toxic and at very high concentrations the toxicity of the compound may reduce the numbers of degrading microorganisms in the immediate environment and thus an increase the acclimation phase becomes evident (Alexander 1973; Wiggins and Alexander 1988; Wiggins, Jones and Alexander 1987; Lappin, Greaves and Slater 1985; Rossin, Lester and Perry 1982; Boethling and Alexander 1979; Atlas and Bartha1972).

The concentration of compounds in nature may be too low to support growth of the degrading microorganisms and the length of the acclimation phase may be the result of the slow growth of these microorganisms on the very low concentrations of a compound (Wiggins and

Alexander 1988, Schindler 1971). As the concentration of the compound increases, so does the growth rate. The more rapid proliferation at these higher concentrations would be reflected in a shorter acclimation phase (Wiggins and Alexander 1988). The concentration of nitrogen (N), phosphorus (P), or both is also important in some environments. Some natural waters have low concentrations of these nutrients which may limit microbial growth and increase the acclimation phase (Alexander 1994).

The dependence of microbial growth on temperature is also well documented. If microbial growth is inhibited by temperature the length of the acclimation phase will increase (Ludzack and Kinkead 1956, Atlas and Bartha1972). Thus, the time of the year and the water temperature at which samples are taken will affect the acclimation periods (Wiggins, Jones and Alexander 1987).

Protozoa have been reported to affect the growth of degrading microorganisms and therefore consequently affect the length of the acclimation phase (Atlas and Bartha 1972, Wiggins and Alexander 1988). They have been shown to regulate the numbers of bacteria in wastewaters (Pike and Curds 1971), soil (Habitue and Alexander 1977) and aquatic environments (Fenchel and Jorgensen 1977). They may also affect nutrient availability and turnover in many ecosystems (Fenchel and Jorgensen 1977, Gude 1985, Stout 1980, Wiggins and Alexander 1988).

The acclimation phase has also been reported to be affected when the microorganism present becomes adapted to the compound being degraded (Pfaeder and Shimp 1985; Spain, Prichard and Bourquin 1980; Ventullo and Larson 1986). Adaptation can be separated into fate and effect components. The fate adaptation refers to a process whereby the rate of biodegradation of a compound is significantly increased as a result of prior exposure to that compound. Under certain conditions when repeated application of the same compounds are made to the same site, microbial populations can develop to the point that succeeding applications will have progressively shorter residual times (Ventullo and Larson 1986). In some situations the extent of adaptation achieved is not directly proportional to the pre-exposure concentration of the compound. There seems to be a threshold concentration below, which no detectable adaptation of the community occurs. Other possible explanations for the length of the acclimation phase include an insufficient utilization of other organic compounds before the chemical of interest (Kuiper and Hanstveit 1984, Alexander 1981). Another possible reason for the acclimation phase may be due to the time it takes for the microorganisms to detoxify the chemical of interest allowing the altered molecule to be degraded (Alexander 1994).

Biodegradation of the cyanotoxins

Microcystins are the most comprehensively researched cyanotoxins, providing the majority of literature for review. However, resent work by Smith (2005) has provided an insight into the biodegradation of cylindrospermopsin. This work confirmed Chiswell's (1999) initial observations that natural aquatic bacteria are capable of degrading cylindrospermopsin and for the first time successfully isolated and characterized bacteria responsible for toxin degradation. This occurrence was previously only inferred rather than directly studied. Little work has been done on the biodegradation of other cyanotoxins.

Microcystin

Biodegradation after an acclimation period of a few days or weeks has been the most important means of detoxification of microcystin in laboratory experiments and in the field

(Jones 1990; Jones et al. 1994; Lahti et al. 1997; Lam, Fedorak and Prepas 1995; Rapala et al. 1994; Watanabe et al.1992; Welker, Steinberg and Jones 2001).

When the degradability of microcystins was tested experimentally using natural microbial communities, persistence of toxins ranged from a period of a few days to several weeks, probably mainly dependent on the origin of the heterotrophic inocula (Watanabe et al.1992).

Studies by Jones et al. (1994) demonstrated that biodegradation of microcystin-LR in natural waters occurred after an induction phase of three days to three weeks. Once degradation was initiated, degradation was >95% complete within an additional three to four days. Re-addition of microcystin-LR to induced water samples led to immediate microcystin-LR degradation with no acclimation phase. This was proposed to be due to induction of degradative competent bacteria during the first exposure to microcystin-LR (Jones 1990).

A study conducted by Jones and Orr (1994) in Lake Centenary after algicide treatment noted a biphasic degradation of microcystin-LR. A rapid initial phase, during which time 90-95% of the original microcystin was degraded, was followed by a slower phase, which was interrupted by a flash flood. The slower phase corresponded to a decimal reduction time of about 15 days, which is the same as the rate detected in the Lake Tuusulanjarvi conducted by Lahti et al. (1997) (Jones and Orr 1994).

A field study conducted by Lahti et al. (1997) showed dissolved microcystin-LR persisted in Lake Tuusulanjarvi for 30 days and in particulate material 15 days. This has been attributed to lower water temperatures (decreasing from 20 to 8°C) and also to lower concentrations of microcystin and co-metabolites that are possibly needed for degradation to occur (Lahti et al. 1997).

Takenaka and Watanabe (1997) tested surface water of a Japanese lake for bacterial degradation of microcystin-LR. The study discovered that the gram-negative bacterium, *Pseudomonas aeruginosa,* degraded this toxin with a lag time of three days [Takenaka, 1997). Microcystin biodegradation was also observed in Lake Suwa, Japan during a toxic *Microcystis aeruginosa* bloom (Park et al. 2001).

Jones et al. (1994) isolated a single bacterial strain capable of degrading microcystin-LR. 16S rRNA sequencing of this organism confirmed it belonged to the genus *Sphingomonas* (Jones et al. 1994).

The enzymes required for microcystin-LR degradation have been reported to be present fortuitously in the *Sphingomonas* sp. and are required for normal cellular growth and metabolism when the bacterium is grown on peptone medium (Bourne et al. 1996). Bourne et al. (1996) postulated that at least three enzymes are involved in the breakdown of microcystin-LR. The most important step is a hydrolysis step, which causes the ring opening of microcystin-LR, and is catalysed by an enzyme named "microcystinase" (MlrA). The initial site of action of the enzyme is at the Adda-Arg peptide bond. The linear microcystin product is 160-fold less reactive towards protein phosphatase than the parent compound. MlrA has been previously characterised as a metalloprotease by limited inhibitor studies and is a 336-residue endopeptidase. Inhibitor studies have similarly characterised the second enzyme in the degradative pathway (MlrB) which cleaves linear microcystin LR to a tetrapeptide degradation product. The enzyme belongs to the "penicillian binding enzyme" family of active site serine hydrolases. The third enzyme (MlrC) has 507 residues and mediates further peptidolytic degradation of the tetrapeptide. Based on further inhibition studies it is suspected to be a metallopeptidases (Bourne et al. 2001).

Nodularin

Lahti et al. (1998) showed that seventeen out of 100 bacterial strains isolated from water and sediment from two Finnish lakes were capable of microcystin degradation. Three strains were also capable of nodularin degradation (Lahti et al. 1998). No further work has been performed on the biodegradation of nodularin.

Cylindrospermopsin

Studies conducted by Chiswell (1999) suggested that biodegradation was an important process in the removal of cylindrospermopsin from contaminated water bodies. In cylindrospermopsin contaminated water treated by either autoclaving, filtering or with antibiotics, no decrease in toxin concentration was reported. However, in untreated dam water cylindrospermopsin degraded rapidly after an initial acclimation period (Chiswell 1999). Chiswell concluded from these studies that biodegradation of the toxin was possible although the action of bacteria had only been inferred rather than directly studied.

Results of a study conducted by Smith (2005) confirmed for the first time that natural aquatic bacteria have the capacity to degrade cylindrospermopsin once the toxin has been released from the cells of *C. raciborskii*. Results from this study also confirmed that biodegradation of the soluble toxin can occur in many different aquatic environments, in the concentration range of 0 to 950 $\mu g\ L^{-1}$. Further studies completed suggested that successful removal appeared to be dependant on either the presence of *C. raciborskii* cells themselves and or previous history of toxic *C. raciborskii* blooms. The loss of toxin via biological mechanisms was thus unlikely to occur if these conditions were not met. If degradation does occur it may be delayed until the endemic populations of microorganisms developed the necessary enzymes for cylindrospermopsin degradation and/or the biomass of cylindrospermopsin degradative bacteria reached a level that the loss of toxin was detectable. The conditions conducive for growth or induction of enzymes responsible for toxin degradation appeared to be supplied by *C. raciborskii* cells or cylindrospermopsin itself. These relationships need to be investigated further. If induction and/or optimum biomass is to be achieved, there is too little information currently available to generalize how long the active population will retain activity, particular in the absence of toxic *C. raciborskii* cells (Smith 2005).

The outcome of experiments conducted in the study also illustrated that repeated cylindrospermopsin exposure reduces the acclimation phase indicating that previous exposure to the toxin may allow rapid growth of cylindrospermopsin degradative bacteria. This results in a shorter acclimation period prior to the commencement of degradation (Smith 2005).

Further laboratory experiments conducted during this study successfully isolated and characterized a bacterial species capable of degrading soluble cylindrospermopsin. This bacterial species was tentatively identified as *Delftia* sp. MS1. This strain was isolated from a Queensland dam, which had a history of toxic *C. raciborskii* blooms. *Delftia* sp. is a member of the Comamondaceae family (Wen et al. 1999) which belongs to the β subclass of the Proteobacteria (Stackebrandt, Murray and Trueper 1988).

Saxitoxins

Jones and Negri (1997) reported that PSP's from *Anabaena circinalis* persisted for approximately three months in river water or irrigation drainage water samples. These results indicated that the bacteria present in the aquatic environment studied were not capable of degrading this cyanotoxin (Jones and Negri 1997).

Anatoxin-a

A *Pseudomonas* sp. has been shown to degrade the neurotoxin anatoxin-a in laboratory experiments (Kiviranta et al. 1991). The fate of anatoxin-a from an *Anabaena circinalis* culture was followed under constant light and temperature for 55 days. At the end of the study no toxin was detected (Kiviranta et al. 1991). The *Pseudomonas* species was reported to degrade the toxin at a rate of 6-30 $\mu g\ mL^{-1}$ per three days. In a study reported by Rapala et al. (1994) anatoxin-a was degraded after an initial acclimation phase. The author reported that up to 2.4 $\mu g\ mL^{-1}$ of toxin was lost in four days. The loss of anatoxin-a occurred equally fast with all sediments and water inoculates, with or without a prior history of bloom events (Rapala et al. 1994).

Anatoxin-a (S)

Literature indicated that no work has been performed on the biodegradation of Anatoxin-a(S).

Kinetics of toxin biodegradation

Knowledge of the kinetics of biodegradation is essential to the evaluation of the persistence of cyanotoxins and to assessing the exposure of these toxins to humans, and animals. Once degradation of the toxin commences, the amount disappearing with time and the shape of the disappearance curve will be a function of the toxin in question, its concentration, the organisms responsible for the degradation, and the effect a variety of environmental factors has on the degradation of the toxin. This information is extremely important because it can help determine the concentration of the toxin remaining at any time. It can also permit prediction of the levels likely to be present at some time in the future, and allows assessment of whether the toxin will be removed before it is transported to a site at which humans and animals may be exposed (Alexander 1994).

The biodegradation order can be calculated using following empirical formulae:

$$\text{Rate of reaction} = d\,[A]/dt \qquad (2.1)$$

where d = change in rate

[A] = concentration of reactant or product

t = time

Using the results obtained from measuring toxin disappearance under a variety of condition and the formulae d[C]/dt, where [C] is the concentration of toxin, the maximum rate of degradation of toxin over a specific concentration range for example can be calculated.

Kinetics of microcystin degradation

Jones and Orr (1994) reported that microcystin degradation was biphasic (study site Lake Centenary). A rapid initial degradation phase, during which time 90-95% of the original microcystin was degraded, was followed by a slower degradation phase .The authors reported that biphasic degradation kinetics had not been observed in their culture studies, although these studies were concluded when degradation was >95% complete. Biphasic degradation kinetics may indicate biodegradation by at least two different stains of bacteria. One of which can metabolize microcystin at relatively high concentrations and the other at low concentrations (Jones and Orr 1994).

When a bacterium capable of degrading microcystins-RR, -YR, and - was added to microcystins present in culture medium, the toxins were completely degraded in four days. The highest degradation rates of microcystins-RR and -LR were 13 and 5.4 mg L^{-1}day, respectively. The degradation rates were strongly dependent on temperature and the maximum rate was at 30 ° C (Park et al. 2001).

Kinetics of cylindrospermopsin degradation

The curve produced from plotting initial concentration of cylindrospermopsin versus the maximum rate of degradation represents zero order kinetics or linear degradation in the concentration range of 0-350 µg L^{-1}. These results indicate that the rate of the reaction is independent of reactant concentration. These experiments were performed using batch cultures of cylindrospermopsin degrading bacteria (Smith 2005).

Conclusions

It has been demonstrated that environmental microorganisms have the capacity to degrade some of the freshwater cyanobacterial toxins in reservoirs and other aquatic systems. In particular, cyclic peptides (the microcystins and nodularin) have been shown to be effectively degraded by microorganisms, with the microcystins receiving the most attention. Cylindrospermopsin, a toxin that has been studied by the research group involved in this project, can also be degraded in reservoirs and under biological filtration conditions. Anatoxin-a is also subject to microbial degradation while anatoxin-a(S) has not been effectively studied to date.

TASTE AND ODOUR COMPOUNDS

The cyanobacteria account for the majority of biogenic taste and odour problems in source waters (Taylor et al. 2006). The two compounds produced by the cyanobacteria that are of primary concern are geosmin and MIB (2-methylisoborneol). Cyanobacteria as sources of taste and odours is well covered in Taylor et al. (2006).

Anabaena and Geosmin

Geosmin concentration is correlated with *Anabaena circinalis* populations in Myponga Reservoir (Figure 2.7) and experimentally (Bowmer et al. 1992). The relationship between cell number and geosmin concentration is strong up to 1000 cells mL^{-1}, but then becomes weaker (Figure 2.8). However, it is evident that at even very low cell numbers geosmin concentrations can exceed the odour threshold of approximately 10 ng L^{-1}.

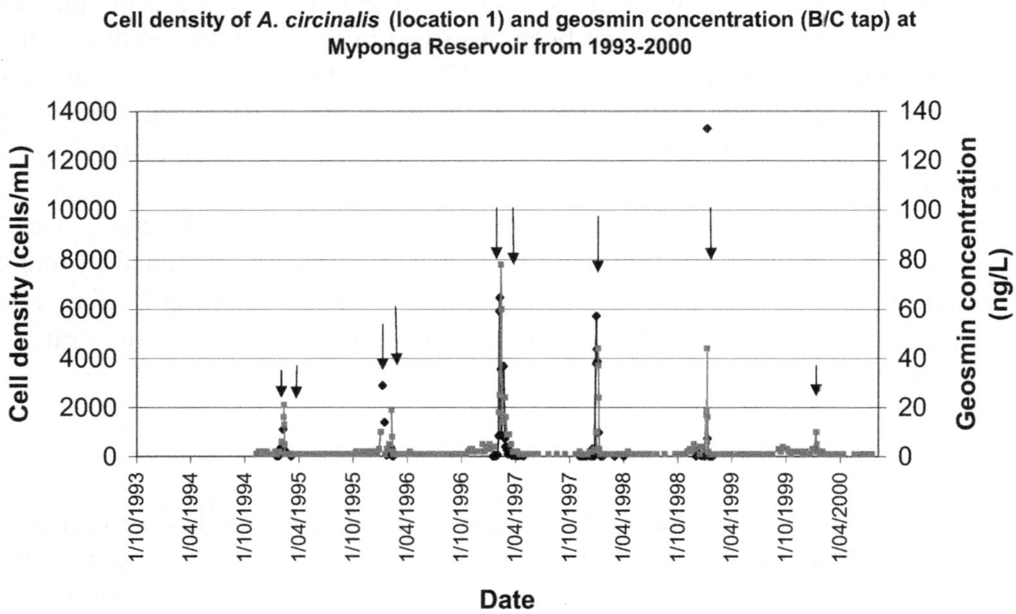

Figure 2.7 *Anabaena circinalis* and Geosmin concentration in Myponga Reservoir. Arrows indicate copper dosing events.

Treatment with copper has been shown to release geosmin contained within *A. circinalis* (Bowmer et al. 1992). Current laboratory procedures used to determine geosmin concentration in water samples would include a proportion of that contained within the cells of *A. circinalis*. The salt stripping technique imparts significant stresses on the phytoplankton within a sample, which would lyse most cells and release the geosmin contained within these cells. Intact cells would normally be flocculated and removed from the water before treatment and not pass into the treatment plant. Therefore the concentration of geosmin that the treatment plant will need to deal with is likely to be an over estimate.

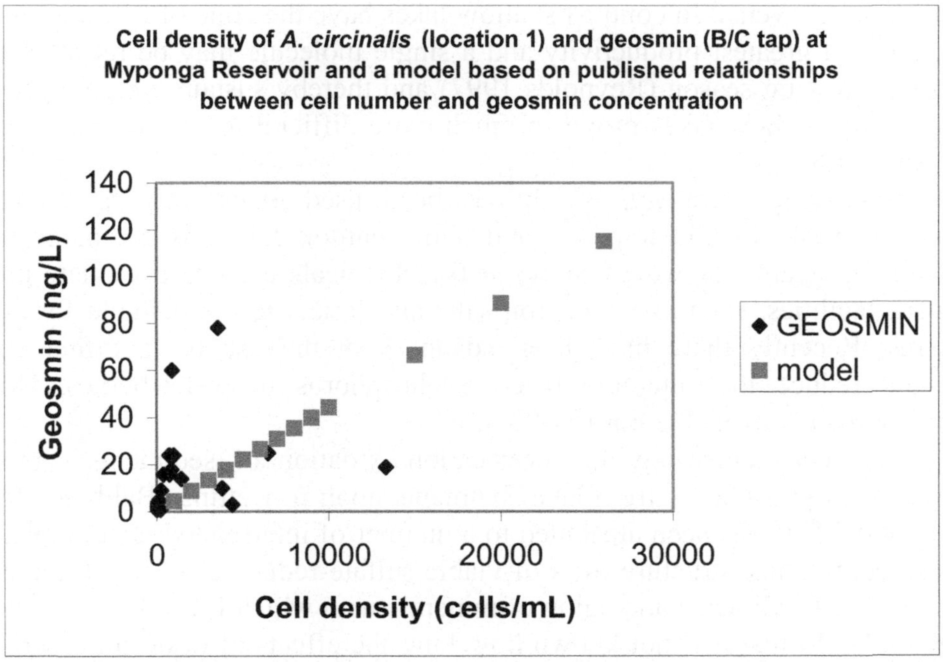

Figure 2.8 Actual and modelled Geosmin concentration by cell number of *A. circinalis* at Myponga Reservoir.

CYANOBACTERIAL CONTROL

Controlling the nutrient load

The major nutrient sources are the catchment, the internal load derived from sediment and in some cases from atmospheric deposition. The sources contributing to elevated nitrogen oxide species in the atmosphere include soil emission, fossil fuel combustion, lightning and biomass burning in forests and savanna (IPPC 1996). These sources, together with nutrients derived from catchment, are difficult for water managers to control. However, strategies to retain phosphorus in the catchment, such as gypsum application (Varcoe et al. 2000) and optimised fertiliser application, will reduce the sustainable algal biomass in lakes and reservoirs. In some instances catchments are naturally high in phosphorus and consequently attempts to reduce phosphorus to limiting levels would be unsuccessful. In these cases alternative strategies to control algae should be sought.

The nutrient source which reservoir managers often do have the ability to control is the internal nutrient load or sediment-derived nutrient load. The internal nutrient load is most often controlled by oxygenation of the hypolimnion either by artificial destratification or by direct oxygen injection. In systems where the internal nutrient load contributes significantly to the total nutrient load, a reduction in nutrient release from sediment can significantly decrease the sustainable algal biomass (Sherman, Whittington and Oliver, 2000).

In a typical phosphorus cycle, phosphorus is remobilised from sediment or decaying organic matter and entrained into the water column, where it is taken up by algae. From there the phosphorus is either passed on to higher levels of the food web or lost to the bottom as the algae sediment. In deep lakes the resolubilisation of P at the sediment is vertically separated from the algae and so each P molecule can only be accessed with entrainment from the hypolimnion to the epilimnion. In strongly stratified deep lakes this may happen only once or twice a year during significant 'over-turn' events. In contrast shallow lakes have the zone of P resolubilisation much closer to the zone of greatest productivity and a single molecule may be recycled a number of times during the growing season (Reynolds 1997) and thereby sustain a high algal biomass for longer. This is why shallow lakes prove so much more difficult to restore than deep ones (Sas 1989, Reynolds 1997)

A lake remediation strategy, which has been used extensively to decrease nutrient concentrations in waste treatment ponds and some natural lakes, is sediment capping with phosphorus binding agents. However, some of the chemicals used to flocculate phosphorus in waste treatment facilities, such as alum, iron salts and lime, are not suitable for application in natural systems. Recently there have been advances in the use of modified clays to bind phosphorus and reduce the filterable reactive phosphorus in water-bodies. However, this technology is not readily available nor in wide use.

There is evidence emerging that desiccation/oxidation of sediments can significantly reduce the release of phosphorus from lake sediments upon re-wetting (Baldwin 1996, Baldwin and Mitchell 2000). This has been attributed to a number of interrelated factors including a shift in the bacterial community structure (loss of viable sulfate-reducing bacteria), increased carbon limitation in the dried sediments and aging of minerals with which P can be associated (Mitchell and Baldwin 1998). Whilst it is not known how long the effects of desiccation will last, drying the lake sediments may be a suitable strategy to reduce P release in some shallow lakes. Many natural shallow lakes may have undergone seasonal wetting and drying before modification of the hydrology to maintain unnaturally high water levels.

The phosphorus binding capacity of sediments depends upon the sediment type and the number of sites available to absorb phosphorus. If the sediment has received phosphorus rich waters for a considerable period then the binding sites may be saturated and removal of this sediment by dredging may be an effective strategy to reduce phosphorus concentrations in the receiving water.

Managing cyanobacteria in source water

Cyanobacteria are difficult to manage once they have become established in a waterbody. Often copper based algicides are used, however, there are environmental issues concerning the addition of heavy metals to aquatic systems. The most environmentally sound methods to control cyanobacterial growth in reservoir is to manipulate the environment to favour other phytoplankton over the cyanobacteria. Previous attempts have included manipulating the physical, chemical and biological components of the reservoir. Manipulation of catchment nutrient sources takes a number of years before any impact is observed. Manipulation of the biological environment to encourage grazing requires rigorous monitoring and is not applicable in some systems (Boon et al. 1994). On the other hand manipulation of the physical environment is a practical means of changing the phytoplankton habitat and also can manipulate the internal nutrient load.

Artificial destratification

During stratification the hypolimnion is effectively separated from the atmosphere and becomes depleted of oxygen. This leads to reducing conditions at depth and contaminants such as iron, manganese and nutrients are released from sediment.

There are basically two types of artificial destratification systems available, bubble plume aerators and mechanical mixers. Both systems weaken stratification so that wind forcing can more readily breakdown stratification and mix the reservoir. Bubble plume aerators operate by pumping air through a diffuser hose near the bottom of the reservoir. As the small bubbles rise to the surface they entrain water such that the plume has a unique temperature and density and will plunge to the level of equivalent density and an intrusion will propagate at that depth. As the intrusion moves through the reservoir there is return flow above and below the intrusion and these circulation cells facilitate exchange between the surface and the hypolimnion.

The role of bubble plume aerators is to weaken stratification and work synergistically with wind to mix the reservoir and to oxygenate the hypolimnion. To control contaminant resolubilisation the hypolimnion must receive sufficient oxygen to satisfy the sediment oxygen demand. Artificial destratification has been relatively successful at controlling the release of contaminants from sediments (Brookes, Burch and Tarrant 2000) but has been less successful in controlling cyanobacteria (McAuliffe and Rosich 1990). Each waterbody has unique physical and ecological characteristics and so the success of destratification may vary, however, there are notable examples of success for cyanobacterial control (Visser et al. 1996; Lewis, Antennuci and Brookes 2001). The inability of destratifiers to mix the stratified surface layers, outside the immediate influence of the plume or mixer, has meant there is still a habitat for buoyant cyanobacteria to exploit (Sherman, Whittington and Oliver 2000).

The production and degradation of cyanobacterial toxins is the culmination of a suitable hydrodynamics, biogeochemistry, cyanobacterial and microbial growth. The factors contributing to cyanobacterial growth are predictable and can be modelled. The factors contributing to cell lysis, toxin release and degradation are less well known and this project has addressed some of these knowledge gaps and incorporated them into the ecological CAEDYM.

CHAPTER 3: ECOLOGY OF *CYLINDROSPERMOPSIS RACIBORSKII*

ABSTRACT

North Pine Dam is a monomictic subtropical reservoir in southeast Queensland, Australia, artificially mixed with a destratification unit in the summer months. It is dominated by cyanobacteria, including the toxic species *Cylindrospermopsis raciborskii* which reaches densities >30,000 cells mL^{-1} during the warmer months. *C.raciborskii* is a species most commonly observed in tropical and subtropical reservoirs and lakes, however it is undergoing a range expansion into temperate regions across the world. Previous studies have shown that *C. raciborskii* may gain a competitive advantage over other phytoplankton due to its ability to: grow under low light conditions; fix atmospheric nitrogen under low dissolved nitrogen concentrations; and a superior phosphate scavenging ability.

This study involved both an examination of the historical water quality data for North Pine Dam, and a series of *in situ* and laboratory-based process studies to determine the contribution of nutrients and light to *C. raciborskii* dominance in North Pine Dam. The study showed that for both *C. raciborskii* in culture and the phytoplankton community in North Pine Dam, there was adaptation to low light levels, consistent with other findings in the literature. The ability of *C. raciborskii* to regulate buoyancy ensured that it remained within the euphotic zone but avoided high light conditions, and hence photoinhibition. *C. raciborskii* also remained active for long periods (28 days) in the dark, giving it an additional advantage over other cyanobacteria. However, contrary to other findings, the dominance of *C. raciborskii* did not appear to be due to nitrogen fixation. Analysis of five years of historical data in North Pine Dam showed that nitrogen availability is generally high throughout the year, with nitrate being present in higher concentrations than ammonium. This is confirmed by nitrogen uptake studies which showed that nitrate and ammonium were preferentially used by the phytoplankton community over nitrogen fixation. Phosphate concentrations were generally low in both the surface and bottom waters. This phosphate regime is likely to enhance the growth of *C. raciborskii* since it has a phosphorus uptake and storage capacity superior to other phytoplankton species.

INTRODUCTION

Ecology of Cylindrospermopsis

Cylindrospermopsis is considered an emerging species in reservoirs as it appears to be occurring more frequently and expanding its geographic distribution. Padisák (1997) provides an excellent review of the worldwide distribution and spread of *Cylindrospermopsis raciborskii.* Although previously considered a subtropical/tropical species it is now found across a wide range of latitudes and has been shown to tolerate a wide range of climatic conditions (Briand, Leboulanger and Humbert 2004). Extensive blooms have been reported in tropical regions in Australia (Fabbro and Duivenvoorden 1996, Harris and Baxter 1996, McGregor and Fabbro 2000, Bormans et al. 2004); North and South America (Branco and Senna 1994, Chapman and Schelske 1997, Huszar et al. 2000, Marinho and Huszar 2002, Dobberfuhl 2003); and temperate regions of Europe (Dokulil and Mayer 1996, Padisák 1997, Padisák and Reynolds 1998, Briand et al. 2002, Fastner et al. 2003) and New Zealand (Ryan, Hamilton and Barnes 2003). *Cylindrospermopsis* is part of the photoadapting solitary filamentous cyanoprokaryotes-functional classification group S (Reynolds et al. 2002). It has been separated into S_N from S_1 (*Planktothrix, Limnothrix, Pseudanabaena*) and S_2 (*Spirulina*, *Arthrospira*, *Raphidiopsis*) and has a preference for warm mixed layers and tolerates light and nitrogen deficient conditions, but is sensitive to flushing.

The species can grow under diverse conditions and has been recorded in shallow and deep lakes and reservoirs, as well as large lowland rivers such as the Nile, Volga and Murray (Padisák 1997). High water temperature is a major factor promoting its growth. In subtropical-tropical regions it is found in deep stratified lakes and can persist all year round, while in temperate regions it is limited to warmer periods and shallow water bodies. The wide distribution of *Cylindrospermopsis* and appearance in newly constructed water bodies, such as Lake Balaton in Hungary (Padisák 1997), indicates that it is invasive and has a large degree of ecophysiological adaptability. It has several attributes that make it a successful invading species including, buoyancy regulation, preference for low light regimes, high affinity for and ability to store phosphorus, efficient ammonium uptake, possession of heterocytes to enable atmospheric nitrogen fixation, and resistance to zooplankton grazing.

Cylindrospermopsis is recognised as a species which dominates the phytoplankton community under thermally, stratified conditions (Hawkins and Griffiths 1993, Fabbro and Duivenvoorden 1996, Harris and Baxter 1996, Huszar et al. 2000, Marinho and Huszar 2002, Bormans et al. 2004). It may be absent under artificially mixed conditions (Hawkins and Griffiths 1993) or during periods of high wind mixing (Dokulil and Mayer 1996), although it persists in North Pine Dam, a reservoir with an artificial destratification unit (Antenucci et al. 2005). Compared to several cyanobacterial species, *Cylindrospermopsis* has a preference for high temperatures (Dokulil and Teubner 2000). Under cultured conditions an increase in temperature from 24°C to 26-28°C results in a doubling of the growth rate from 0.4-0.5 d^{-1} to 0.8-1.1 d^{-1}, respectively (Saker, Neilan and Griffiths 1999). The highest growth rate of 1.6 d^{-1} was observed at 30°C (Shafik et al. 2001). During adverse conditions such as low temperatures, specialized reproductive cells, known as akinetes, may form. Phosphorus is known to be an essential requirement for full development of the akinetes (Moore et al. 2003). These akinetes remain dormant in the sediment until growth conditions improve. Temperatures of 22-23.5°C have been shown to germinate akinetes (Padisák 1997). This requirement has implications for the likelihood of its occurrence within a growth season and its distribution in temperate regions. In shallow water bodies sediment temperatures are likely to reach 22°C in a warm summer, whereas in deep temperate lakes the species is absent as akinetes are unlikely to germinate (Padisák 1997). Seasonal temperatures will influence the timing and magnitude of *Cylindrospermopsis* populations by affecting growth rates and recruitment from sediments. For example, high temperatures were recorded in the sediment of the Fitzroy River (Queensland, Australia) before the onset of a bloom (Fabbro and Duivenvoorden 1996).

Cylindrospermopsis is considered a low light adapted species (Dokulil and Mayer 1996) and based on its shade tolerance with an light compensation point (I_k) of 20 µmol m^{-2} s^{-1} (Padisák 1997), has been given a separate algal functional assemblage of S_n compared to other heterocystous cyanobacteria- H (Padisák and Reynolds 1998). It is often observed to form subsurface cell maxima under stratified conditions or occur in water bodies with high light attenuation. Saker, Neilan and Griffiths (1999) found the greatest concentrations were recorded at 4 m depth where temperatures approximated 30.5°C in the Solomon Dam (Queensland, Australia). A decrease in the euphotic depth by 50% caused the loss of *C. raciborskii* in the Paranoa Reservoir in Brazil (Branco and Senna 1994). In Lake Julius (Queensland, Australia), the highest concentrations were recorded at 2 m depth while the euphotic (Z_{eu}) and mixed depths (Z_{mix}) approximated 0-4.5 and 6 m, respectively (McGregor and Fabbro 2000). The development of sub surface maxima poses a problem for water quality monitoring programs as there is a lack of visual cues, such as surface scums, as is common with *Microcystis* and *Anabaena*.

Low light transparency is another factor contributing to the proliferation of *Cylindrospermopsis*. In the Waikato lakes of New Zealand the species was recorded when the euphotic zone was shallow with secchi disk depths ranging between 0.41-2.6 m (Ryan, Hamilton and Barnes 2003), while in the Francs-Pecheurs pond in France and the Paranoa Reservoir in Brazil Secchi disk depth approximated 0.67 m (Branco and Senna 1994, Briand et al. 2002). The preference for low light has also been explored under laboratory conditions. Dokulil and Mayer (1996) found that photosynthetic rates became light saturated at light intensities of 30-60 µmol m^{-2} s^{-1}. In contrast, maximum growth rates have been recorded at higher light intensities of 118-120 µmol m^{-2} s^{-1} using cultured isolates (Saker, Neilan and Griffiths 1999; Shafik et al. 2001). Briand, Leboulanger and Humbert (2004) found that ten strains of *Cylindrospermopsis* from around the world can display growth at a wide range of light intensities, i.e. 30 to 400 µmol m^{-2} s^{-1}. In samples isolated from the Solomon Dam, Saker, Neilan and Griffiths (1999) recorded growth of 0.92 and 1.23 d^{-1} for straight and coiled trichomes, respectively. Growth was also recorded at 433 µmol m^{-2} s^{-1} whereas Shafik et al. (2001) reported an inhibition in growth at 200 µmol m^{-2} s^{-1}.

An examination of the relationship between mixed depth, Secchi disk depth (Z_{eu} +3 x Z_{secchi}) and cyanobacterial dominance by Dokulil and Teubner (2000) indicated that *Cylindrospermopsis* prefers regimes where $Z_{mix}/Z_s < 10$ equivalent to a Z_{mix}/Z_{eu} ratio of 3. This ratio is greater than the finding of Bormans et al. (2004) who reported on the interaction between inflow, turbidity, Z_{eu}, Z_{mix}, nutrient concentrations and cyanobacterial concentrations in the Fitzroy River. They found that cyanobacteria including *Cylindrospermopsis* were light limited when the Z_{mix}/Z_{eu} ratio was between 3-4.5. However after flooding, stratification enabled the settling of particles, which decreased turbidity and increased Z_{eu}. The Z_{mix}/Z_{eu} ratio decreased to 1 enabling the growth of cyanobacteria and *C. raciborskii* to dominate for several months until the next flooding event. A light intensity of 120 µmol m^{-2} s^{-1} equates to a daily light dose of 5.184 mol d^{-1}, assuming a light dark cycle of 12:12 h. Based on the maximum growth rates observed by Saker, Neilan and Griffiths (1999) and Shafik et al. (2001) at 120 µmol m^{-2} s^{-1}, *Cylindrospermopsis* would display maximum growth with a Z_{mix}/Z_{eu} ratio varying between 1-2, consistent with the findings of Bormans et al. (2004). Fabbro and Duivenvoorden (1996) also reported a bloom in the Fitzroy River. They found that with higher surface irradiances and a clearing of sediment from the water column caused the bloom to deepen in the water column.

Although *Cylindrospermopsis* has been reported to form subsurface maxima and its presence or absence has been reported under different mixing regimes, there are no detailed investigations into how it regulates it buoyancy in relation to light dose and nutrients, or is entrained within the water column. Subsurface maxima suggest that it maintains near neutral buoyancy similar to *Planktothrix rubescens* (Walsby et al. 2004). Evidence of buoyancy regulation includes: its occurrence in lakes with strong nutrient gradients and the ability of the species to persist by absorbing nutrients from deeper waters (Branco and Senna 1994, Ramberg 1987) dense populations do not seem to collapse; and that different coloured filaments vary in their floating ability with green filaments forming a scum and red filaments sinking in culture flasks (Padisák 1997).

The interaction between wind mixing, light penetration, phytoplankton cell distribution, photosynthesis and growth is complex (Brookes, Regel and Ganf 2003; Regel, Brookes and Ganf 2004; Walsby et al. 2004). *Planktothrix rubescens* is another cyanobacterium which forms subsurface cell maxima and has low light requirements. Walsby et al. (2004) examined its buoyancy in relation to daily light dose in Lake Zurich and Blelham Tarn. They reported that buoyancy status varied during a 24 h light dark cycle and was influenced by small changes in light intensity. For example in low light (2.5 µmol m^{-2} s^{-1}), close to 50 % of filaments remained buoyant whereas at a higher light intensity (22 µmol m^{-2} s^{-1}) buoyancy was lost in most filaments and recovered slowly in the dark phase. They found daily light doses of 0.28 mol m^{-2} and 0.51 mol m^{-2} were required to maintain neutral buoyancy under stratified conditions in Lake Zurich and Blelham Tarn, respectively. The neutral buoyancy depth at which these doses occurred was also found to match the depth at which maximum concentrations were recorded. Under stratified conditions, cell distribution would be governed by buoyancy related responses to cloud cover and light attenuation whereas when the mixed depth exceeded the neutral buoyancy depth, filaments would become entrained in the surface mixed layer. The influence of light dose on the buoyancy regulation of *Cylindrospermopsis* needs to be explored to assess the applicability of techniques such as artificial mixing.

Cylindrospermopsis has been recorded in lakes ranging from oligotrophic (Ramberg 1987) to hypertrophic (Chapman and Schelske 1997; Dobberfuhl 2003; Ryan, Hamilton and Barnes 2003). A lack of nitrate is considered to be the main reason for the development of heterocystic species such as *Cylindrospermopsis*. *Cylindrospermopsis* is not only able to fix atmospheric nitrogen but has a preference for ammonium compared to nitrate and is able to absorb ammonium at low concentrations (Marinho and Huszar 2002, Bormans et al. 2004, Sprőber et al. 2003). Saker, Neilan and Griffiths (1999) showed that growth with ammonium addition (0.86, 1.23 d^{-1}) was higher compared to nitrate addition (0.7, 1.02 d^{-1}). In addition, low percentages of heterocytes have been reported in trichomes from some reservoirs with low nitrate concentrations (Branco and Senna 1994, Bouvy et al. 1999, Présing et al. 1996) indicating a preference for ammonium before nitrogen fixation. Branco and Senna (1994) found that when ammonium concentrations were high in the Paranoa Reservoir during the wet season, few heterocytes were observed in *Cylindrospermopsis* trichomes. Conversely, when ammonium decreased, the percentage of filaments with heterocytes increased to 30 %. This was also observed in the Juturnaiba Reservoir in Brazil (Huszar et al. 2000). The high affinity for ammonium means that it is also able to compete with other heterocystic cyanobacteria under low ammonium concentrations when the other species are already fixing atmospheric nitrogen (Présing et al. 1996, Padisak 1997).

The occurrence of *Cylindrospermopsis* is less predictable in relation to phosphorus concentration and loading. However, a recent study by Istvanovics et al. (2000) examined the phosphorus-use efficiency of the species under continuous and pulsed phosphorus supply and regarded the species as phosphorus (P) opportunistic. Phosphorus uptake affinities were higher (U_{max}/K = 800-4000) compared to many algal species and the species also had an efficient storage capacity as determined by pulsed P supply experiments (Q_{ex}= 95 µg P mg C^{-1}). Such an adaptation enables the species to be competitive when exposed to low ambient concentrations or waterbodies that have large fluctuations in phosphorus. The difference between its minimal and maximal P cell quota can satisfy the demand of three subsequent cell divisions in the case of a lack of P in the medium (Istvanovics et al. 2000). Its ability to maintain a lower cell quota of surplus P compared to other cyanobacteria (Istvanovics et al. 2000) enables it to produce more biomass per unit phosphorus. This adaptation makes lake restoration efforts difficult when the primary aim is to reduce phosphorus loading (Dobberfuhl 2003). However, in contrast, increases in phosphorus may cause *Cylindrospermopsis* to lose competitive advantage. Dobberfuhl (2003) found that *C. raciborskii* density was negatively correlated to total phosphorus (TP) in three central Florida lakes. They found that as TP increased above 0.02 mg L^{-1}, *Cylindrospermopsis* concentrations decreased while *Microcystis* maintained its cell density.

Another factor that contributes to the occurrence of Cylindrospermopsis is its general resistance to zooplankton grazing (Padisak 1997). For example, Cladocera decreased when *C. raciborskii* increased in Solomon Dam (Hawkins 1988). In Lake Balaton the community changed from *Daphnia* and *Eudiaptomus* species to cyclopoids (Toth and Zankai 1985). *C. raciborskii* filaments have also been reported to reduce filtering rates of *Daphnia* species (Hawkins and Lampert 1989). In contrast, the rotifer *Brachionus* sp. has been observed to ingest short straight filaments but not coiled filaments (Fabbro and Duivenvoorden 1996).

Mixing, algal entrainment and photosynthesis

The interception and utilisation of light by phytoplankton are among the key processes determining aquatic productivity in the pelagic zone. These processes also contribute to phytoplankton species succession and the abundance of higher trophic-order organisms.

The availability of light is determined by solar insolation and the optical properties of the water body. Light interception is partially a function of vertical position of the phytoplankters in the water column, which is determined by the mixing characteristics and the migratory behaviour of the phytoplankton. In stratified waterbodies, where turbulent mixing is restrained, the vertical velocity of a phytoplankter, such as a colony of *Microcystis aeruginosa*, will be governed by its density and size according to Stokes' Equation:

$$\nu=2gr2(\rho'-\rho)/9\eta\theta \qquad (3.1)$$

where ν = the terminal velocity

g = gravitational acceleration

r = the radius of a sphere of volume equivalent to that of the organism

ρ' = the density of the organism

ρ = the density of the medium

η = the viscosity of the medium

θ = the form resistance defined as $\nu/\nu s$, where vs is the terminal velocity of a sphere of equal volume and density to that of the organism

Cyanobacterial colonies are not densely packed with cells but also contain mucilage that contributes to the effective diameter of the colony and changes the overall density of the colony. To account for colony composition, Equation 3.1 may be multiplied by a ratio (A) that accounts for colony composition, i.e. the ratio of cell volume to colony volume (Kromkamp and Walsby 1990).

The buoyancy of *Microcystis* plays a significant role in maximising light interception and growth. However, the benefit of buoyancy is dependent upon the nature of the turbulent mixing regime (Oliver and Ganf 2000). The entrainment of phytoplankton cells within turbulent eddies is a function of the floating or sinking velocity of the phytoplankton (v) and the velocity of the turbulent eddies (u) that transport cells. The quotient (ψ), proposed by Humphries and Imberger (1982) to determine the extent of entrainment, is given by:

$$\psi=15|v|/u \qquad (3.2)$$

If $\psi>1$, the intrinsic velocity of the phytoplankton cells plays a role in phytoplankton distribution, but when $\psi<1$, phytoplankton distribution is dominated by mixing processes (Humphries and Lyne 1988, Oliver and Ganf 2000). Reynolds (1994) expanded this concept and provided a model of the sinking (or floating) velocities of common freshwater phytoplankton and the shear velocity (u^*) necessary to entrain them. Oliver and Ganf (2000) modified the model of

Reynolds (1994) to include more genera of cyanobacteria and characteristic u^* of different habitats (Figure 3.1).

When the lake is not losing heat and wind speeds are low, the characteristic velocity of turbulence (u) within the diurnal surface layer can be equated to the water friction velocity u^* estimated from the wind speed (Denman and Gargett 1983):

$$u^{*2}=\rho_a c U_{10}^2/\rho \qquad (3.3)$$

where ρ_a = the density of air

ρ = water density

c = the dimensionless drag coefficient (1.3×10^{-3})

U_{10} = the wind speed measured 10 m above the surface

For typical densities (ρ_a = 1.2 kg m^{-3}: ρ = 1000 kg m^{-3}) the calculation of u^* simplifies to $0.001U_{10}$ (Oliver and Ganf 2000). U_{10} can be calculated from the wind speed at 2 m (U_2) using the Equation 3.4.

$$U_{10} = U_2 \times \ln(10/Z_o)/\ln(2/Z_o) \qquad (3.4)$$

where Z_o = 0.000115

In order to assess the validity of the Humphries and Lyne (1988) model of entrainment for the cyanobacterium *Microcystis aeruginosa,* it was necessary to identify a feature of phytoplankton that was sensitive to light and that could be used to assess the previous light history of the cells and give insight into entrainment, turbulent mixing and light capture.

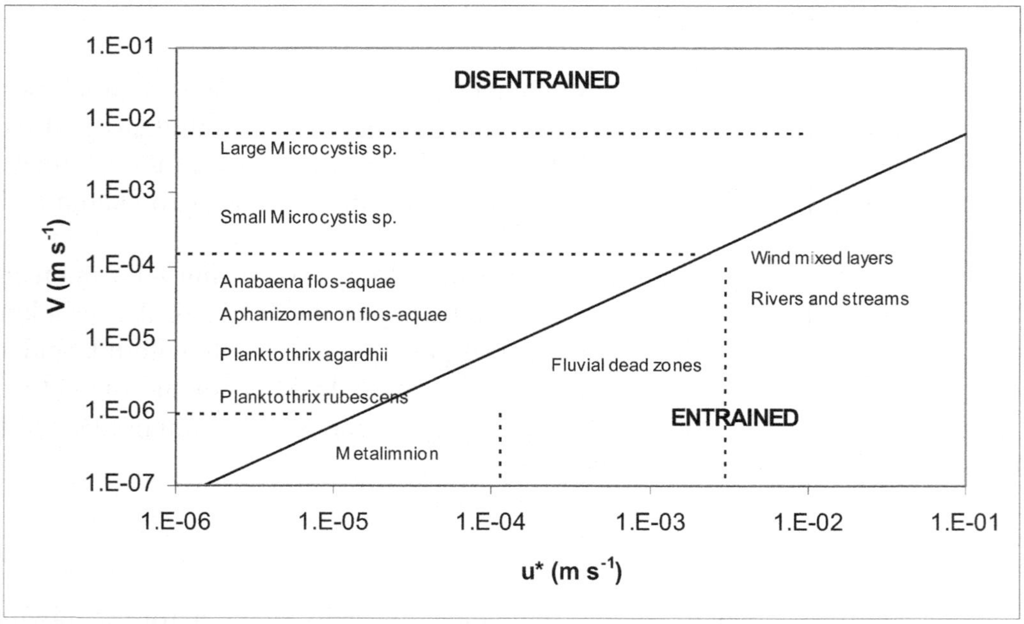

Figure 3.1 A comparison of floating and sinking velocities (v) of selected cyanobacteria with the velocity of turbulence (u*). The solid line separates regions of particle entrainment and disentrainment. Values of u in different habitats provide perspective (after Reynolds 1994). Figure 3.1 text copied from Oliver and Ganf (2000).

Harris (1980) identified a 'fluorescence ratio' which was the DCMU-sensitive chlorophyll *a* fluorescence divided by the *in vivo* chlorophyll fluorescence. This ratio displayed a vertical gradient during stratified conditions as the fluorescence ratio of surface populations became depressed. Since the time-scales of mixing and phytoplankton photochemical processes are similar (Denman and Gargett 1983), Harris (1986) concluded that the fluorescence ratio could be used as an indicator of mixing in surface waters. With the improved fluorometers now available, the physiological basis for Harris's observations are better understood (Oliver and Whittington 1998) and variable fluorescence can be more readily assessed.

Photons are absorbed by the light harvesting pigments, which then enter an excited state. The return to ground state may be by transfer of the energy along an electron transport chain via a series of photochemical reactions, transfer of energy to adjacent pigments, heat dissipation or by the emission of a fluorescence photon (Campbell et al. 1998). Fluorescence emission, primarily from PSII, changes with intensity of the excitation light as excited electrons participate in photochemical and non-photochemical pathways.

The difference between the maximum fluorescence yield (F_m), when all reaction centres are momentarily closed by a saturating light pulse, and the minimum fluorescence yield (F_o: reaction centres open) provides considerable information on the photochemical state of PSII. The variable fluorescence ($F_v = F_m - F_o$) is usually standardised to F_m and the ratio F_v/F_m is closely related to the maximum yield of photosynthesis (Genty, Briantais and Baker 1989). F_v/F_m can be used as a quantitative indicator of photo-inhibition (Ibelings 1996, Whittington et al. 2000).

Because F_v/F_m is sensitive to photon flux density it shows promise as a parameter to offer insight into the immediate past light-history of phytoplankton. This will, in turn, yield information on water mixing, entrainment of phytoplankton and vertical migration.

Brookes, Regel and Ganf (2003) used the variable fluorescence ratio and high quality hydrodynamic and meteorological data to validate the entrainment model. The variable to maximal chlorophyll fluorescence ratio (F_v/F_m) was used to determine light history of *M. aeruginosa* colonies in a shallow, stratified lake. As insolation increased, the lake stratified and *M. aeruginosa* colonies displayed a depression in F_v/F_m that became less severe with depth. In the afternoon, wind speed increased, entraining colonies and disrupting the discrete depth variable F_v/F_m response. The point where the photochemical response became homogenised allowed determination of the water friction velocity necessary to entrain colonies *of M. aeruginosa* ($u^* = 0.003$ m s^{-1}). This fitted the entrainment model proposed by Humphries and Lyne (1988). Rates of F_v/F_m depression were dependent upon the photon flux density, whereas the rate of recovery of F_v/F_m was dependent upon the historical photon density. A model was presented which modelled F_v/F_m response to five different mixing scenarios.

North Pine Dam

North Pine Dam, a drinking water reservoir in southeast Queensland has had severe summer blooms of *C. raciborskii* since the 1980s with resultant concentrations of the toxin, cylindrospermopsin, being detected. An analysis of this data in the mid 1990s suggested that *C. raciborskii* blooms dominated during drier years and when the water column was more stable (Harris and Baxter 1996). It was concluded that vertical stratification and low nitrogen inputs were likely to provide the competitive advantage for *C. raciborskii* over other algal species. However, the installation of a destratification unit in NPD in the mid 1990s designed to reduce stratification and hence reduce *C. raciborskii* dominance had little or no effect. Blooms now commence earlier and are more sustained (Antenucci et al. 2005).

Managing blooms of the toxic cyanobacteria, *Cylindrospermopsis raciborskii* requires not only an understanding of the chemical and physical factors promoting its growth but also the interactions between *C. raciborskii* and other competing algal species. Research on the interactions between algal assemblages has focused on taxonomic grouping of species and/or genera. However taxonomic similarities do not necessarily translate into functional similarities. Functional groupings for species and genera have been established based on a suite of characteristics of algae in temperate freshwater systems (Reynolds et al. 2002). In a management context, understanding the tolerances and sensitivities of particular functional groups to variables such as light, grazing, nutrients etc. can aid in developing techniques appropriate to the management of dominant functional groups.

This study focussed on the importance of light, nutrients and phytoplankton species dynamics on *C. raciborskii* growth in North Pine Dam. This involved a combination of analysis of the historical water quality data, field process studies and laboratory culture studies.

MATERIALS AND METHODS

Historical water quality data

The analysis of the historical water quality data involved a comparison of the physico-chemical environment in artificially mixed North Pine Dam and two adjacent naturally mixed reservoirs, Somerset and Wivenhoe. North Pine, Somerset and Wivenhoe Dams are located in subtropical, southeast Queensland, Australia, and provide drinking water to the city of Brisbane and a number of regional towns and cities (1.2 million people). The catchments for these reservoirs are typically unprotected, and are dominated by cattle grazing pasture (>50%), with approximately 20% natural vegetation.

North Pine Dam (Lake Samsonvale, 27°15′S, 152°55′E) is the smallest of the three reservoirs with an area of 21.8 km^2 215,000 ML full reservoir volume (Table 3.1, Figure 3.2). An artificial destratification system was installed and began operating in 1995. The system is typically switched on in mid-September and runs continuously until the end of April each year. It consists of two diffusers, made of polyethylene pipe, each being approximately 550 m long. Each diffuser discharges at least 47 L s^{-1} air at the depth of the diffuser. It was installed 1 m above the sediment surface. Somerset Dam (27°08′S, 152°32′E) has a surface area of 42.1 km^2 and a volume of 360,000 ML when full. It is located upstream of Wivenhoe Dam. Wivenhoe Dam (27°24′S, 152°37′E) is the largest reservoir with a surface area of 107.5 km^2 and a full reservoir volume of 1,165,000 ML.

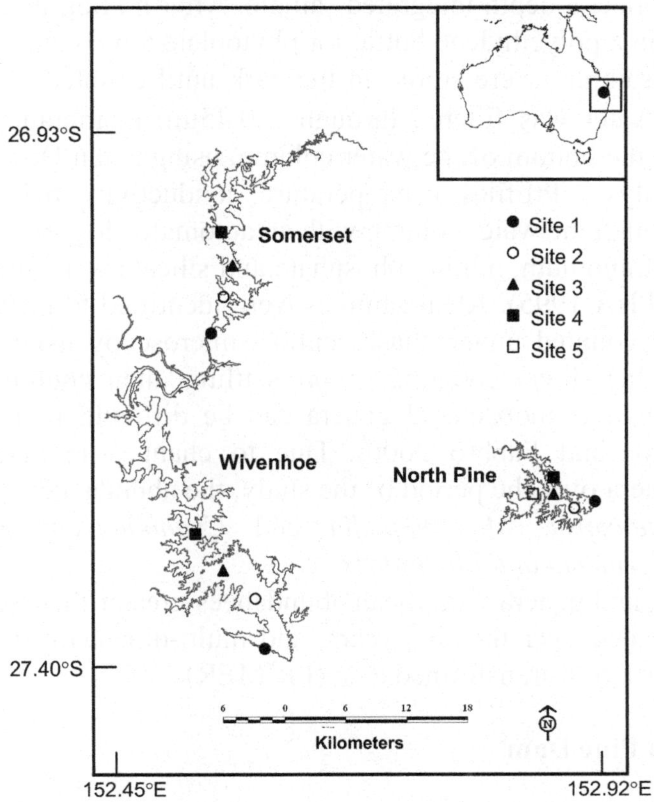

Figure 3.2 Map of the three water reservoirs, North Pine, Somerset and Wivenhoe Dams in southeast Queensland.

Table 3.1
Physical attributes of the three water reservoirs. Annual rainfall and evaporation are averages for the period 1997-2002

Parameter	North Pine	Somerset	Wivenhoe
Reservoir area (km^2)	21.8	42.1	107.5
Full reservoir volume (ML)	215,000	380,000	1,165,000
Maximum depth at wall (m)	35	39	44
Catchment area (km^2)	347	1330	7020
Rainfall (mm y^{-1})	1060	924	712
Evaporation (mm y^{-1})	1375	1775	1872

Rainfall data was collected by the Australian Bureau of Meteorology (BOM) using rain gauges located at all three reservoirs. Maximum and minimum ambient temperatures were measured at the nearest BOM sites, i.e. Brisbane airport for North Pine Dam, Beerburrum for Somerset Dam, and Amberley for Wivenhoe.

Water samples were collected 100 m from the dam wall, biweekly for phytoplankton counts and monthly for nutrients from July 1997 until June 2002. A 3 m long hosepipe sampler, 5 cm dia., was used to collect a depth-integrated sample. After mixing in a bucket, a subsample of 250 ml was collected in a polyethylene bottle for phytoplankton counts. 1 ml of 100% Lugols solution was added and samples were stored in the dark until counted. A subsample was also taken for nutrients and water was filtered through a 0.45μm membrane filter and frozen. A sample was also taken at the bottom of the water column, using a van Dorf sampler, for nutrient analyses. for nutrient analyses. Profiles of temperature, conductivity, pH, oxygen and turbidity were also determined through the water column with an automated logger.

In the laboratory, ammonium, nitrate, phosphate and silicate were analysed using standard colorimetric methods (APHA 1995). Algal samples were identified to genus level, and in some cases, species level, and counted under phase-contrast microscopy using a Sedgewick Rafter counting chamber by either direct counting or pre-settling, then counting of fixed samples. Taxonomic identification of cyanobacterial genera can be difficult, particularly for the order Chroococcales (McGregor and Fabbro 2000). Due to changes in taxonomic naming and resolution for various genera over the period of the study, the abundance of the following genera were combined: *Aphanocapsa, Merismopedia* and *Cyanodictyon*; *Monoraphidium* and *Ankistodesmus*; and *Geitlerinema and Limnothrix.*

Data for all genera, and genera with mean abundances greater than 1000 cells ml^{-1} in each reservoir were then averaged over the five years, and multi-dimensional scaling (MDS) plots were generated using fourth root-transformed data (PRIMER).

Process studies at North Pine Dam

The effect of light and nitrogen on algal growth in North Pine Dam was examined over a one year period.

Effect of light on phytoplankton growth

Measurements of primary productivity were done six times over a one year period (3 Feb'04, 5 Feb'04, 9 Jun'04, 12 Jan'05, 13 Jan'05, 9 Feb'05) in North Pine Dam. Water samples were collected using a 3 m depth-integrated sampler at the dam wall site (10001) at North Pine Dam. At the same time, light profiles through the water column were measured using a 4 pi sensor (Licor). Water was transported to a laboratory at the bank and used to fill acid-washed plastic bottles. For the first three sampling occasions, 500 ml polycarbonate bottles were used, thereafter 350 ml PET drink bottles were used. There were 5 treatments for the first three sampling occasions, 100, 50, 14, 5 and 0% light, increasing to 6 treatments, i.e. 100, 50, 25, 14, 5 and 0% light for the last three sampling occasions, in order to more fully characterize the photosynthesis-irradiance curve. All bottles for shading were placed in bags made of shade cloth of the appropriate light intensity. There were three replicates for each treatment. 20 μCi ^{14}C-bicarbonate (specific activity of 1.11 to 2.22 GBq $mmol^{-1}$ (30 to 60 mCi $mmol^{-1}$)) was added to the bottles.

Samples were incubated by securing them on a mesh rack floating in the reservoir for 4 h. The ambient light intensity was logged every 10 min during the incubation. The water temperature at the location of the bottles was also monitored using a temperature logger (Etemperature, OnSolutions). Bottles were placed in black bags to stop the incubations and known volumes of water were filtered onto 0.2 μm polycarbonate filters. 0.2 ml 6 N hydrochloric acid was added to each filter in scintillation vials. Samples were then processed to determine ^{14}C uptake (Rothlisberg et al. 1994).

An additional ^{14}C-incubation was done in February 2005 to determine whether the addition of nutrients would stimulate primary productivity. The protocols were the same as previous incubations however there was only one light level, 100% light. There were four treatments: no nutrient addition, phosphate addition (0.23 mg L^{-1} KH_2PO_4), ammonium addition (1.51 mg L^{-1} NH_4Cl), and ammonium plus phosphate addition. There were three replicate bottles for each treatment. The nutrients added were designed to be a ten times increase over background nutrient concentrations based on data from February in previous years.

Subsamples were taken for chlorophyll *a* analysis and alkalinity. For chlorophyll analysis, known volumes of water were filtered through glass fibre filters. The filters were then frozen until returned to the laboratory. Filters were extracted by sonicating for 1 min in cold 100% acetone, adjusting the acetone concentration to 90%, then measuring the extract spectrophotometrically, when there was sufficient chlorophyll, and using a spectrofluorometer, when chlorophyll values were low (Jeffrey and Welshmeyer 1997). The spectrophotometric values were then used to calibrate the fluorescence readings. Water for alkalinity analysis was stored in plastic vials on ice until analyzed in the laboratory. Alkalinity was analyzed by the method of the American Public Health Association (1995).

Subsamples were also taken for measures of photosynthetic activity using a PHYTOPAM (Pulse Amplitude Modulation device, Walz) on 9 Jun '04, 28 Aug '04, 10 Nov '04, 12 Jan'05, 13 Jan'05, 9 Feb'05. Samples were pre-adapted to low light conditions for a minimum of 20 min prior to undertaking measurements. Yield measures were taken for the three algal groupings – diatoms and dinoflagellates, green algae, and cyanobacteria at light levels ranging from 1 to 764 $\mu mol\ m^{-2}\ s^{-1}$.

Subsamples were also collected for total algal and *C. raciborskii* cell counts, and fixed with Lugol's solution. Samples were settled on Sedgewick Rafter cells and counted using a phase contrast microscope at 400 times magnification by Queensland Health Services.

Phytoplankton nitrogen uptake and fixation

Nitrogen uptake and nitrogen fixation rates by phytoplankton in North Pine Dam were determined using ^{15}N-nitrogen uptake and acetylene reduction methods. Sampling for ^{15}N-nitrogen uptake experiments was conducted six times over a one year period (3 Feb'04, 7 Apr'05, 9 Jun '04, 28 Aug '04, 10 Nov '04, 13 Jan'05). Water samples were collected using a 3 m depth-integrated sampler at the dam wall site (10001) at North Pine Dam. Water was transported to a laboratory at the bank and used to fill acid-washed 500 ml polycarbonate bottles. Two concentrations of ^{15}N-ammonium chloride or potassium nitrate were added (0.01 and 0.001 mg N L^{-1}) in bottles incubated at 0 and 100% light. Once background ammonium and nitrate concentrations had been determined, the appropriate ^{15}N-addition rate was used to ensure that addition rates were in the range of 10% of background. Bottles were incubated by attaching them to a mesh rack floating in the reservoir for 1 h. To detect the effect of irradiance, three replicates at five light levels, 100 %, 50%, 30%, 10% and 0%, were examined on two occasions.

Following the incubation, known volumes from each bottle was filtered using 25mm pre-combusted GF/F filters and frozen. In the laboratory, filters were dried and the $^{15}N/^{14}N$ isotopic ratio and %nitrogen was determined using a mass spectrometer (GV Isoprime, Manchester UK). Uptake rates and relative preference indices were calculated using the equations of Glibert et al. (1991). Relative preference indices were calculated using the equations of McCarthy, Taylor and Taft (1977).

Water was also used to fill 500 ml glass bottles for nitrogen fixation measurements. Acetylene gas, prebubbled through water, was bubbled into through rubber septa in the top of the bottles into the water for 5 min to ensure saturation of the water with acetylene. Triplicate bottles were incubated at five light levels, 100, 50, 30, 10 and 0% light for 24 h. 3 ml subsamples were withdrawn from the bottles into venoject vials containing 0.2 ml 20% trichloroacetic acid at 0. 4 and 24 h. Samples were stored at 4°C until analyzed. A gas chromatograph with a flame ionization detector was used to determine the ethylene content of the samples. Nitrogen fixation was determined from the conversion of acetylene to ethylene (Knowles 1990).

Subsamples were also taken for determination of ammonium, nitrate/nitrite and phosphate concentrations. Water was filtered through 0.45 μm cellulose filters and stored on ice until returned to the laboratory. Nutrients were analyzed at Queensland Health Services using standard methods (APHA 1995).

C. raciborskii culturing studies

To determine how *Cylindrospermopsis raciborskii* responds to light intensity, a log phase culture with an initial concentration of 11,167 trichomes mL^{-1} was placed in triplicate under 7 different light intensities. The light intensities ranged from darkness to 10, 20, 30, 70, 175 and 475 $\mu mol\ m^{-2}\ s^{-1}$. The cultures were set up as batch cultures in Erlenmeyer flasks. Growth rate was determined by changes in trichome concentration. Measurements of average trichome length were also made to determine whether or not light intensity may have implications on floating and sinking velocities associated with Stokes law.

Cylindrospermopsis raciborskii (strain 020A) was obtained from the Australian Water Quality Centre. The strain was originally isolated from Palm Island (Queensland). Stock and experimental cultures were grown in ASM media (Gorham et al. 1964) on a 12:12 light-dark cycle in a cabinet at a constant temperature of 25°C. The experimental cultures were grown in 100 mL Erlenmeyer flasks and were illuminated from below with fluorescent tubes. A LI-COR LI-1000 light meter was used to measure photon irradiance (E) at the base of the flasks.

Trichome concentration was determined using a Sedgewick Rafter counting chamber and an Olympus BH-2 microscope at 40x. At least 100 trichomes or the entire chamber was counted. Prior to counting subsamples were preserved with Lugol's iodine solution (1 %) and their gas vesicles collapsed at 800 KPa.

C. raciborskii growth rate was determined by the change in trichome concentration over the first 0-6 days. Measurements on individual trichomes were made using an image analysis system. Images were acquired via a Panasonic model GP-KR222 black and white ½" format interline transfer CCD camera mounted on an Olympus BX50 microscope equipped with DIC optics.

RESULTS

Historical water quality data

The climate of southeast Queensland, where the three reservoirs are situated, is warm (monthly mean max. temperature 28.4-31.4°C) and wet in the summer (December-February), and mild (monthly mean max. temperature 20.8-23.3°C) and dry in the winter (May-July) (Table 3.2). North Pine and Somerset Dams typically had more rain (1060 and 924 mm y^{-1} respectively) than Wivenhoe Dam (712 mm y^{-1}) (Table 3.1).

There was little difference between average monthly surface and bottom temperatures in North Pine Dam in autumn and winter with surface temperatures a few degrees higher (Figure 3.3). The pattern was similar in autumn and winter at Wivenhoe and Somerset Dams, however in spring and summer differences between the surface and bottom were up to 8°C. Oxygen concentrations in the bottom waters were low in summer at all three reservoirs, 1.8 ± 2.0, 0.8 ± 1.3 and 2.1 ± 2.9 mg L^{-1} in North Pine, Somerset and Wivenhoe Dams respectively. Overall, temperatures were lowest in Wivenhoe and highest in North Pine Dam in winter. The reverse was true in summer.

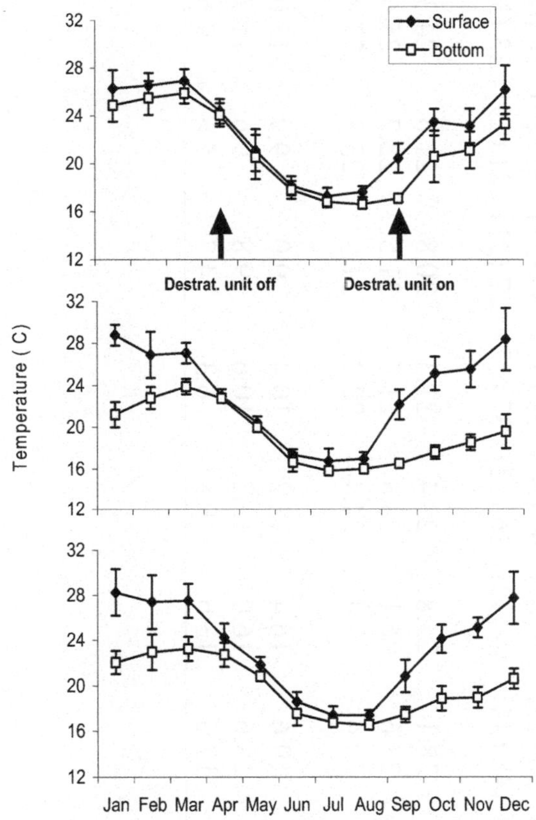

Figure 3.3 Mean (± SD) monthly surface and bottom water temperatures (°C) from July 1997 to June 2002 in (a) North Pine, (b) Somerset, and (c) Wivenhoe dams. Months when destratification unit is turned on and off are indicated for North Pine Dam.

Table 3.2

Mean (± SD) monthly rainfall (mm), and mean monthly maximum and minimum air temperatures (°C) of the three water reservoirs from 1997 to 2003. Nearest BOM sites used for temperature were: Brisbane airport for North Pine Dam, Beerburrum for Somerset Dam and Amberley for Wivenhoe Dam.

Month **PARAMETER**	**JAN**	**FEB**	**MAR**	**APR**	**MAY**	**JUN**	**JUL**	**AUG**	**SEP**	**OCT**	**NOV**	**DEC**
RAINFALLL (MM)												
North Pine	87 (53)	189 (148)	90 (45)	81 (43)	78 (41)	47 (41)	36 (38)	39 (40)	33 (35)	77 (46)	110 (51)	135 (38)
Somerset	64 (66)	173 (140)	50 (28)	58 (61)	67 (50)	39 (31)	35 (37)	32 (29)	36 (49)	66 (32)	88 (64)	138 (35)
Wivenhoe	53 (36)	110 (65)	48 (34)	44 (32)	36 (22)	37 (27)	27 (21)	21 (19)	33 (36)	72 (42)	71 (53)	104 (35)
Temperature (°C max)												
North Pine	28.7	28.8	28.1	25.8	23.4	21.3	20.8	21.6	24.3	25.4	26.1	28.4
Somerset	30.3	29.2	28.6	26.1	23.8	21.7	21.5	22.7	25.6	27.6	26.9	29.4
Wivenhoe	31.5	30.6	30.4	27.4	24.4	22.2	21.9	23.3	27.2	28.3	28.8	31.2
Temperature (°C min)												
North Pine	20.6	20.9	19.6	16.4	13.2	10.4	9.0	9.9	13.0	15.6	17.3	20.0
Somerset	19.4	19.3	18.6	16.0	13.2	10.6	9.8	9.6	11.7	14.8	15.9	18.1
Wivenhoe	18.7	19.2	17.5	13.6	9.8	6.3	4.7	5.6	9.6	13.0	15.6	18.0

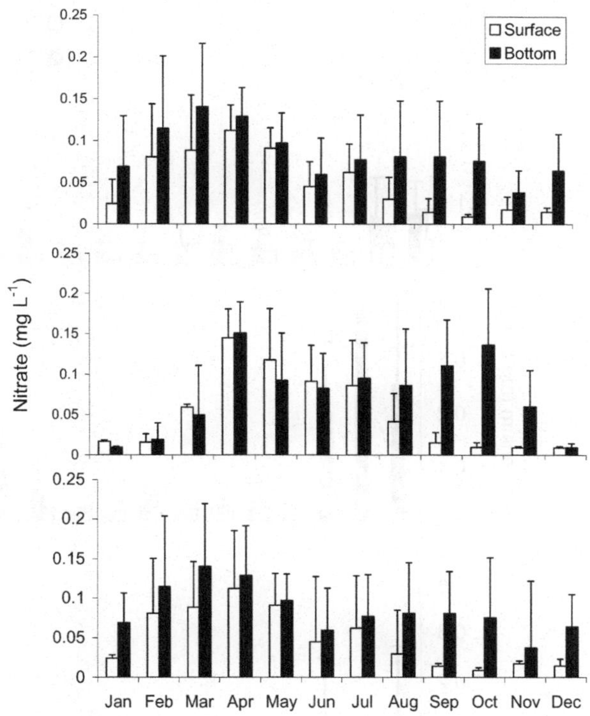

Figure 3.4 Mean (± SD) monthly surface and bottom water nitrate concentrations (mg L^{-1}) from July 1997 to June 2002 in (a) North Pine, (b) Somerset, and (c) Wivenhoe dams.

Mean monthly nitrate concentrations were higher at the bottom than surface, and highest in late summer through winter in all three reservoirs with concentrations up to 0.15 mg L^{-1} (Figure 3.4). In contrast, ammonium concentrations were generally below 0.05 mg L^{-1} with the exception of higher concentrations in bottom waters during summer in all three reservoirs (Figure 3.5). The differences between the surface and bottom waters were more marked at Somerset and Wivenhoe Dams than at North Pine Dam. Phosphate concentrations were higher overall in Wivenhoe and Somerset Dams compared with North Pine Dam (0.007, 0.008 and 0.003 mg L^{-1} respectively) (Figure 3.6). Summer peaks were only detected in Wivenhoe and Somerset Dams. Phosphate concentrations in surface and bottom waters were generally similar at all three reservoirs. Mean monthly silicate concentrations were generally higher in Wivenhoe and Somerset Dams (6 mg L^{-1}) than North Pine Dam (3 mg L^{-1}).

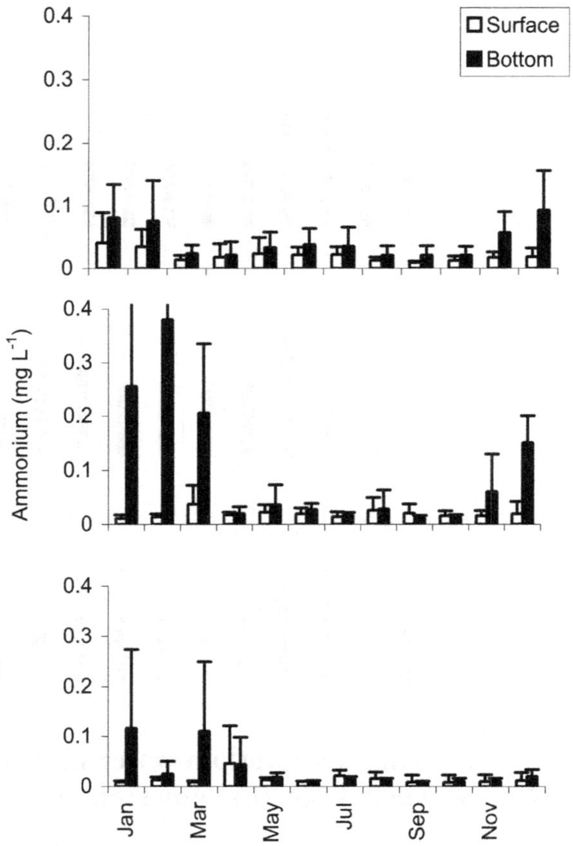

Figure 3.5 Mean (± SD) monthly surface and bottom water ammonium concentrations (mg L^{-1}) from July 1997 to June 2002 in (a) North Pine, (b) Somerset, and (c) Wivenhoe dams.

Total algal densities peaked in the summer each year in all three reservoirs (Figure 3.7). Summer densities were generally higher in Somerset and Wivenhoe Dams than North Pine Dam. *C. raciborskii* densities also peaked each summer in all three reservoirs (Figure 3.7). In some summers, *C. raciborskii* dominated the algal community. *C. raciborskii* blooms in Somerset and Wivenhoe Dams were similar in intensity (10^5 cells ml^{-1}) and timing (summer). Blooms in north Pine Dam, the reservoir with the destratification unit installed, commenced earlier than the other two reservoirs. During the winter months, densities were higher in North Pine Dam than in Somerset or Wivenhoe Dams.

C. raciborskii densities were relatively low below 23-25°C in all three reservoirs, but as the temperature increased, densities increased rapidly (Figure 3.8). However, the increase was more dramatic in Somerset and Wivenhoe Dams than in North Pine Dam.

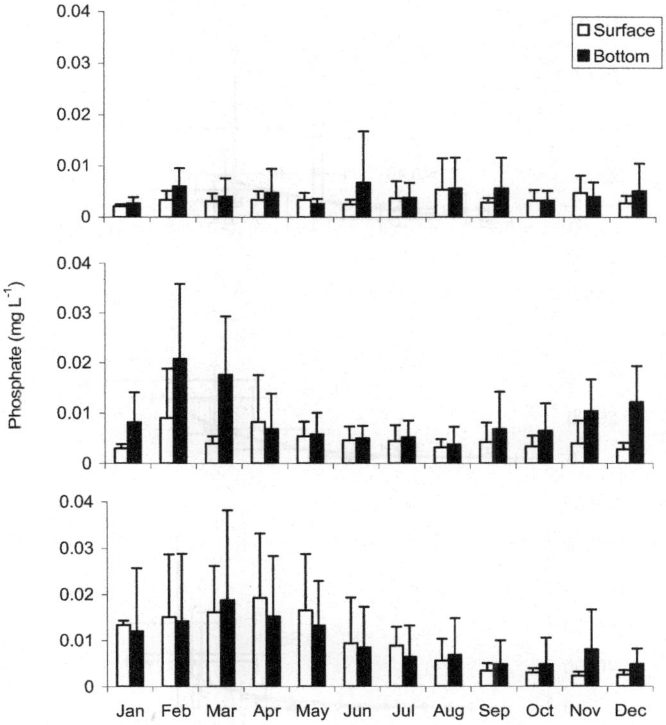

Figure 3.6 Mean (± SD) monthly surface and bottom water phosphate concentrations (mg L^{-1}) from July 1997 to June 2002 in (a) North Pine, (b) Somerset, and (c) Wivenhoe dams.

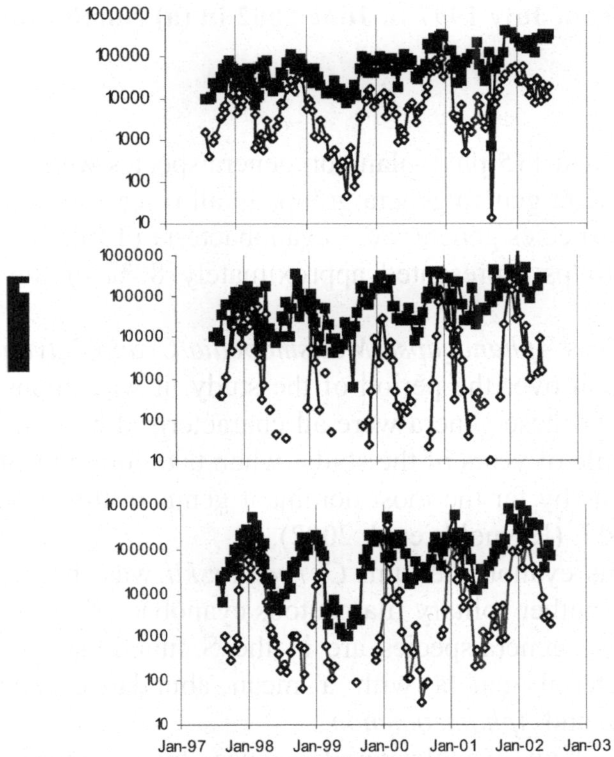

Figure 3.7 Total algal and C. raciborskii cell densities in 3 m depth-integrated surface waters from July 1997 to June 2002 in (a) North Pine, (b) Somerset, and (c) Wivenhoe dams.

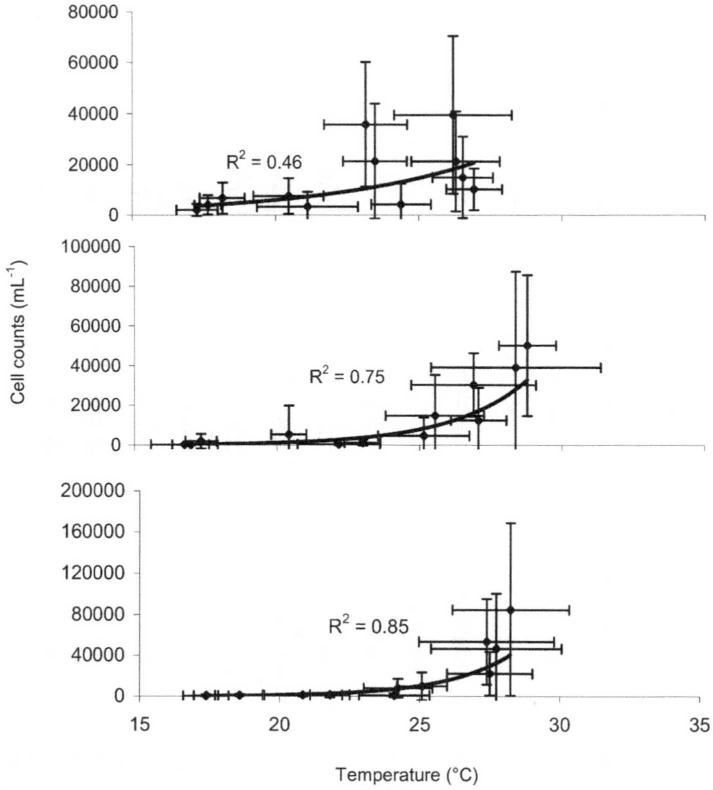

Figure 3.8 Mean (± SD) monthly temperature (°C) versus mean (± SD) monthly *C.raciborskii* cell densities from July 1997 to June 2002 in (a) North Pine, (b) Somerset, and (c) Wivenhoe dams.

A total of between 104 and 115 phytoplankton genera/species were identified in the North Pine Dam, however the dominant genera/genera groups in all reservoirs were cyanobacteria. Six of the eight most abundance species/genera were cyanobacteria (Table 3.3). The top four most abundance species/genera groups represented approximately 80% of the total phytoplankton abundance.

Most dominant group was *Aphanocapsa/Merismopedia/Cyanodictyon* (Table 3.3). Due to change in taxonomic resolution over the period of the study, it was impossible to differentiate between these species however these genera were all characterised by mucilaginous colonies of small cells. Over the last couple of years of the study, when taxonomic resolution had improved, *Aphanocapsa* was identified as by far the most dominant genus in this group. These genera are functionally grouped as K and L (Reynolds et al. 2002).

The solitary filamentous cyanobacterium, *C. raciborskii* was the second most dominant species in North Pine Dam. Another solitary filamentous cyanobacterium, *Planktolyngbya,* had a similar abundance. Both these genera/species are in the S functional group (Reynolds et al. 2002). The other cyanobacterial genera with a mean abundance >1000 cells ml^{-1} were *Pseudanabaena*, *Aphanothece* and *Aphanizomenon*.

In addition to the dominant cyanobacterial species, two grouped chlorophyte genera, *Monoraphidium/Ankistodesmus,* were also present in substantial numbers (Table 3.3). The diatom, *Achnanthidium* and the chrysophyte, *Chroomonas* were also present at abundances >1000 cells ml^{-1}.

Table 3.3

Mean abundance (± SD) (cells ml^{-1}) of algal genera in North Pine Dam. Only genera with cell counts >1000 ml^{-1} are included. CYAN = cyanobacteria, CHLO = chlorophyte, CHRYS = chrysophyte, DIAT = diatom. Functional groups also given in brackets (Reynolds et al. 2002).

Genus	Cell abundance (ml^{-1})
Aphanocapsa/Merismopedia/ Cyanodictyon (CYAN) (K, L_o, L_m)	59300 (48800)
Cylindrospermopsis raci (CYAN) (S_n)	28000 (23700)
Planktolyngbya (CYAN) (S_i)	15400 (14900)
Monoraphidium/ *Ankistodesmus* (CHLO) (X_i, S_i)	4700 (13900)
Pseudanabaena (CYAN) (S_i)	3800 (5600)
Microcystis (CYAN) (L_M)	3000 (4900)
Aphanothece (CYAN) (K)	1930 (5500)
Aphanizomenon (CYAN) (H_i)	2400 (6200)
Achnanthidium (DIAT)	1750 (4100)
Chroomonas (CHRYS) (Y)	800 (600)
Aulacoseira (DIAT) (C)	400 (400)

Multidimensional scaling was used to assess the dissimilarity between the dominant genera/species (Figure 3.9). The *Aphanocapsa/Merismopedia/Cyanodictyon/Microcystis* group was distinctly different from all other genera in all reservoirs. *C. raciborskii* and *Planktolyngbya* appear to have similar patterns, however the patterns of abundance for the other solitary filamentous cyanobacterium, *Pseudanabaena* was distinctly different from these two genera. Groupings were not obviously along taxonomic or functional lines (Reynolds et al. 2002).

The abundance of the dominant *Aphanocapsa/Merismopedia/Cyanodictyon/Microcystis* group decreased in the dry winter months but only for about 3 months (Figure 3.10). In contrast, *C. raciborskii* decreased in abundance much earlier, i.e. in late summer. *Planktolyngbya* also decreased in late summer but continued to decrease during the winter months to reach a low in August.

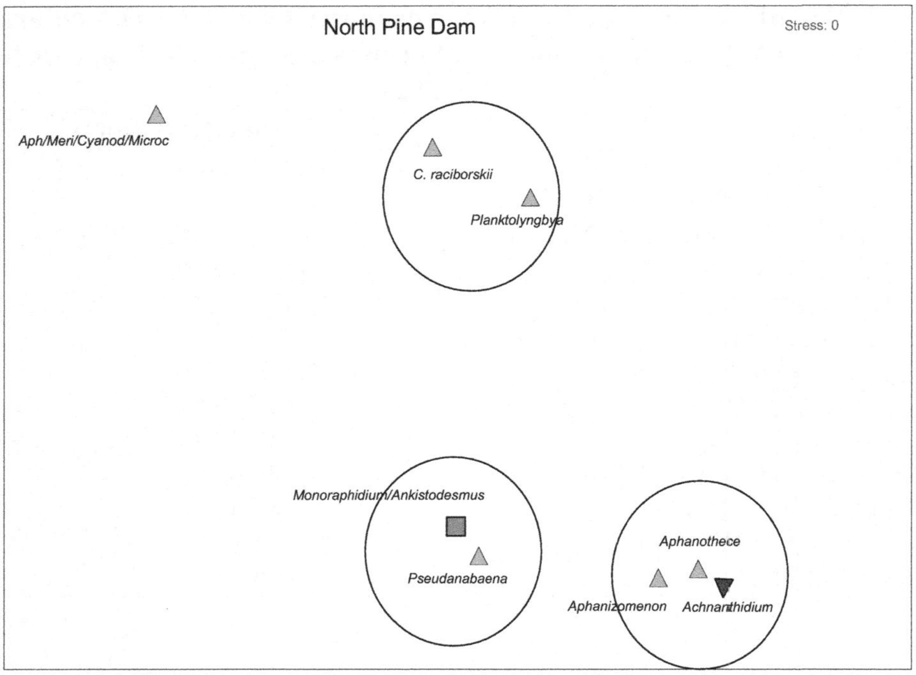

Figure 3.9 Multidimensional scaling (MDS) plot of the dominant algal species in North Pine Dam across 5 sites and 5 years. Triangle = cyanobacteria, square = chlorophyte, inverted triangle = diatom.

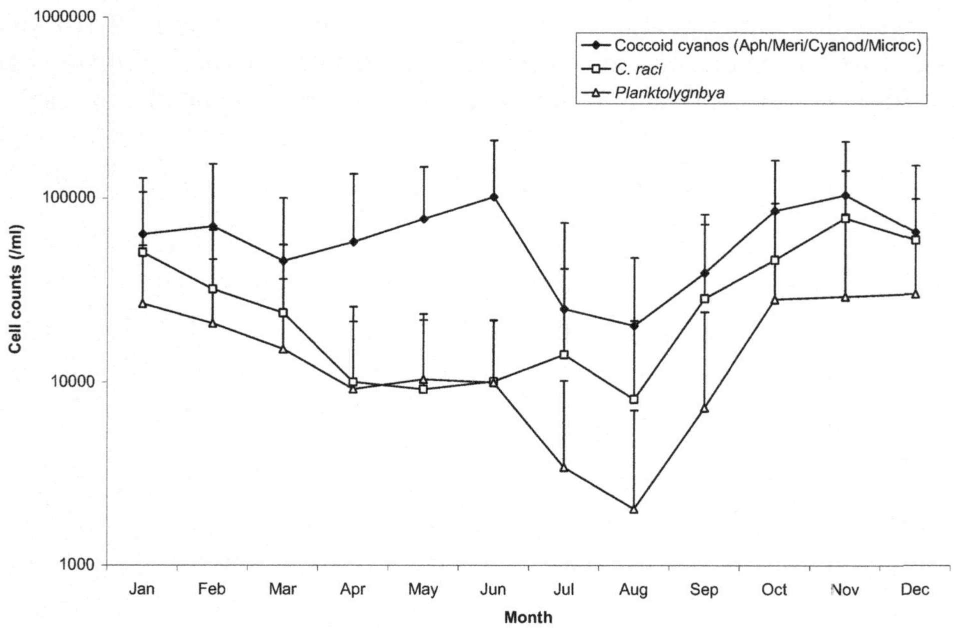

Figure 3.10 Mean monthly algal cell abundances (ml^{-1}) of the three most dominant algal genera/genera groups in North Pine Dam over 5 years

Process studies – North Pine Dam

Effect of light on phytoplankton growth

Primary productivity studies were conducted in North Pine Dam over a one year period. During summer, *C. raciborskii* was present in high numbers (> 20,000 cells/ml) although it only represented on average approximately 15% of the total cell densities (Table 3.4). *Aphanocapsa*, a small colony-forming cyanobacterium, is the most dominant genus. The algal community in general was found to be highly tolerant of low light conditions, with a light compensation point (i.e. light for optimal growth, I_K) for ^{14}C-productivity of 218 ± 80 mmol m^{-2} s^{-1} (Figure 3.11). There was evidence of photoinhibition at 100% light levels on most sampling occasions. The photosynthetic rate was also determined using a PHYTOPAM (pulse-amplitude modulation) device and a I_K for photosynthetic yield was found to be 105 ± 63 mmol m^{-2} s^{-1}. The tolerance of the algal community to low light conditions is consistent with laboratory studies for *C. raciborskii*, although this species had an even lower I_K (22-26 mmol m^{-2} s^{-1}) than the general algal community (Shafik et al. 2001).

Mean areal primary productivity rates in North Pine Dam were 0.71 ± 0.55 g m^{-2} d^{-1} and chlorophyll *a* concentrations were 11.53 ± 3.19 µg L^{-1} (Table 3.4).

Table 3.4
Primary productivity and photosynthetic activity for phytoplankton at the dam wall site in North Pine Dam. P_{max} is max primary productivity, α is slope of max productivity, Chl is chlorophyll a.

Parameter	Mean	SD
Depth-int. prim. prod. (g m^{-2} d^{-1})	0.71	0.55
P_{max} (g m^{-3} d^{-1})	0.33	0.20
α	0.0005	0.0003
Chl spec. P_{max} (g m^{-3} d^{-1} mg chl^{-1})	0.126	0.172
Light comp. point (µmol m^{-2} s^{-1})	218	80
Yield	0.547	0.067
PHYTOPAM light comp. point (mmol m^{-2} s^{-1})	105	63
Chlorophyll *a* (µg L^{-1})	12.38	3.82
C. raciborskii cell density (mL^{-1})	17,448	117,412
Total phytoplankton density (mL^{-1})	117,412	45,068
Euphotic depth (m)	4.0	1.1

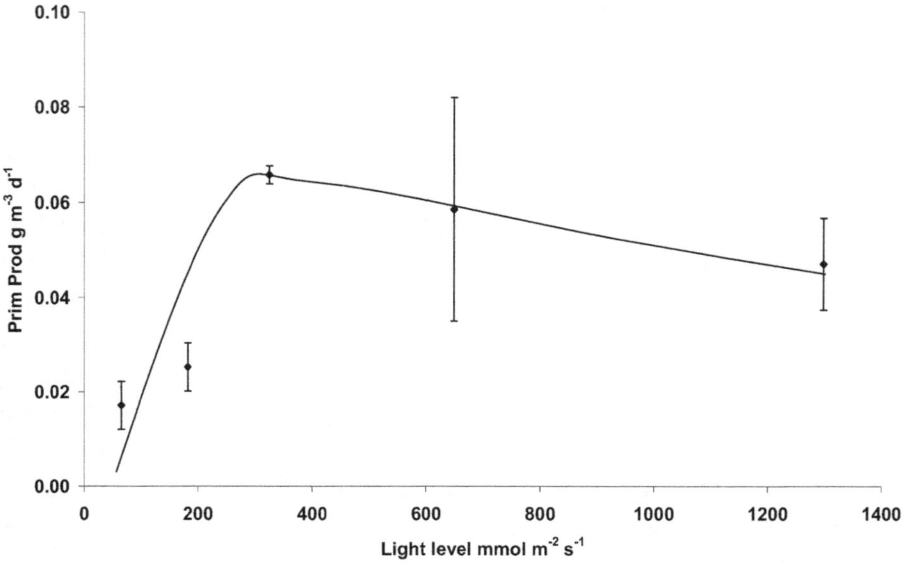

Figure 3.11 Example of primary productivity (g C m^{-3} d^{-1}) versus light (mmol m^{-2} s^{-1}) curve showing the light compensation point at the point of inflexion of the curve.

There was no evidence that areal primary productivity rates were lower in winter than (water temperatures 19°C) than in summer (28°C) during the incubations (Figure 3.12). However, areal primary productivity rates were correlated with the euphotic depth (R^2= 0.75) (Figure 3.13). This suggests that light, rather than temperature, was the key factor affecting primary productivity rates in the reservoir.

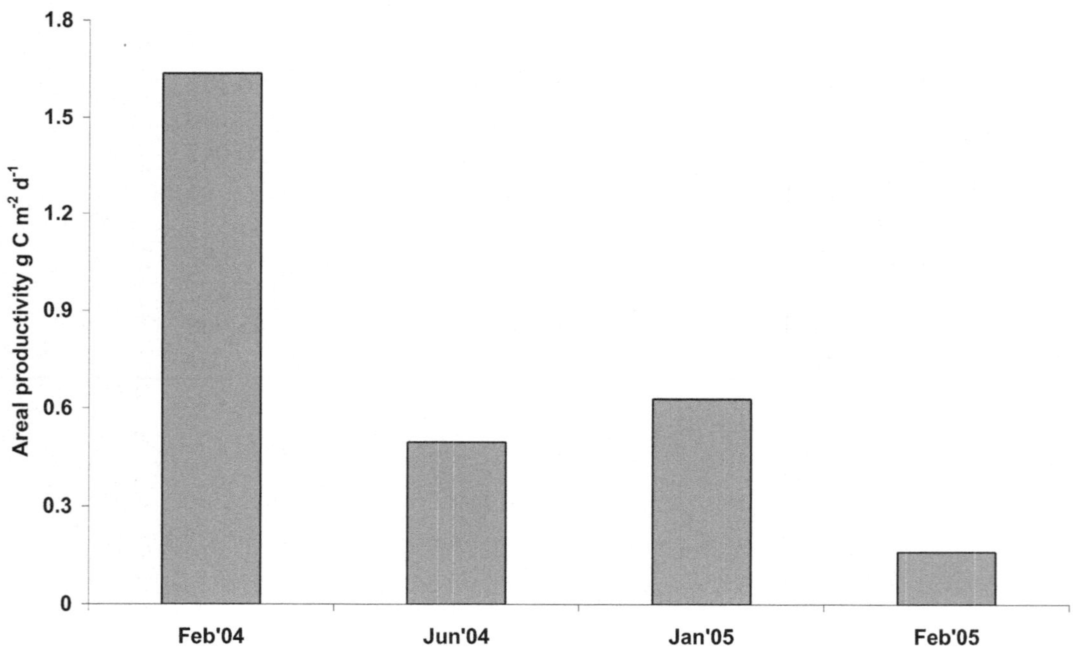

Figure 3.12 Depth-integrated areal primary productivity (g C m^{-2} d^{-1}) over the course of the study at the dam wall site at North Pine dam.

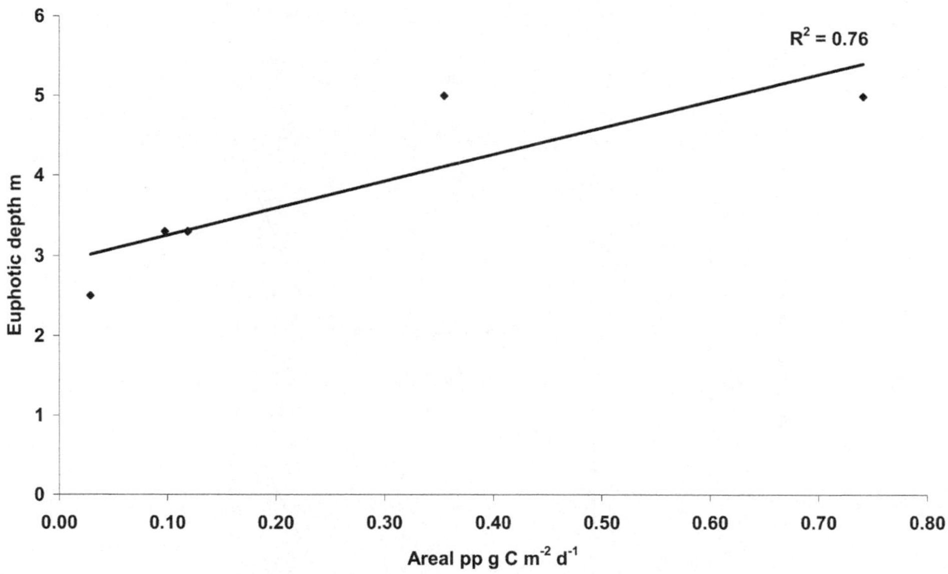

Figure 3.13 Correlation between depth-integrated areal primary productivity (g C m^{-2} d^{-1}) and euphotic depth (m) at the dam wall site at North Pine dam.

Nutrients are another factor that could be limiting primary productivity and the effect of dissolved inorganic nitrogen will be examined in more detail in the next section. An additional experiment was done in summer 2005 to determine if nutrient addition (phosphate and ammonium) would increase primary productivity rates, as measured by the ^{14}C- method. There was no evidence of an increase in primary production rates with nutrient addition (Table 3.5). In fact rates were lower than the control.

Table 3.5
Effect of nutrient addition on algal growth at North Pine Dam in February 2005.

Treatment	Mean primary productivity (g C m^{-3} d^{-1})	sd	n
Control	0.025	0.001	3
Phosphate addition	0.016	0.006	3
Nitrate/ammonium addition	0.017	0.002	3
N + P addition	0.016	0.003	3

Effects of nitrogen on phytoplankton growth

Maximum ammonium uptake rates ranged from 0.17 to 4.66 µM h^{-1} over the year of the study and there appeared to be a seasonal effect with lower rates in winter months (Figure 3.14). Nitrate uptake rates were lower, 0.04 to 0.40 µM h^{-1} although over winter and spring rates were similar to ammonium uptake rates. Both ammonium and nitrate uptake rates were lower in the dark than the light (39 and 88% respectively of uptake rates in the light).

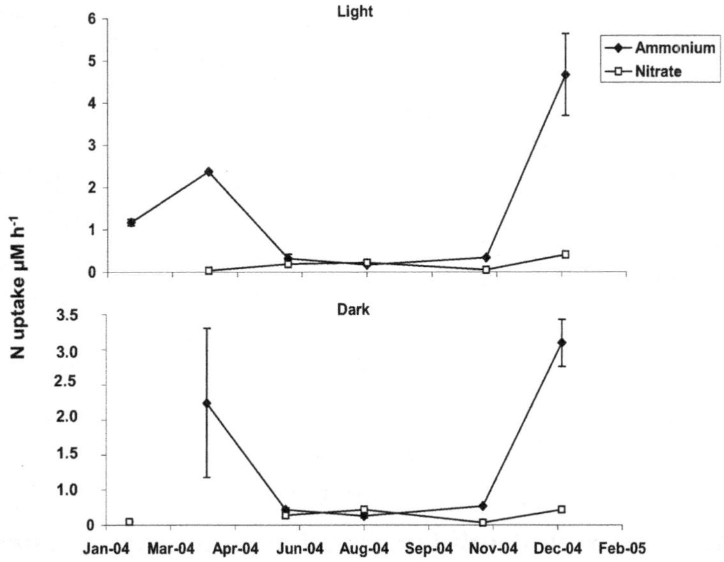

Figure 3.14 Ammonium and nitrate uptake (µM h^{-1}) at 100% light and dark from February 2004 to January 2005.

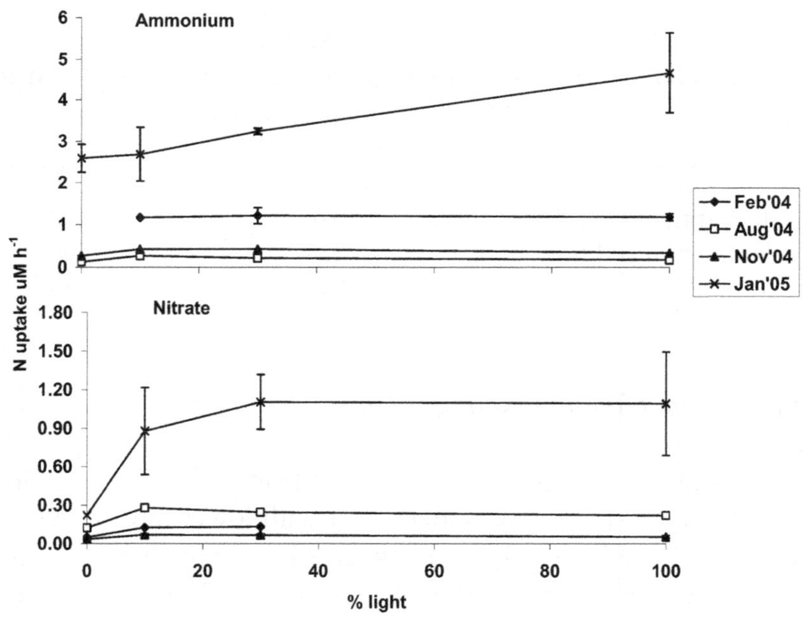

Figure 3.15 Ammonium and nitrate uptake (µM h^{-1}) over a range of light levels (% of surface light) in February, August, November 2004 and January 2005.

In February, August, November 2004, and January 2005, the effect of a range of light levels on nitrate and ammonium uptake was determined (Figure 3.15). For both nutrients, there was generally little change in uptake rates over a range of light levels, from 10 to 100%. This contrasts with the primary productivity vs. irradiance curves which showed a maximum primary productivity rate between about 220 to 500 mmol m^{-2} s^{-1} (10-30% of surface light), with lower rates at higher and lower light intensities.

The mean turnover rate for ammonium in the water column was 1.5 h, reflecting low background concentrations of ammonium coupled with a relatively high phytoplankton uptake (Table 3.6). In contrast, the mean nitrate turnover rate was 59 h, reflecting both a relatively high background concentration of nitrate/nitrite coupled with lower uptake rates. Ammonium was the preferred source, as determined by the relative preference indices for ammonium and nitrate (a higher value denotes a higher level of preference).

Table 3.6

Background ammonium and nitrate/nitrite concentrations (μM), turnover of ammonium and nitrate (h^{-1}) by phytoplankton, and the relative preference indices (RPI) for ammonium and nitrate. *Nitrate/nitrate concentrations for this month were inferred from mean April data for North Pine Dam for the last 5 y.

Month	NH_4 conc (uM)	NO_3 conc (uM)	NH_4 turnover (h^{-1})	NO_3 turnover (h^{-1})	RPI NH_4	RPI NO_3
Feb'04	6.71	3.71	5.7	-	-	-
Apr'04	1.93	8.57*	0.8	218.0	5.3	0.1
Jun'04	0.28	8.57	0.9	45.8	19.9	0.4
Aug'04	0.14	3.50	0.8	15.8	11.3	0.6
Nov'04	0.12	0.64	0.6	12.2	5.5	0.2
Jan'05	1.00	4.21	0.21	3.8	4.8	0.1

Nitrogen fixation rates were also measured five times over the year however values were generally below the detection limit of the method. Highest rates were in August and November 2004 but did not exceed 0.03 μM h^{-1}. This coincided with a higher proportion of heterocysts on the trichomes in November but not August 2004 (Table 3.7). Rates were generally <0.005 μM h^{-1}. These rates are low compared with ammonium and nitrate uptake rates. Therefore, phytoplankton in North Pine Dam during this study preferentially used ammonium, followed by nitrate, followed by nitrogen fixation. This is consistent with a study of *C. raciborskii* in culture which showed the same order of preferences (Sprőber et al. 2003).

Table 3.7

Density of C. raciborskii heterocysts and trichomes (# mL^{-1}) at the dam wall site in North Pine Dam

Date	Trichomes	Heterocysts	% heterocysts
7 Apr '04	<100	0	0
9 Jun '04	<100	0	0
26 Aug '04	<100	0	0
10 Nov '04	111	33	30

Culturing studies

The influence of light intensity on *C. raciborskii* growth is expressed as the change in trichome density over time (Figure 3.16) and growth rate versus light dose (Figure 3.17). When placed in darkness, *C. raciborskii* maintained a relatively constant density over the first 28 days ranging between 11,167 mL^{-1} to 30,515 mL^{-1}. After 28 days, trichome density decreased and on day 44, the density approximated 450 mL^{-1}. Positive growth was recorded in the 10-30 μmol m^{-2} s^{-1} light treatments for the entire 44 days. The maximum growth rate of 0.179 d^{-1} was obtained at 30 μmol m^{-2} s^{-1} or the equivalent of 1.3 mol m^{-2} d^{-1}. At light intensities of 70-475 μmol m^{-2} s^{-1}, positive growth was recorded only in the first 6 days (0.161-0.185 d^{-1}).

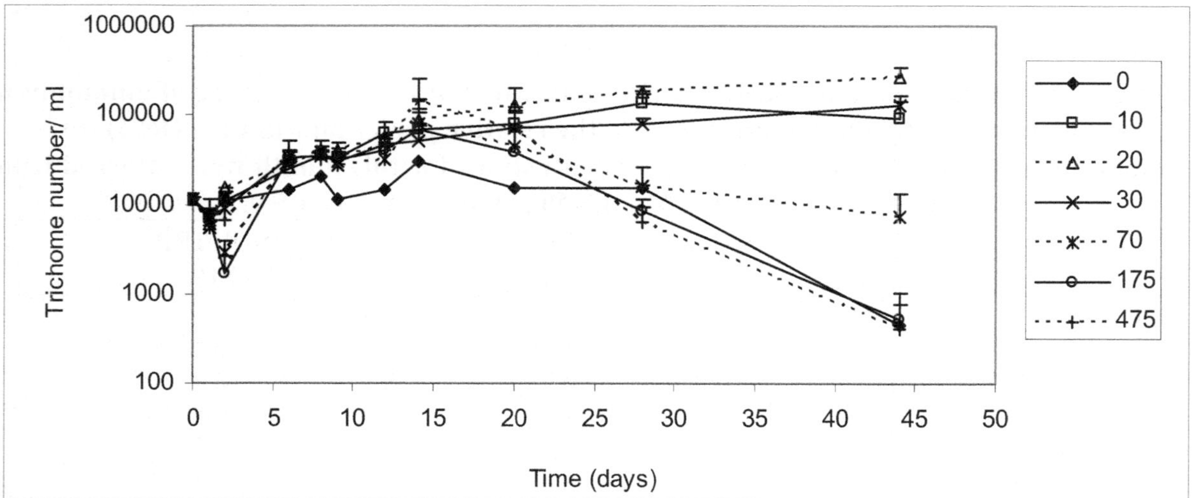

Figure 3.16 Changes in Cylindrospermopsis density over time at different light intensities (μmol m^{-2} s^{-1}).

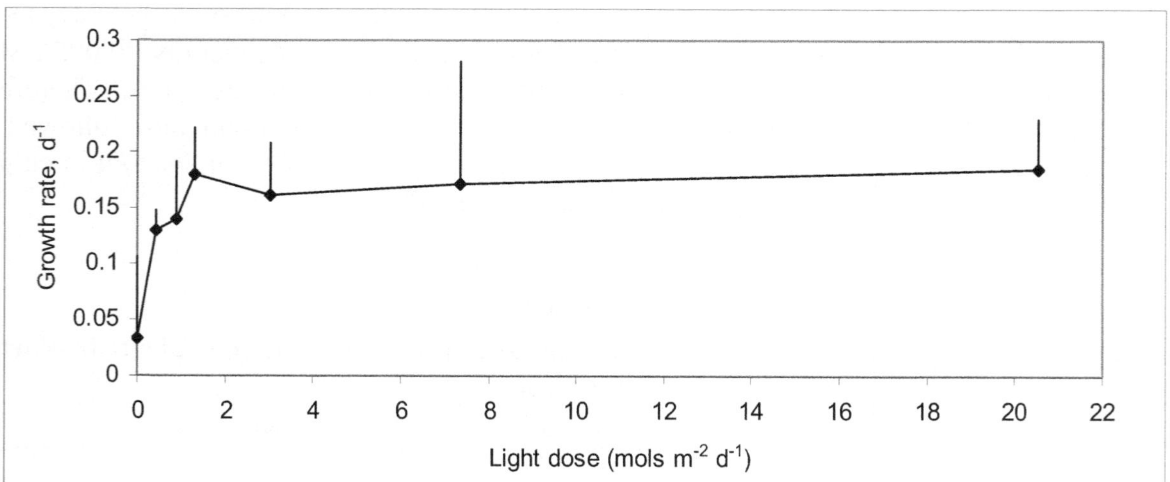

Figure 3.17 Cylindrospermopsis growth rate light dose curve. Growth rate was determined by changes in trichome number over 0-6 days from data in Figure 3.16.

C. raciborskii trichome length decreased in all treatments over time but the effect was greatest in at the highest light intensities (Figure 3.18). These also treatments had the lowest growth rates.

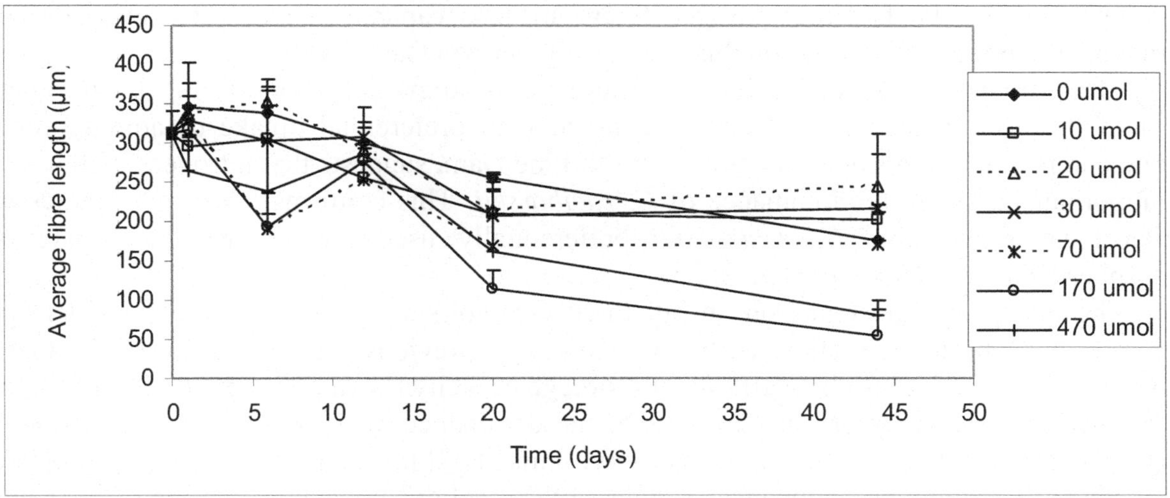

Figure 3.18 Changes in Cylindrospermopsis trichome length over time at different light intensities (µmol m^{-2} s^{-1}).

DISCUSSION

North Pine Dam is typically dominated by cyanobacteria with *Aphanocapsa/Merismopedia/Cyanodictyon*, *Planktolyngbya* and *C. raciborskii* representing about 80% of the algal numbers. Blooms of *C. raciborskii* occurred every summer in the five years examined in this study in both the artificially mixed reservoir, North Pine Dam and the adjacent naturally mixed reservoirs, Wivenhoe and Somerset. North Pine, Somerset and Wivenhoe Dams have similar temperature and rainfall patterns, with wet warm summers and dry mild winters. Wivenhoe and Somerset Dams can be characterized as monomictic with stratification during the summer months and turnover of the water mass in autumn, reflected in the lack of thermal stratification. North Pine Dam was also a monomictic system until the establishment of the destratification unit in 1995 but continued to have a thermal and oxygen gradient (Antenucci et al. 2005).

The dominance of *C. raciborskii* in the three water reservoirs in this study is consistent with many other water reservoirs in tropical and subtropical Australia, and Brazil (Bouvy et al. 2000, McGregor and Fabbro 2000). In addition, there is evidence that *C. raciborskii* is becoming established in an increasing number of waterbodies, including in temperate regions (Padisák 1997; Saker et al. 2003; Ryan, Hamilton and Barnes 2003). A number of factors have been suggested as possible causes of *C. raciborskii* dominance in reservoirs, including nitrogen fixing capacity, a high phosphate affinity and intracellular storage, the presence of a thermally stratified water column, water temperatures > 25 C, shade tolerance, long hydraulic residence times, and resistance to grazing (Padisák 1997, McGregor and Fabbro 2000).

The issue of phosphorus versus nitrogen limitation to algal growth in freshwater systems is complex and it is not possible to generalise about which nutrient is likely to be limiting in a particular waterbody (Oliver and Ganf 2000). In this respect, the use of nitrogen:phosphorus ratios to infer limitation is largely outdated (Downing, Watson and McCauley 2001). Most often nitrogen plus phosphorus addition is more likely to stimulate algal growth than either nutrient alone. The situation becomes more complex when considering nitrogen-fixing cyanobacteria, which are not reliant on a dissolved form of nitrogen for growth. It this case, it would be assumed that these organisms are phosphorus limited rather than nitrogen limited. However, this is complicated by the fact that nitrogen fixers are also capable of using dissolved forms of nitrogen, and indeed grow faster on this sources (Oliver and Ganf 2000).

The presence of detectable levels of nitrate and ammonium (>0.005 mg L^{-1})throughout most of the year in the three reservoirs coupled with preferential uptake of ammonium and nitrate compared with nitrogen fixation in North Pine Dam, suggests that nitrogen fixation is not an important factor in the dominance of *C. raciborskii*. This confirms laboratory studies with which have shown that *C. raciborskii* preferentially used ammonium and nitrate over atmospheric nitrogen (Sprőber et al. 2003).

The phosphate concentrations in all three reservoirs were near the limits of detection, therefore phosphorus limitation is likely. However, previous studies have shown that *C. raciborskii* has a high phosphate affinity and storage capacity (Istvànovics et al. 2000) compared with other species. Therefore, it may be that the dominance of *C. raciborskii* is explained by their enhanced adaptation to low phosphate environments. If this is the case, phosphate additions may reduce *C. raciborskii* dominance. Studies of the algal dynamics in North Pine Dam suggest that *C. raciborskii* gained a competitive advantage from low phosphate availability prior to the implementation of the destratification unit but light became a more important factor once the unit was operational (Antenucci et al. 2005).

Phytoplankton in North Pine Dam were preadapted to the low light levels and became photoinhibited at high light levels. Additionally culture studies with *C. raciborskii* showed that maximum growth occurred at low light levels, i.e. 30 µmol m^{-2} s^{-1}. Shafik et al. (2001) also found a maximal growth rate at 30 µmol m^{-2} s^{-1} and growth inhibition at 200 µmol m^{-2} s^{-1}. The primary productivity rates (0.71 ± 0.55 g m^{-2} d^{-1}) and chlorophyll *a* concentrations (11.53 ± 3.19 µg L^{-1}) were comparable with a number of reservoirs in USA and a reservoir in Hungary (Knoll, Vanni and Renwick 2003; Présing et al. 2001). Yield values were consistent with studies in other Australian reservoirs and reflect an actively photosynthesizing algal community (Brookes, Regel and Ganf 2003; Oliver et al. 2003).

C. raciborskii tolerated prolonged darkness for up to 28 days. This is substantially longer than *Microcystis aeruginosa* which decreased to 1 % of its initial cell number after 20 days (Furusato, Asaeda and Manatunge 2004). Fallon and Brock (1979) also found that after 10 days of darkness, *Aphanizomenon*, *Anabaena* and *Phormidium* decreased and reached a minimum corresponding to 1 % of initial abundance.

The positive growth rates of *C. raciborskii* in low light and its tolerance to prolonged darkness suggest that the use of artificial mixing to control *C. raciborskii* may be difficult and depend highly on the depth of light penetration and the achievable mixing depth. The laboratory growth results indicate that growth would be reduced at irradiances above 3 mol m^{-2} d^{-1} which is the equivalent of Z_{eu}/Z_{mix} ratios of 1.3-2 (Table 3.8).

Conversely, a *C. raciborskii* trichome would experience a low photon dose when the euphotic depth is shallow and the mixed depth is deep. Considering a euphotic depth of 4-5 m and a mixed depth of 7 to 10 m, the daily photon dose a theoretical trichome circulating through the layer would experience, approximates 0.86 to 0.26 mol m^{-2} d^{-1}. This value is in the range for positive growth observed in cultures grown at 10-30 µmol m^{-2} s^{-1}. For artificial mixing to decrease the photon dose below that of 10 µmol m^{-2} s^{-1} or 0.26 mol m^{-2} d^{-1}, the Z_{eu}/Z_{mix} (e.g. 4 m/10m) ratio would have to be smaller than 0.4. Furthermore, considering the tolerance of *C.*

raciborskii to prolonged darkness, the mixing would have to be persistent and able to overcome strong stratification.

An examination of the relationship between mixed depth and secchi disc depth (Z_s = 3x Z_{eu}) and cyanobacterial dominance indicated that *Cylindrospermopsis* prefers regimes where Z_{mix}/Z_s equivocal to a Z_{eu}/Z_{mix} ratio of 0.33 (Dokulil and Teubner 2000). Bormans et al. (2004) also found that cyanobacteria including *C. raciborskii* in the Fitzroy River, Australia, were light limited when the Z_{eu}/Z_{mix} ratio was between 0.22-0.33. After flooding, stratification enabled the settling of particles, which decreased turbidity and increased Z_{eu}. The Z_{eu}/Z_{mix} ratio increased to 1 enabling the growth of cyanobacteria and *C. raciborskii* to dominate for several months until the next flooding event.

Table 3.8

Modelled mean daily light dose (mol m^{-2} d^{-1}) experienced by phytoplankton circulating through various mixed depths in North Pine Dam with different euphotic depths and a sub-surface light dose of 11.93 mol m^{-2} d^{-1}. Shaded light grey cells represent conditions where *C. raciborskii* would be negatively buoyant. White cells represent conditions where *C. raciborskii* would display positive growth.

	Euphotic depth (m)					
Mixed depth (m)	1	2	3	4	5	6
1	2.56	5.16	7.77	10.36	12.95	15.54
2	1.29	2.46	3.79	5.11	6.42	7.71
3	0.86	1.49	2.33	3.22	4.09	4.96
4	0.65	1.00	1.53	2.14	2.79	3.43
5	0.52	0.72	1.04	1.45	1.91	2.39
6	0.43	0.54	0.73	0.98	1.29	1.62
7	0.37	0.42	0.53	0.68	0.86	1.07
8	0.32	0.34	0.39	0.47	0.57	0.69
9	0.29	0.28	0.30	0.34	0.38	0.44
10	0.26	0.23	0.23	0.24	0.26	0.28

C. raciborskii trichome length decreased by approximately 50% after 14 days in response to light intensity. Differences in the length of trichomes have also been reported for *Planktothrix agardhii* at different depths in a shallow fishpond (Hašler and Poulíčková 2003). Long filaments dominated the surface with an average length of 167 µm whereas at 50 cm depth they were 150 µm and at the bottom which was only 1 m deep they were 135 µm in length.

The size of an algal cell or colony or trichome has implications on its floating and sinking velocity through the water column. Following from Stokes Law, the velocity of a particle, $v = 2gr^2(\rho'-\rho)9\eta\theta$ where g = gravitational acceleration, r = radius of a sphere of volume equivalent to that of the organism, ρ' = density of the organism, ρ = density of the medium, η = viscosity of the medium, θ = form of resistance defined as v/v_s where v_s is the terminal velocity of the a sphere of equal volume and density of the organism. Therefore, for *C. raciborskii* a 50 % difference in trichome length will influence its radius and floating velocity. Trichome length was longer in low light and consequently trichomes would have a faster floating or sinking velocity. Along with buoyancy regulation this may act to move trichomes away from high light to low light or conversely from darkness to low light. Brookes et al. (1999) reported that the greatest axial linear dimension (GALD) of *Anabaena circinalis* suspended in bottles in Chaffey Dam

(New South Wales, Australia) increased with decreasing irradiance from 70 to 30 to 1 % of surface irradiance.

In conclusion, the examination of the historical water quality data in North Pine Dam coupled with measures of *in situ* processes and laboratory studies highlighted factors which appear to play the most important role in *C. raciborskii* dominance in North Pine Dam – the ability to grow under low light conditions and regulate their buoyancy to achieve this outcome, and the ability to utilize low phosphorus concentrations.

CHAPTER 4: BIODEGRADATION OF THE CYANOTOXIN M-LR IN A SAND FILTER AND NATURAL WATERS AND ASSOCIATED DETECTION OF A MOLECULAR MARKER (*MLRA* GENE) FOR DEGRADATION

INTRODUCTION

Extracellular microcystins are chemically stable in water (Jones and Orr 1994, Tsuji et al. 1994) and have been documented to be recalcitrant to conventional water treatment processes of flocculation and filtration (Hoffman 1976, Keijola et al. 1988, Himberg et al. 1989, Lahti and Hiisvirta 1989). Activated carbon adsorption and ozone oxidation have been shown to be successful in their mitigation from drinking water (Jones et al. 1993, Rositano et al. 2001, Newcombe et al. 2003, Ho 2004). However, the presence of natural organic material (NOM) reduces the effectiveness of both of these treatment processes for the removal of microcystin (Lambert et al. 1996, Rositano et al. 2001, Newcombe et al. 2003, Ho 2004). For activated carbon, NOM decreases the adsorption capacity for these metabolites through competitive adsorption and/or pore blockage mechanisms, while NOM can also act to consume ozone thereby reducing its oxidation potential.

Biological filtration systems for the removal of problematic contaminants, such as microcystin toxins, are becoming more attractive to water suppliers as they are generally of low technology, requiring little maintenance and infrastructure. In addition, biological filtration systems are able to remove contaminants without the addition of other chemicals that may have the potential to produce undesirable by-products such as organochlorides.

Many studies have reported biological degradation of microcystin in natural lakes and reservoirs (Jones and Orr 1994; Rapala et al. 1994; Cousins et al. 1996; Christoffersen, Lyck and Winding 2002). Only a few studies have implicated degradation of microcystin in biologically active sand filters. Lahti and Hiisvirta (1989) found up to 86 % removal of microcystin was achieved in a pilot scale slow sand filter. Studies by Sherman, Tully and Gibson (1995) and Grützmacher et al. (2002) also implicated biological degradation of microcystin in slow sand filters. However, these studies were unable to conclusively prove that the removal of microcystin was through biological degradation rather than physical processes, such as adsorption to sand. Furthermore, only slow sand filtration studies have been documented and to date no removal of microcystin has been reported under rapid sand filtration conditions.

Perhaps the most important aspect of any biological degradation processes is the identification of the degrading bacteria. Only a few strains of the genus *Sphingomonas* have been reported to degrade microcystin (Bourne et al. 1996; Bourne et al. 2001; Park et al. 2001; Saito et al. 2003; Harada et al. 2004; Ishii, Nishijima and Abe 2004). This diminutive population of bacteria responsible for the degradation of microcystin suggests that specific genes are involved in the degradation processes. Bourne et al. (1996, 2001) identified a gene cluster in a *Sphingomonas* species which was responsible for the degradation of microcystin-LR. Recently, Saito et al. (2003) developed a polymerase chain reaction (PCR) test for the detection of one of these microcystin degrading genes, *mlrA*. This gene is important as it encodes the enzyme which is responsible for cleaving the cyclic structure of microcystin.

The objective of this study was to determine whether biologically active sand filters were capable of effectively removing microcystin toxins under slow and rapid sand filtration conditions. Furthermore, an additional objective was to confirm that the removal was primarily through biological degradation rather than any physical processes. This will be conducted using molecular techniques, such as the PCR developed by Saito et al. (2003), to detect the microcystin degrading gene *mlrA*. A second objective was to examine the potential for microcystin degradation within a range of natural water bodies that experience *Microcystis* blooms. This was firstly examined by determining degradation rates of microcystin in two water sources. It was also examined by screening a wide range of waters for the presence of the microcystin-degrading gene, *mlrA*. It was considered that the presence of this gene would provide strong evidence for the presence of microorganisms that have the ability to degrade microcystin toxins. All water sources chosen were experiencing a bloom of *Microcystis* sp. at the time of this study, or had previously experienced *Microcystis* blooms.

METHODS

Materials and reagents

All reagents used were analytical grade or better unless otherwise stated. Solutions were prepared using Milli-Q water (Millipore Pty Ltd, USA) and autoclaved (121°C for 15 minutes) when possible, otherwise filter sterilised through 0.2 μm cellulose nitrate membranes. All glassware and equipment was autoclaved (121°C for 15 minutes).

Microcystin-LR (m-LR) and -LA (m-LA) were isolated from a natural bloom of *Microcystis aeruginosa* that occurred in the Torrens Lake in South Australia during the summer of 1998-99. The isolation procedure involved freeze-thawing the bloom material in water and methanol, followed by preparative reverse phase flash chromatography and preparative high performance liquid chromatography (HPLC).

Sample water was collected from the Myponga Reservoir, South Australia. Sand was obtained from the filter beds at the Morgan Water Treatment Plant (WTP) where no pre-chlorination is practiced.

Laboratory column experiments

Morgan WTP filter sand was packed into three separate glass columns. Each column was fed with Myponga Reservoir water (MRW) spiked with m-LR and m-LA at target concentrations of 20 μg/L each. Samples were taken in duplicate at regular intervals for microcystin analysis. The conditions of the column experiments are listed in Table 4.1. Experiments were conducted at room temperature (20±2°C).

Table 4.1
Laboratory column conditions

Column Conditions	Parameters
Internal Diameter	2.5 cm
Sand Bed Height	15 cm
Empty Bed Contact Time (EBCT)	7.5-30 minutes
Flow Rate	2.5-10 mL/min

At the completion of the column experiments, sand was removed from a column (where complete removal of microcystin was observed) and the biofilm detached by periodic vortexing for 15 minutes. The supernatant, containing the biofilm, was collected and any carryover of sand was pelleted by slow speed centrifugation at 1,000 x g for 30 seconds.

Toxin degradation in natural waters

Toxin degradation rates were assessed by conducting batch experiments. For the purposes of this study, two microcystin analogues were investigated; m-LR and m-LA. Approximately 10 L of raw water was obtained from the Torrens Lake, South Australia, which receives a high organic loading, and approximately 5 L was obtained from the Myponga Reservoir, South Australia. Half of the Torrens Lake water was sterilised by filtration through 0.2μm filters. Unsterilised Torrens Lake water, sterilised Torrens Lake water and unsterilised Myponga Reservoir water were added to three bottles. These bottles were spiked with toxin extracted from a natural bloom of *Microcystis aeruginosa* from a lake in Gippsland, Victoria, Australia. Initial samples (90 mL) were collected for determination of m-LR and m-LA concentrations, with each treatment having initial concentrations of approximately 20 $\mu g\ L^{-1}$. In order to determine degradation rates, m-LR/LA concentrations were monitored 1-2 times each day over a period of 10 days. This involved collecting 90 mL samples from each batch and analysing them for m-LR/LA using HPLC.

Presence of mlrA in natural waters

Water samples were collected by various groups from a range of sources throughout Australia that contained *Microcystis* blooms. These sources were:

- Three samples from Gippsland, Victoria, supplied by SSG Environmental Services
- Two sites in Gum Lagoon, Victoria, supplied by Goulburn-Murray
- Site 12 - Abattoir
- Site 13 - Outfall
- Chichester Dam in Hunter Valley, New South Wales, supplied by Hunter Water
- Torrens Lake in South Australia
- Two sites within Sydney Brick Pit, New South Wales, supplied by Sydney Water

DNA from microorganisms within the waters was extracted by freeze-thawing the cells. The extracted DNA was then used as a template in the PCR analysis, targeting the *mlrA* gene using two specific oligonucleotide primers, MF and MR (Table 4.2) (Saito et al. 2003). Dilutions of the DNA of each sample were also analysed using the same PCR (1:10 and 1:100). This was conducted because high concentrations of DNA have been shown to inhibit PCR tests.

Analysis of microcystin

Prior to HPLC analysis, microcystins were concentrated from sample waters by C18 solid phase extraction according to the methods described by Nicholson, Rositano and Burch (1994). A HPLC system consisting of a 600 pump controller, 717plus autosampler and 996 photodiode array detector (Waters Pty Ltd, Australia) was employed. Sample volumes of 50 μL were injected into a 150 x 4.6 mm Luna C18 column (Phenomenex, Australia) at a flow rate of 1 mL/min. Two mobile phases were used for the gradient run (30 % acetonitrile/0.05 %

trifluoroacetic acid and 55 % acetonitrile/0.05 % trifluoroacetic acid). Concentrations of m-LR were determined by calibration of the peak areas (at 238 nm) with that of external standards (Sapphire Bioscience Pty Ltd, Australia). All m-LA concentrations were expressed in terms of m-LR equivalents. The HPLC method has a detection limit of 0.025 μg/L. Microcystin recoveries were >95 % with a relative precision of 10 %.

Molecular techniques

DNA, from the column biofilm, was extracted and prepared using the InstaGene™ Matrix DNA Preparation Kit according to the manufacturer's instructions (Bio-Rad, Australia). The DNA was then used as a template for a PCR test using two 20 base pair (bp) oligonucleotide primers, MF and MR (Table 4.2). These primers target the *mlrA* gene which has been shown to initiate the degradation of microcystin. The PCR test employed in this study was modified from one which was previously developed by Saito et al. (2003). Amplifications were performed in 50 μL volumes, containing 1 pmol of each primer, 2.5 mM $MgCl_2$ (Perkin Elmer, Australia), 1 x PCR Buffer II (Perkin Elmer, Australia), 200 μM deoxynucleoside triphosphates (dNTPs) (Roche Diagnostics Pty Ltd, Australia), 1.25 units of Ampli *Taq* Gold DNA Polymerase (Perkin Elmer, Australia) and 1-10 ng of template DNA. A GeneAmp® 2400 PCR System (Perkin Elmer, Australia) was used for the amplifications under the following conditions: 94°C for 10 minutes; 35 cycles of 94°C for 20 seconds, 60°C for 10 seconds and 72°C for 30 seconds; and 72°C for 10 minutes. Five μL of the PCR products were separated by gel electrophoresis with the resultant DNA bands visualised on a Dark Reader illuminator (Clare Chemical Research, USA).

Table 4.2
Oligonucleotide primer sequences

Primer	Sequence (5' – 3')
MF	GACCCGATGTTCAAGATACT
MR	CTCCTCCCACAAATCAGGAC

Source: "Reprinted from *Water Research*, 40, Ho, L., Meyn, T., Keegan, A., Hoefel, D., Bookes, J.D., Saint, C., and Newcomb, G., Detection of the *mlr*A Gene in a Biologically Active Sand Filter Responsible for the Degradation of Microsystin Toxins. 768-774, Copyright (2006), with permission from Elsevier"

The PCR products were cloned using the TOPO TA Cloning® Kit (Invitrogen™, USA) and the resulting plasmid DNA purified using the QIAprep® Miniprep Kit (QAIGEN, USA). The purified DNA then underwent a sequencing reaction using the Big Dye v2.0 Terminator Cycle Sequencing Kit (PE Applied Biosystems, Australia). Sequences of DNA were aligned using the SeqMan™ II and EditSeq™ software programs (DNASTAR, USA). Identification of the amplified DNA fragment was performed online using the Nucleotide-nucleotide BLAST® (Basic Local Alignment Search Tool) network service on the NCBI GenBank® database (http://www.ncbi.nlm.nih.gov/BLAST).

RESULTS AND DISCUSSION

Biological filtration of microcystin

Three laboratory sand columns were employed for the biological filtration of m-LR and m-LA. All three columns contained sand sourced from the filters at the Morgan WTP; however each were commissioned differently. Column A had been in operation for approximately 12 months prior to the commencement of this study, during which time it was continuously fed with Myponga Reservoir water (MRW) with periodic spikings of m-LR and m-LA. During this time column A was able to effectively remove both microcystin analogues (unpublished data). Prior to the commencement of this set of experiments, column A had not been exposed to either of the microcystin analogues for a period of 6 months. The purpose of this column was to determine whether it still had the ability to remove microcystin after a prolonged break from microcystin exposure. This simulates a realistic situation since microcystin episodes are usually of a transient nature.

Column B was commissioned at the commencement of this study using filter sand directly obtained from the Morgan WTP. Therefore, the sand had an existing biofilm which was not acclimatised to the new experimental regime. In contrast to columns A and B, the Morgan WTP sand in column C was autoclaved prior to the commencement of the experiments to inactivate the existing biofilm.

Each column was fed with MRW spiked with m-LR and m-LA at target concentrations of 20 μg/L each. However, different empty bed contact times (EBCTs) were employed for each column. For column C, a constant EBCT of 30 minutes was used for the 5 day experiment, whereas for columns A and B the EBCT was varied as shown in Table 4.3. Results of the biological filtration of microcystin through each of the columns are shown in Figures 4.1-4.3. Only m-LR results are shown as the trends for m-LA were identical. During the initial stages of the experiment it was difficult to maintain consistent microcystin influent concentrations due to degradation of the compounds in the influent water. This was resolved by cleaning the influent water reservoir and tubing at regular intervals to minimise any biological activity occurring prior to the sand filters.

Table 4.3
Empty bed contact times (EBCTs) employed for columns A and B

EBCT (minutes)	Time Employed (days)	
	Column A	Column B
30	0-6, 28-29, 35-39	0-20, 28-29, 35-36
15	7-20	21-27
10	21-27	30-34
7.5	30-34	37-39

Source: "Reprinted from *Water Research*, 40, Ho, L., Meyn, T., Keegan, A., Hoefel, D., Bookes, J.D., Saint, C., and Newcomb, G., Detection of the *mlr*A Gene in a Biologically Active Sand Filter Responsible for the Degradation of Microsystin Toxins. 768-774, Copyright (2006), with permission from Elsevier"

No microcystin was detected in the effluent of column A at any time during the experiment, even with the low EBCTs employed (Figure 4.1). Furthermore, no microcystin was detected in samples obtained from a sampling port approximately mid way up the column bed. Samples from this port effectively halve the EBCT of the sand bed; consequently, the lowest EBCT of approximately 4 minutes still yielded no toxin breakthrough. This low EBCT corresponds to a filtration rate of approximately 4 cm/min which is similar to conditions of a rapid sand filter. To date, no data has been published in reference to the degradation of microcystin under rapid sand filtration conditions. This is a significant finding as it indicates that removal of microcystin can occur through rapid sand filters, which are commonly used in Australian WTPs.

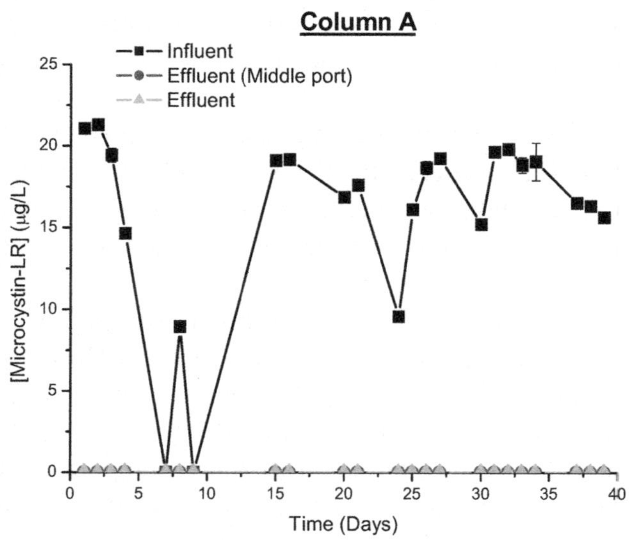

Figure 4.1 Biological filtration of microcystin-LR through laboratory sand column A. Source: "Reprinted from *Water Research*, 40, Ho, L., Meyn, T., Keegan, A., Hoefel, D., Bookes, J.D., Saint, C., and Newcomb, G., Detection of the *mlr*A Gene in a Biologically Active Sand Filter Responsible for the Degradation of Microsystin Toxins. 768-774, Copyright (2006), with permission from Elsevier"

The results from column A also imply that all the degradation of microcystin occurred in the top half of the column. This is because no microcystin was detected in samples taken from the middle port of the column at any time during the experiment. Similarly, Holst et al. (2003) found that the order of microcystin degradation in a sand infiltration basin was greater at the upper layers of the filter.

In contrast to column A, breakthrough of microcystin was observed in for only 3 days in the effluent of column B (Figure 4.2). However, after this point of time no microcystin was detected in the effluent for the duration of the experiment, even with the low EBCTs employed. Similar results were observed for column C which employed a constant EBCT of 30 minutes (Figure 4.3). The results suggest that the time required for a biofilm to establish and acclimatise to microcystin in this set of experiments is quite short. This is in contrast to other studies which have documented lag periods of up to 16 days prior to degradation of microcystin commencing (Miller and Hallowfield 2001). Furthermore, the initial breakthrough of microcystin in columns

B and C suggests that a majority of the removal of microcystin through the columns was through biological degradation and not any other physical processes such as adsorption to the sand or glass columns.

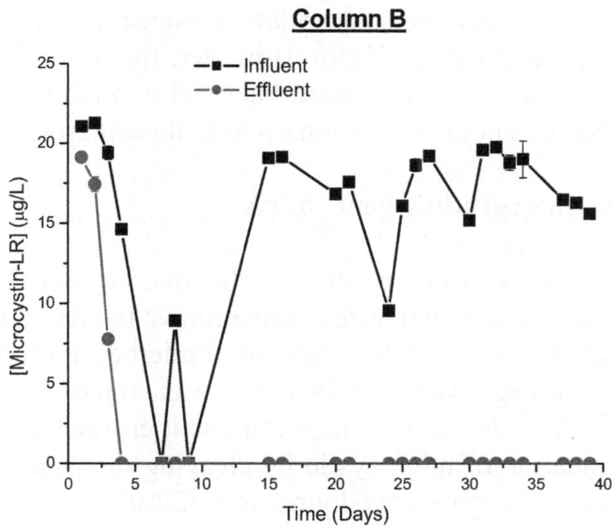

Figure 4.2 Biological filtration of microcystin-LR through laboratory sand column B.
Source: "Reprinted from *Water Research*, 40, Ho, L., Meyn, T., Keegan, A., Hoefel, D., Bookes, J.D., Saint, C., and Newcomb, G., Detection of the *mlr*A Gene in a Biologically Active Sand Filter Responsible for the Degradation of Microsystin Toxins. 768-774, Copyright (2006), with permission from Elsevier"

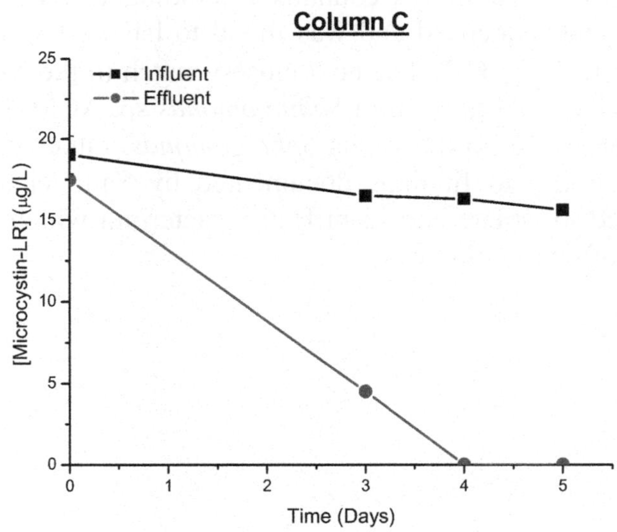

Figure 4.3 Biological filtration of microcystin-LR through laboratory sand column C.
Source: "Reprinted from *Water Research*, 40, Ho, L., Meyn, T., Keegan, A., Hoefel, D., Bookes, J.D., Saint, C., and Newcomb, G., Detection of the *mlr*A Gene in a Biologically Active Sand Filter Responsible for the Degradation of Microsystin Toxins. 768-774, Copyright (2006), with permission from Elsevier"

The immediate removal of microcystin in column A suggests that pre-exposure of the biofilm to microcystin was able to eliminate the lag period prior to degradation commencing, since breakthrough was initially observed in column B. Similarly, Rapala et al. (1994), Christoffersen, Lyck and Winding (2002) and Holst et al. (2003) have shown that lag periods in microcystin degradation studies could be substantially reduced, in some circumstances removed, when bacteria were pre-exposed to microcystin. However, the immediate removal in column A could also have been due to the more advanced bacterial community in column A, compared with that of column B which had not been acclimatised to the experimental conditions.

Detection of the microcystin degrading gene, mlrA

Molecular techniques were used to identify the microcystin degrading gene, *mlrA*, in column A which was previously shown to effectively remove microcystin. In addition, molecular techniques would conclusively prove that the removal of microcystin through the laboratory sand filters was through biological degradation. A PCR test was employed using the MF-MR primer set which target this gene. The *mlrA* gene is important as it encodes the enzyme which has been shown to initiate the degradation of microcystin by cleaving the Adda-arginine peptide bond in microcystin, opening up its cyclic structure (Bourne et al. 2001, Saito et al. 2003).

DNA was extracted from the biofilm in column A and used as a template in the PCR test. This results in a PCR product of approximately 800 bp in size. Figure 4.4 shows results of the *mlrA* detection by the PCR. Lane 2 corresponds to the DNA from *Sphingomonas* sp. ACM-3962. This was used as a positive control for the PCR test since the primers, MF and MR, were initially designed from the *mlrA* gene sequence from this bacterium (Saito et al. 2003). Lanes 3 and 4 correspond to bacteria which were previously shown to be closely related to *Sphingomonas* sp. MBIC4193 and *Sphingomonas* sp. KT-1 based on 99 % sequence similarities to NCBI accession numbers AB023290.1 and AB22601.1, respectively. A band was evident in lane 5 which indicates that the biofilm from column A contains a homologous *mlrA* gene. The PCR product from lane 5 was cloned and sequenced and was found to have a 98 % similarity to a putative *mlrA* gene from *Sphingopyxis* sp. C-1. The next closest match in the NCBI GenBank® database was a 91 % similarity to the *mlrA* gene from *Sphingomonas* sp. ACM-3962. These results imply that the *mlrA* gene is not unique to the genus *Sphingomonas*, rather it appears to be unique to microcystin degraders, similar to findings documented by Saito et al. (2003). Attempts are currently being conducted to isolate and identify the bacterium which contains this *mlrA* gene homologue within the biofilm of column A.

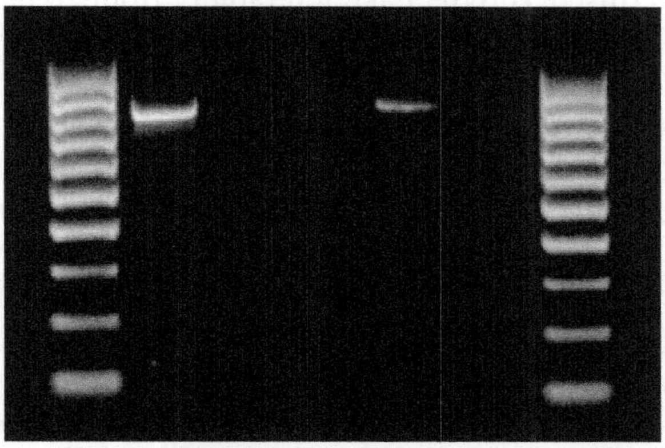

Figure 4.4 Detection of *mlrA* gene by PCR using the MF-MR primer set: Lane 1 - DNA reference marker (100 bp ladder); Lane 2 - *Sphingomonas* sp. ACM-3962; Lane 3 - *Sphingomonas* sp. MBIC4193; Lane 4 - *Sphingomonas* sp. KT-1; Lane 5 - Column A biofilm DNA; Lane 6 - Milli Q water; Lane 7 - DNA reference marker (100 bp ladder).
Source: "Reprinted from *Water Research*, 40, Ho, L., Meyn, T., Keegan, A., Hoefel, D., Bookes, J.D., Saint, C., and Newcomb, G., Detection of the *mlr*A Gene in a Biologically Active Sand Filter Responsible for the Degradation of Microsystin Toxins. 768-774, Copyright (2006), with permission from Elsevier"

To date, no other studies have reported detection of the *mlrA* gene in biological sand filters. The detection of the homologous *mlrA* gene in this study indicates that there is potential for this PCR test to be used to assess biologically active sand filters for their microcystin degrading capability. This would be a valuable tool to evaluate sand filters prior to an impending episode of microcystin.

Biological sand filtration was determined to be an effective treatment process for the complete removal of two microcystin analogues, m-LR and m-LA. Furthermore, complete removal of the microcystins was evident under rapid sand filtration conditions, which has not been previously reported.

The removal of microcystin through the sand filters was shown to be primarily through biological degradation by the detection of the *mlrA* gene, a microcystin degrading gene, in the biofilm of one of the filters. This is the first study to report detection of this gene in sand filters which were actively removing microcystin. Consequently, there is potential for this PCR test to be used to assess the microcystin degrading ability of biologically active sand filters.

Toxin degradation in natural waters

In the sterilised Torrens Lake water, both m-LR and m-LA dropped to approximately 15 $\mu g\ L^{-1}$ during the experimental period, but were approximately 20 $\mu g\ L^{-1}$ at the completion of the experiment (Figure 4.5). This indicated undetectable levels of degradation. In contrast, in unsterilised water sources (Torrens Lake and Myponga Reservoir) m-LR and m-LA concentrations fell with time indicating degradation of both analogues of microcystin. In unsterilised Torrens Lake water, m-LR concentrations fell rapidly from 18.8 $\mu g\ L^{-1}$ on day 0 to 11.2 $\mu g\ L^{-1}$ on day 2 (Figure 4.5). This degradation continued to day 5 when concentrations reached 3.6 $\mu g\ L^{-1}$ and concentrations were undetectable by day 7. Over the experimental period, the degradation rate of m-LR in Torrens Lake water was 0.411 day^{-1} (Table 4.4). The degradation rate of m-LA in unsterilised Torrens Lake water was slightly lower. As with m-LR concentrations of m-LA in unsterilised Torrens Lake fell rapidly from 20 μg L^{-1} on day 0 to 14.6 μg l^{-1} on day 2 (Figure 4.5). By day 5 concentrations had reached 4.8 μg L^{-1} and reached undetectable concentrations by day 7.

Although, there was no sterilised control Myponga Reservoir water, it appeared that there was degradation of m-LR and m-LA during the experimental period (Figure 4.5). Unlike Torrens Lake water, in Myponga Reservoir water there was little difference in m-LR and m-LA degradation rates (Table 4.4). In addition, the observed degradation rates were not as rapid as those in the Torrens Lake water (Table 4.2, Figure 4.5). At the beginning of the experiment, the m-LR concentration was 19.4 μg L^{-1} in Myponga Reservoir water. By day 2 it had fallen to 15.4 μg L^{-1} and was undetectable by day 8 (Figure 4.5). Microcystin-LA concentrations were initially 21.0 μg L^{-1} and fell to 16.9 μg L^{-1} by day 2 and had reached undetectable concentrations by day 8 (Figure 4.5).

Table 4.4
Microcystin-LR and microcystin-LA degradation rates in Torrens Lake and Myponga Reservoir water.

Toxin	Degradation rate (day^{-1})	
	Torrens water (unsterilised)	Myponga water
Microcystin-LR	0.411	0.302
Microcystin-LA	0.386	0.296

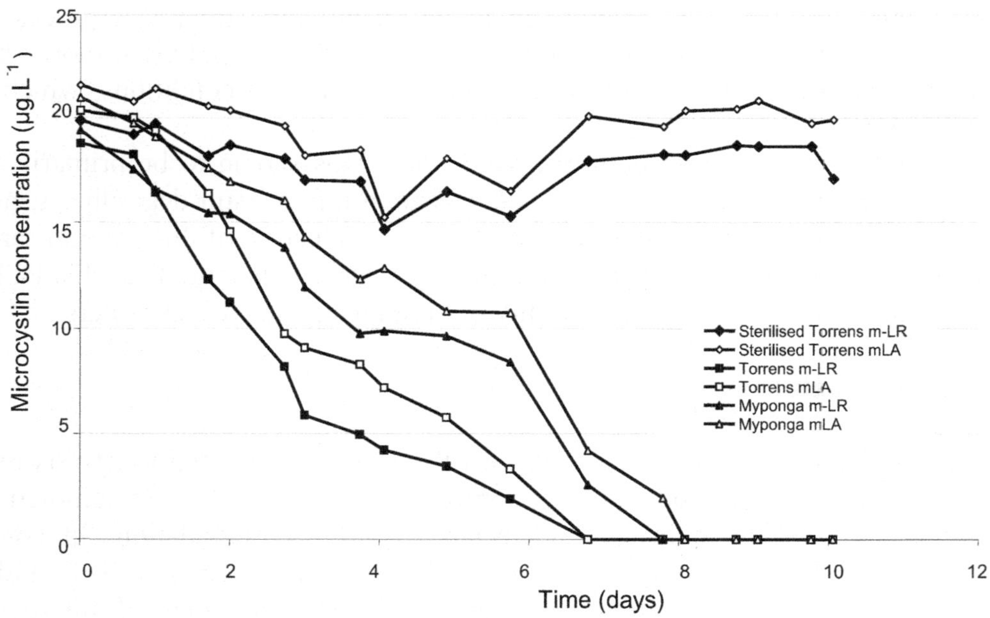

Figure 4.5 Changes in microcystin-LR and microcystin-LA concentrations over time in a batch reactors containing unsterilised and sterilised Torrens Lake water and Myponga Reservoir water.

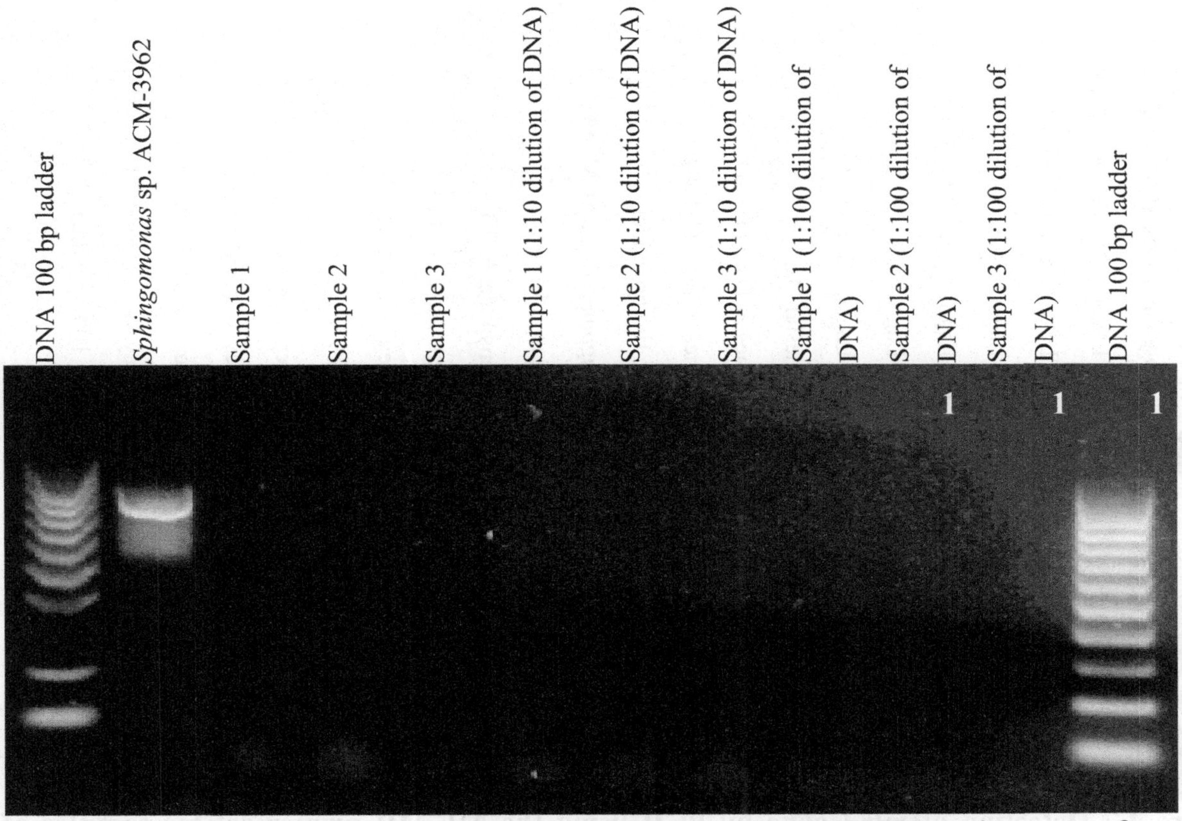

Figure 4.6 Detection of *mlrA* gene by PCR using the MF-MR primer set for three Gippsland-Murray water samples (3-11), including 1:10 and 1:100 dilutions with DNA 100 bp ladder used as a template and *Sphingomonas* sp. ACM-3962 used as a control.

Presence of mlrA in natural waters

In the PCR analyses a band was obtained for *Sphingomonas* sp. ACM-3962 (Figures 4.6 and 4.7), indicating that the PCR functioned properly. However, no bands were detected with the PCR for all samples collected from Gippsland (Figure 4.6), Gum Lagoon (Goulburn Murray), Chichester Dam (Hunter Valley) and Torrens Lake (Adelaide) (Figure 4.7) at any dilutions. This suggests that the *mlrA* gene was not present in these samples and that the microorganisms in these water bodies would be incapable of degrading microcystin toxins using the *mlrA* degradation mechanism. However, bands were evident in the 1:10 dilution of the Sydney Water Brick Pit 1 sample and both dilutions of the Sydney Water Brick Pit 2 sample (Figure 4.7). This suggests that *mlrA* was present within both Sydney Water Brick Pit samples and that these samples contained micrcoorganisms that have the ability to degrade microcystin toxins (using the *mlrA* degradation pathway).

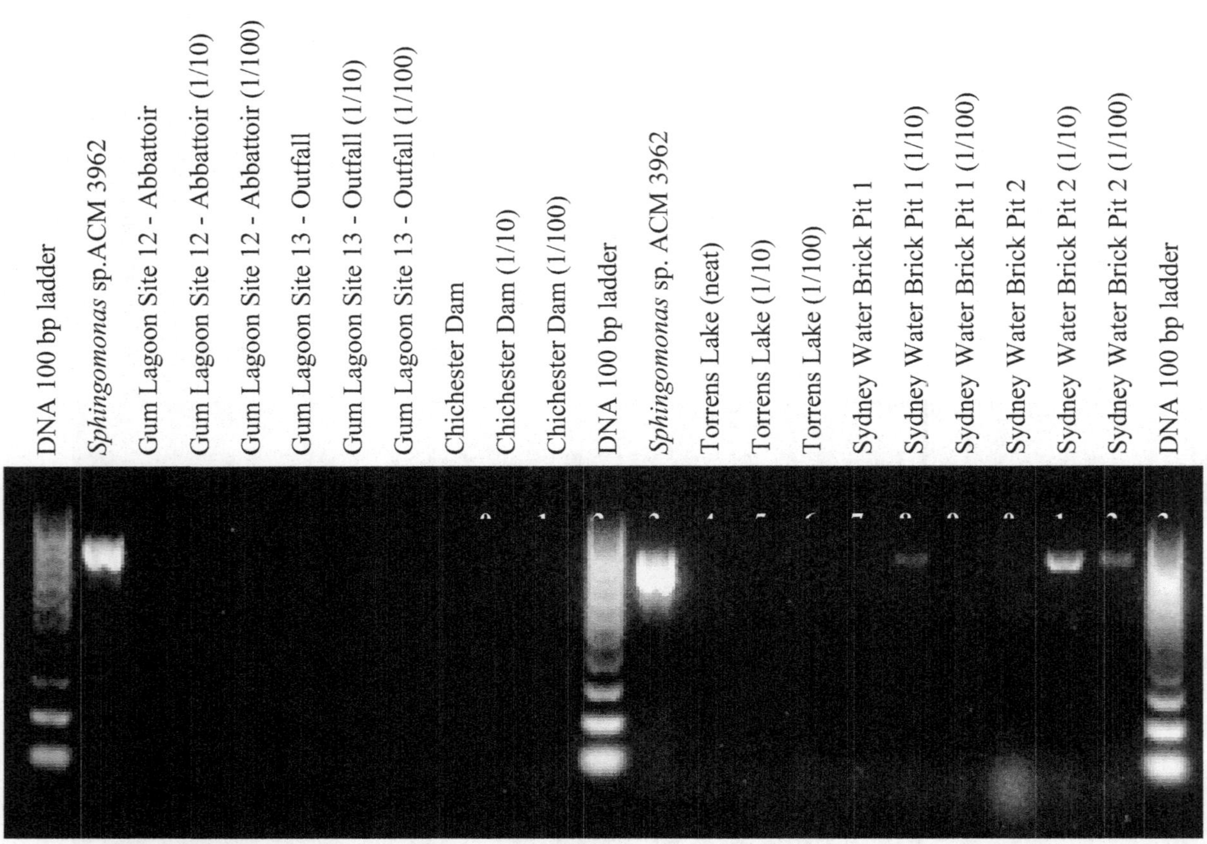

Figure 4.7 Detection of *mlrA* gene by PCR using the MF-MR primer set for samples of Gum Lagoon (Goulburn Murray), Chichester Dam (Hunter Valley), Sydney Water Brick Pit (Sydney) and Torrens Lake (Adelaide), including 1:10 and 1:100 dilutions, with DNA 100 bp ladder used as a template and *Sphingomonas* sp. ACM-3962 used as a control.

The results demonstrated that microcystin degradation within two water natural water sources that experience *Microcystis* blooms was reasonably rapid, with microcystin concentrations reaching undetectable levels after 6-8 days from an initial concentration of 20 μg L^{-1}. However, these rates were slower than those within sand-filters which degrade microcystin toxin from 20 μg L^{-1} to undetectable concentrations with 4 minutes (Ho et al. 2006).

Interestingly, although microcystin degradation occurred within a sample collected from the Torrens Lake, a gene that has been identified to be responsible for microcystin degradation, *mlrA*, was absent from these samples suggesting that there are other mechanisms or other genes responsible for microcystin degradation. In addition, *mlrA* was absent from a number of water samples that were collected from a number of Australian water sources that experience *Microcystis* blooms.

The presence of *mlrA* was detected in two samples from the Sydney Water Brick Pit, despite no microcystin toxins being detected by HPLC in either of these samples. It is unclear whether this was due to biodegradation of the toxins due to the presence of microorganisms that can degrade microcystin or that no microcystin was produced by the *Microcystis* blooms.

Further work is required to substantiate the theory that the microorganisms containing the *mlrA* gene are able to biodegrade microcystin toxins. This would involve isolating these mircoorganisms using conventional culturing techniques. Once isolated, the microorganisms could then be identified and used as pure cultures in batch experiments to verify that they could degrade microcystin toxins.

CHAPTER 5: MICROBIAL DEGRADATION OF CYLINDROSPERMOPSIN LABORATORY STUDIES

INTRODUCTION

Exposure of humans to hazardous cylindrospermopsin (CYN) levels through drinking water is mainly expected in regions with little or no choice of drinking water resources or where treatment is rudimentary or absent (Welker, Steinberg and Jones 2001). Understanding and therefore estimating the persistence of CYN under natural conditions is essential in assisting authorities in making decisions regarding CYN residence times in aquatic environments.

In *Cylindrospermopsis raciborskii* batch culture studies conducted by Griffiths and Saker (2003) 80-90% of CYN remained in the cyanobacterial cells during the active or log period of growth. However, during the stationary period, up to 50% of total CYN was found to be extracellular (Griffiths and Saker 2003). Chiswell et al. (1999) found that at different stages of a *C. raciborskii* bloom extra cellular CYN varied between 19-98% of total CYN. Shaw et al. (1999) also reported that during a bloom of *Aphanizomenon ovalisporum* 85% of total CYN was extracellular.

Once CYN has been released from the cell into the surrounding water several mechanisms may determine its fate in the environment. These mechanisms may include; dilution with non-contaminated water, adsorption to suspended solids and sediments, incorporation into aquatic plants and animals, degradation via photolysis and biodegradation by natural aquatic organisms (Welker, Steinberg and Jones 2001).

Biodegradation is probably the most common process involved in the removal of toxins from raw waters. Several conditions must be satisfied for biodegradation to take place in a particular environment. The degrading organism/s must exist in the environment containing the compound in question, it must have the necessary enzymes to bring about degradation of the compound in question (Daubaras and Chakrabarty 1992, Alexander 1994) and the compound must be accessible to the organism. A compound that is not readily accessible to such microorganisms or their extracellular enzymes will not be degraded (Alexander 1973).

Often before biodegradation occurs, a period of time is noted in which no destruction of the compound is evident. This time interval is commonly known as the lag period. It is the length of time between the addition or entry of the compound into the environment and evidence of its loss (Alexander 1994). The length of the lag period can vary enormously. It may be a matter of hours to many months and the duration can vary among compounds and environments.

A number of explanations have been proposed for the length of the lag period. Firstly, the length can be attributed to the time needed for enzymes participating in the initial periods of degradation of the compound to be induced (Alexander 1973; Torstensson, Stark and Goransson 1975; Stephenson, Lester and Perry 1984). The time required for enzyme induction is usually minutes to hours (Richard 1968) although the period for enzymes to become fully induced may be longer, especially at low substrate concentrations (Button and Robertson 1985). If such a compound does not induce the appropriate enzyme/s needed for the degradation and if there be no other inducers in the immediate vicinity, the compound will not be degraded. (Alexander 1973).

Mutation or the appearance of new genotypes by genetic exchange may also account for the length of the lag period (Walker and Newman 1956; Smidt, Hellwig and Knackmuss 1983). It has been suggested that lag periods that have reproducible duration among replicates are not the result of mutations (Fournier, Codaccioni and Soulas 1981; Spain et al. 1984).

The lag period can often be a reflection of the time it takes for small populations of organisms capable of degrading compounds of interest to reach cell densities sufficiently high to give detectable loss of the compound (Horvath 1972; Spain, Prichard and Bourquin 1980; Ventullo and Larson 1986). However, there may be constraints that prevent the low density of these active organisms from proliferating and these may vary from one environment to another. The presence of toxins, the lack of adequate supply of essential nutrients, temperature or predation by protozoa may delay the increase in cell numbers or activity of the degrading organism/s (Wiggins, Jones and Alexander 1987). For compounds that exhibit lag periods longer than one month, it is unlikely that the growth of cells initially present in the sample can account for the lag time (Wiggins and Alexander 1988).

The concentration of compounds in nature may be too low to support growth of the degrading organisms and the length of the lag period may be the result of the slow growth of the these organisms on the very low concentrations of a compound (Schindler 1971; Goldstein, Mallory and Alexander 1985; Wiggins and Alexander 1988). As the concentration of the compound increases, so does the growth rate. The more rapid proliferation at these higher concentrations would be reflected in a shorter lag period (Wiggins and Alexander 1988). The concentration of nitrogen (N), phosphorus (P), or both is also important in some environments. Some natural waters have low concentrations of these nutrients, which may limit microbial growth and increase the lag period (Alexander 1994).

The dependence of microbial growth on temperature is also well documented (Farrell and Rose 1967). If microbial growth is inhibited by temperature the length of the lag period will increase (Ludzack and Kinkead 1956, Atlas and Bartha 1972). Thus, the time of the year and the water temperature at which samples are taken will affect the lag periods (Tiedje 1977; Wiggins, Jones and Alexander 1987).

Other possible explanations for the length of the lag period include an insufficient utilization of other organic compounds before the chemical of interest (Alexander 1981; Kuiper and Hanstveit 1984; Goldstein, Mallory and Alexander 1985) and the time it takes for the microorganisms to detoxify the chemical of interest allowing the altered molecule to be degraded (Alexander 1994).

Studies conducted by Chiswell (1999) suggested that biodegradation was an important process in removal of CYN from water bodies once the toxin was released from the cells. Studies conducted by Smith (2005) confirmed Chiswell's (1999) initial observations that natural aquatic organisms are capable of degrading CYN. Smith (2005) showed that biodegradation of soluble CYN occurred in many different environments, in the concentration range of 4 to 950 $\mu g\ L^{-1}$, that were contaminated with toxic *C. raciborskii* cells. However, loss of toxin was initially delayed (i.e. lag period) before loss of toxin was detected. The study also illustrated that after repeated exposure to soluble CYN the lag period was reduced and the time taken for degradation shorter.

Although studies conducted by Smith (2005) and Chiswell (1999) provided valuable insight into the ability of natural aquatic bacteria to degradation CYN, additional investigation into CYN degradation in other water bodies will provide a larger pool of knowledge from which authorities can confidently make decisions on the effectiveness of CYN removal from contaminated waters and appropriate withholding times.

The aims of this study were to: isolate CYN degrading organisms from North Pine Dam (NPD) waters; determine the effect of CYN concentration on the biodegradation of CYN; determine the effect of re-addition of CYN to induced samples from NPD; determine the effect of sample site and temperature on biodegradation of CYN; investigate the possibilities of inducing CYN degrading organisms in water bodies with no previous history of toxic *C.*

raciborskii blooms; and investigate the effect of addition of copper based compounds to toxic *C. raciborskii* blooms on CYN degradation.

MATERIALS AND METHODS

Cylindrospermopsis raciborskii Isolate

A toxic culture of *C. raciborskii* (AWT 205) was obtained from Dr Peter Hawkins, Sydney Water, Sydney, New South Wales, Australia. Purification of the strain was performed.

The purification process involved the streaking of an aliquot of *C. raciborskii* (AWT 205) culture onto a 10 mL agarose plate and placing this plate under a light microscope. A well-isolated trichome was located and cut from the agarose plate using a sterile scalpel. The cut section of agarose was viewed under the microscope to ensure a single trichome was isolated. The single trichome and agarose were placed into 10 mL of sterile Jaworski's Medium (JW) and incubated at room temperature with 24 hour illumination. The isolate was checked regularly for evidence of growth. Once growth occurred, the culture (AWT 205/1) was regularly maintained by sub culturing into sterile JW and incubation at room temperature under 24 hour illumination. This stock culture was then used for all future research work involving *C.raciborskii* cells and/or toxin.

Preparation of purified Cylindrospermopsin

C. raciborskii culture (AWT205/1) was grown in batches of 4 to 6 L over a 10 to12 week period in JW. The cells were removed by filtration (Glass fibre filter, fine grade). The resultant solution was then passed through a carbograph (1000 mg of graphitized carbon) solid period extraction cartridge (Alltech) which had been previously washed with 5% Formic Acid in Methanol (10 mL) followed by milliQ water (10 mL). Once all crude CYN solution has been passed through (about 4 to 6 hours), the carbograph cartridge was washed 3 times with 5% Formic Acid in Methanol (10 mL). The Eluents were pooled and evaporated to dryness under reduced pressure (Rota-Vac). The contents were dissolved in 10 mL of milliQ water, filtered through a 0.45 μm filter and purified on a C18 reversed period column (Waters Prep Nova-Pak 100 x 25 mm Prep Pak cartridge) using 5% methanol (isocratic flow) on a Shimadzu LC system and a diode array detector (Shimadzu SPD-M10) set at 262 nm. The CYN fraction was identified and freeze dried over night. This was then taken up in 5 mL of milliQ water, filtered using a 0.45 μm filter and purified on a C18 reverse period column (Alltech Econosil C18 250 x 22 mm preparative column) using 10%methanol as eluent. CYN purity was checked by HPLC/MS/MS using an API300 mass spectrometer (Perkin Elmer Sciex Instrument, Thornhill, Ontario, Canada) using previously calibrated standards. Dr Wasa Wickramasinghe, National Research Centre Environmental Toxicology, Queensland, Australia performed purification of CYN.

Preparation of cell free extract – Cylindrospermopsin (CFE-CYN)

An appropriate amount of freeze-dried *C. raciborskii* culture (AWT 205/1) was weighed into a 50 mL sterile centrifuge tube and sonicated for 10 minutes (Branson Sonifier 250) with 35 mL of milliQ water. The supernatant was filtered through a 0.22 μm sterile filter and the CYN content determined by HPLC/MS/MS. Preparation of CFE-CYN was performed by Dr Wasa Wickramasinghe, National Research Centre Environmental Toxicology, Queensland, Australia.

Cylindrospermopsin identification and analysis

CYN was analysed by HPLC combined with MS/MS. The HPLC used an Altima C18 (250 x 4.6 mm, 5 µm) column at 40°C. A gradient of 1% to 60% methanol over five minutes, with a final isocratic stage holding at 60% methanol of one minute was used. The mobile period was buffered with 5 mM-ammonium acetate. The total flow rate was 1.1mL min^{-1}. Post column splitting was used to submit 20% of the column effluent only to the MS/MS interface. The transition from the M+H ion (416 m/z) to the 194 m/z fragment was monitored using multiple reactant monitoring (MRM) mode. Analytical standards were run against samples allowing quantitation of CYN (Eaglesham et al. 1999). Geoff Eaglesham, Queensland Health Scientific Services, Queensland, Australia performed CYN analysis.

Site descriptions

Figure 5.1 indicates the geographical location of each sample site investigated within North Pine Dam (NPD), Brisbane, QLD, Australia.

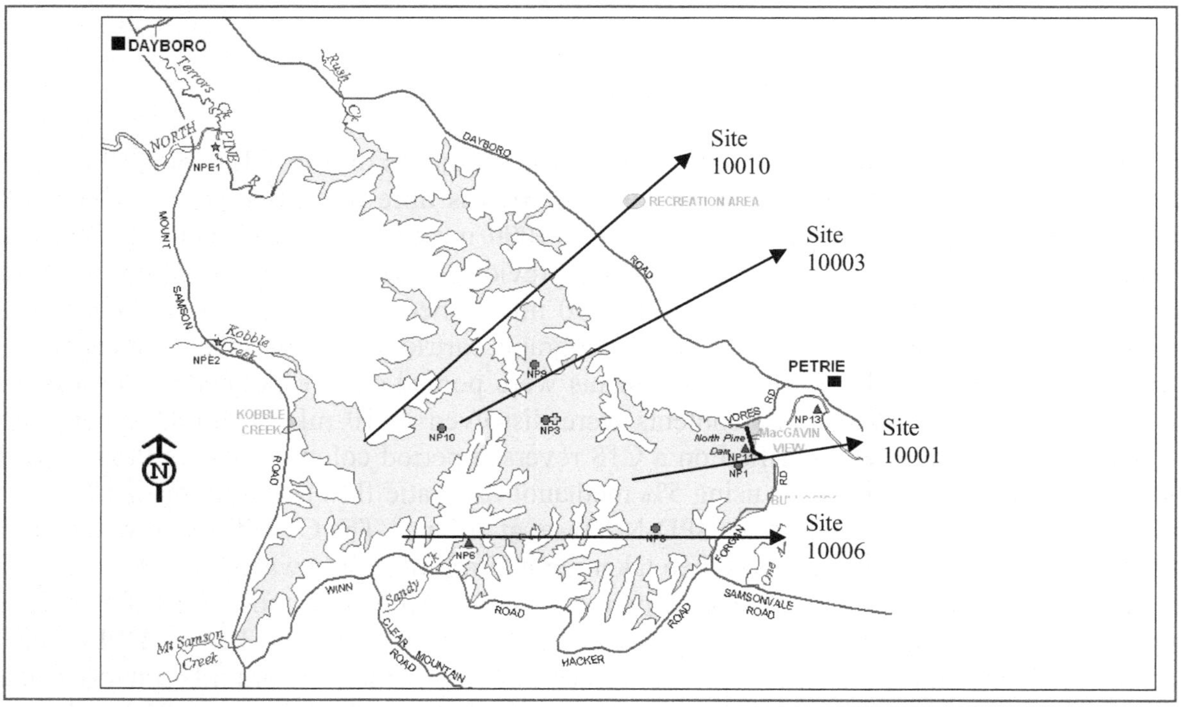

Figure 5.1 Map of NPD, Brisbane, QLD, Australia indicating locations of sample sites investigated. Approximate scale ____1km.

Enrichment of aquatic organisms capable of degrading CYN from NPD, QLD, Australia, for use in laboratory culture kinetic studies using purified CYN – site 10003

A surface water sample (site 10003) from NPD, QLD, was spiked with approximately 20 μg L^{-1} purified CYN. A control sample containing 50 mL of autoclaved (121°C, 15 minutes) NPD water was also spiked with approximately 20 μg L^{-1} purified CYN. The samples were incubated at 25°C in the dark. At intervals a sub-sample (1 mL) of the extracts were filtered through a 0.22 μm sterile filter and the CYN concentration determined using HPLC/MS/MS.

Enrichment of aquatic organisms capable of degrading CYN from NPD, QLD, Australia, for use in laboratory culture kinetic studies using CFE-CYN – site 10006

Surface water samples (2x50 mL) from NPD (site 10006) were spiked with approximately 20 μg L^{-1} of CFE-CYN. A control sample containing 50 mL of autoclaved (121°C, 15 minutes) NPD water was also spiked with approximately 20 μg L^{-1} CFE-CYN. The samples were incubated at 25°C in the dark. At intervals a sub-sample (1 mL) of the extracts were filtered through a 0.22 μm sterile filter and the CYN concentration determined using HPLC/MS/MS. Once the concentration of CYN reached a level below 1μg L^{-1} the entire extract was centrifuged, supernatant decanted and the bacterial pellet washed several times with sterile NPD water. The washed pellet was re-suspended in 10 mL sterile JW medium. The two suspensions were combined. An aliquot of this solution (1 mL) was added to sterile NPD water (50 mL) containing approximately 20 μg L^{-1} CFE-CYN and incubated at 25°C in the dark. This procedure was repeated with two more consecutive sub-cultures to ensure that the enrichment of organisms capable of degrading CYN occurred. These two subcultures were added to sterile JW medium. At regular intervals, an aliquot was taken from these extracts to determine CYN concentration. On the forth enrichment an aliquot was streaked onto solidified JW containing 20 μg L^{-1} CFE-CYN and incubated at 25°C in the dark. Single colonies from these plates were transferred back into sterile JW medium (50 mL) containing approximately 20 μg L^{-1} CFE-CYN and incubated at 25°C in the dark. At regular intervals an aliquot was taken from this extract and CYN degradation monitored to ensure CYN degradative bacteria were isolated.

Effect of CYN concentration on biodegradation of CYN – NPD site 10006

Duplicate tubes (12x50 mL) of NPD water (site 10006) were treated with CFE-CYN at varying concentrations (76 μg L^{-1}, 43 μg L^{-1}, 28 μg L^{-1}, 13 μg L^{-1}, 6.8 μg L^{-1}, 3.0 μg L^{-1} and <1 μg L^{-1}). A control sample containing 50 mL of autoclaved (121°C, 15 minutes) NPD water was spiked with approximately 20 μg L^{-1} CFE-CYN. The samples were incubated at 25°C in the dark. At intervals a sub-sample (1 mL) of the extracts were filtered through a 0.22 μm sterile filter and the CYN concentration determined using HPLC/MS/MS.

Re-addition of CYN to samples previously spiked with CYN at varying toxin concentrations

Once complete degradation of CYN was observed in the NPD samples at the varying concentrations examined, or at day 40 of the experiment if total degradation was not observed (Figure 5.4) CYN-CFE was re-added to the induced samples. The concentration of CYN added to the induced samples approximated the original concentration added. The samples were incubated at 25°C in the dark. At regular intervals a sub-sample (1 mL) of each sample was filtered through a 0.22 µm sterile filter and the CYN concentration determined using HPLC/MS/MS.

Effect of low CYN concentrations on biodegradation of CYN at an alternative site – NPD site 10003

Duplicate tubes of NPD water (site 10003) were treated with approximately 1 µg L^{-1}, 7 µg L^{-1} or 35 µg L^{-1} CFE-CYN. Duplicate control samples containing 50 mL of autoclaved (121°C, 15 minutes) NPD water were spiked with approximately 40µg L^{-1} CFE-CYN. The samples were incubated at 25°C in the dark. At intervals a sub-sample (1 mL) of the extracts were filtered through a 0.22 µm sterile filter and the CYN concentration determined using HPLC/MS/MS.

Effect of sample site on biodegradation of CYN

Water samples from four different sites (10001, 10003, 10006 and 10010) within NPD (refer Figure 5.1) were collected. Duplicate samples (50 mL) were treated with approximately 15 µg L^{-1} CFE-CYN and incubated at 25°C in the dark. At intervals a sub-sample (1 mL) of each sample was filtered through a 0.22 µm sterile filter and the CYN concentration determined using HPLC/MS/MS. Duplicate control samples containing 50 mL of autoclaved (121°C, 15 minutes) NPD water spiked with approximately 15 µg L^{-1} CFE-CYN were also run.

Effect of temperature on biodegradation of CYN

Duplicate samples of NPD water (site 10003) were treated with approximately 35 µg L^{-1} CFE-CYN and incubated at 20°C, 25°C or 35°C in the dark. At intervals a sub-sample (1 mL) of each sample was filtered through a 0.22 µm sterile filter and the CYN concentration determined using HPLC/MS/MS. Duplicate control samples containing 50 mL of autoclaved (121°C, 15 minutes) NPD water spiked with approximately 35 µg L^{-1} CFE-CYN and incubated at either 20°C, 25°C or 35°C in the dark were also run.

Investigations into the inhibition of CYN degradation at NPD when purified CYN was used as the toxin source

Effect of CYN size fractions on CYN biodegradation

Duplicate samples of NPD water (site 10006) were treated with either 15 µg L^{-1} whole CFE-CYN or <10 K CFE-CYN fraction and incubated at 25°C in the dark. At intervals a sub-sample (1 mL) of each sample was filtered through a 0.22 µm sterile filter and the CYN concentration determined using HPLC/MS/MS. Duplicate control samples containing 50 mL of autoclaved (121°C, 15 minutes) NPD water spiked with approximately 15 µg L^{-1} of either whole CFE-CYN or <10K CFE-CYN fraction were also run.

Effect of CYN source on CYN biodegradation at various sites within NPD, QLD, Australia

Water samples from four different sites (10001, 10003, 10006 and Pontoon) within NPD (refer Figure 5.1) were collected. Duplicate samples (50 mL) were treated with approximately 30 µg L^{-1} of either CFE-CYN or purified CYN and incubated at 25°C in the dark. At intervals a sub-sample (1 mL) of each sample was filtered through a 0.22 µm sterile filter and the CYN concentration determined using HPLC/MS/MS. Duplicate control samples containing 50 mL of autoclaved (121°C, 15 minutes) NPD water spiked with approximately 15 µg L^{-1} of CFE-CYN or purified CYN were also run.

Induction of CYN degrading organisms in water bodies with no previous history of toxic C. raciborskii blooms

South Australian water bodies, without a history of CYN contamination and without the presence of toxic *C. raciborskii* cells, at the time of the study and NPD water which has a history of CYN contamination and the presence of *C. raciborskii* cells evident at the time of the study were sampled. These samples were treated with approximately 20 µg L^{-1} CYN-CFE and incubated at 25°C in the dark. At intervals a sub-sample (1 mL) of each sample was filtered through a 0.22 µm sterile filter and the CYN concentration determined using HPLC/MS/MS. Duplicate control samples containing sterile RO water and 25 µg L^{-1} CYN-CFE were also run. Confirmation of the presence or absence of *C. raciborskii* cells in the samples was performed by period contrast microscopy (x 400).

Effect of addition of copper based compounds to toxic C. raciborskii blooms on the biodegradation of CYN

Duplicate samples of NPD water (site 10003) were treated with either 0.5 mg L^{-1} copper as copper sulphate or 0.5 mg L^{-1} copper as COPTROL® at pH 7.2 and pH 8.2. Approximately 10 µg L^{-1} CFE-CYN was then added to all samples and samples incubated at 25°C in the dark. At intervals a sub-sample (1 mL) of each sample was filtered through a 0.22 µm sterile filter and the CYN concentration determined using HPLC/MS/MS. Duplicate control samples containing 50 mL of autoclaved (121°C, 15 minutes) NPD water and duplicate positive control sample containing untreated NPD water spiked with approximately 15 µg L^{-1} CFE-CYN were also run.

Effect of addition of copper based Compounds to toxic C. raciborskii blooms on the biodegradation of CYN – at higher CYN concentrations than decribed above

Samples of NPD water (site 10003) were treated with either 0.5 mg L^{-1} copper as copper sulphate or 0.5 mg L^{-1} copper as COPTROL® at pH 7.2. Approximately 15 to 25 µg L^{-1} CFE-CYN was then added to all samples and samples incubated at 25°C in the dark. At intervals a sub-sample (1 mL) of each sample was filtered through a 0.22 µm sterile filter and the CYN concentration determined using HPLC/MS/MS. Control samples containing 50 mL of autoclaved (121°C, 15 minutes) NPD water were also run.

RESULTS AND DISCUSSION

Enrichment of aquatic organisms Capable of degrading CYN from NPD, QLD, Australia, for use in laboratory culture kinetic studies using purified CYN – site 10003

At site 10003 no reduction of CYN was evident at day 94 of the experimental period (Figure 5.2). The concentration of the control remained at 100% (+/- analytical error) of the initial concentration of CYN at day 94 of the experiment.

These observations differ from studies previously conducted by Smith (2005). Smith (2005) reported that biodegradation of CYN was observed in NPD water by day 13 of the experiment and was completed by day 19 of the experiment (Smith 2005). These results suggest that the conditions required for CYN biodegradation have been previously met by field samples from NPD. The lack of degradation in the present study suggest these conditions where not satisfied during this current experiment.

The inconsistencies between the two studies may have been due to a number of factors that may have prevented a low cell density of CYN degrading organisms from proliferating to densities necessary for a detectable loss of the toxin (Wiggins, Jones and Alexander 1987). Factors that may have contributed to the absence of biodegradation include, the concentration of CYN, the lack of an adequate supply of essential nutrients or growth factors or incubation temperatures (Wiggins, Jones and Alexander 1987).

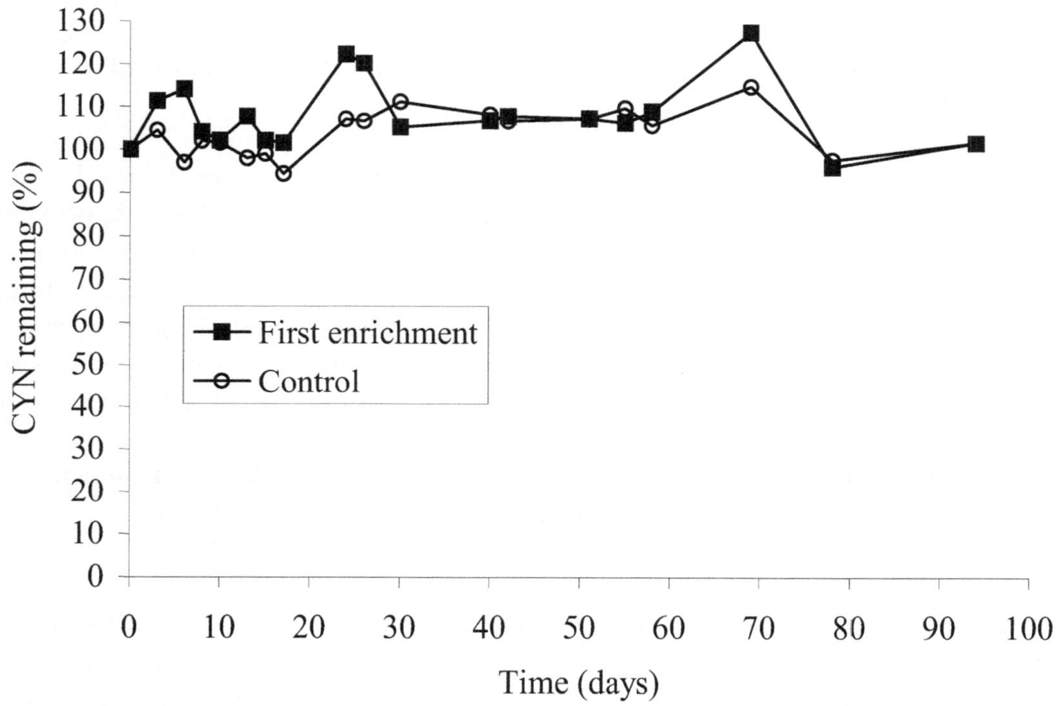

Figure 5.2 Enrichment of CYN degrading organisms from NPD using purified CYN as the toxin source.

Enrichment of aquatic organisms capable of degrading CYN from NPD, QLD, Australia, for use in laboratory culture kinetic Studies using CFE-CYN – site 10006

Duplicate samples (first enrichment A and B) of NPD water spiked with CFE-CYN at approximately 20 μg L^{-1} showed degradation within 17-20 days after a lag period of 6 days. These initial samples were then pooled and subsequently used for two more additional enrichments (second and third enrichment) (Figure 5.3). The lag and time taken for complete degradation of CYN were reduced at each enrichment step. The lag period was reduced from approximately 7 days on the second enrichment to <2 on the third enrichment and the time taken for complete degradation was reduced from 14 days on the second enrichment to 6 days in duration on the third enrichment. The third enrichment served for the isolation of very small, slow growing single colonies. These colonies were placed into JW spiked with approximately 20 μg L^{-1} CYN. At intervals a sub-sample (1 mL) was filtered through a 0.22μm sterile filter and the CYN concentration monitored. No loss of toxin was evident at day 47 of the experimental period, indicating the isolation of CYN degrading organisms from the third enrichment was unsuccessful during this initial isolation attempt.

Previous studies of CYN biodegradation in natural waters have not only implicated the role of bacteria in the removal process but have reported the isolation and identification of bacteria capable of degrading CYN (Chiswell 1999, Smith 2005). Smith (2005) reported the isolation of and identification of a *Delftia* sp. from a water body located At Harvey Bay, QLD Australia. *Delftia* sp. is a member of the Comamondaceae (Wen et al. 1999) family, which belongs to the β subclass of the Proteobacteria (Stackebrandt, Murray and Trueper 1988).

Although this study was not successful in isolating the CYN degrading organisms from the third enrichment culture, the results confirm the ability of aquatic organisms, from NPD, to degrade soluble CYN.

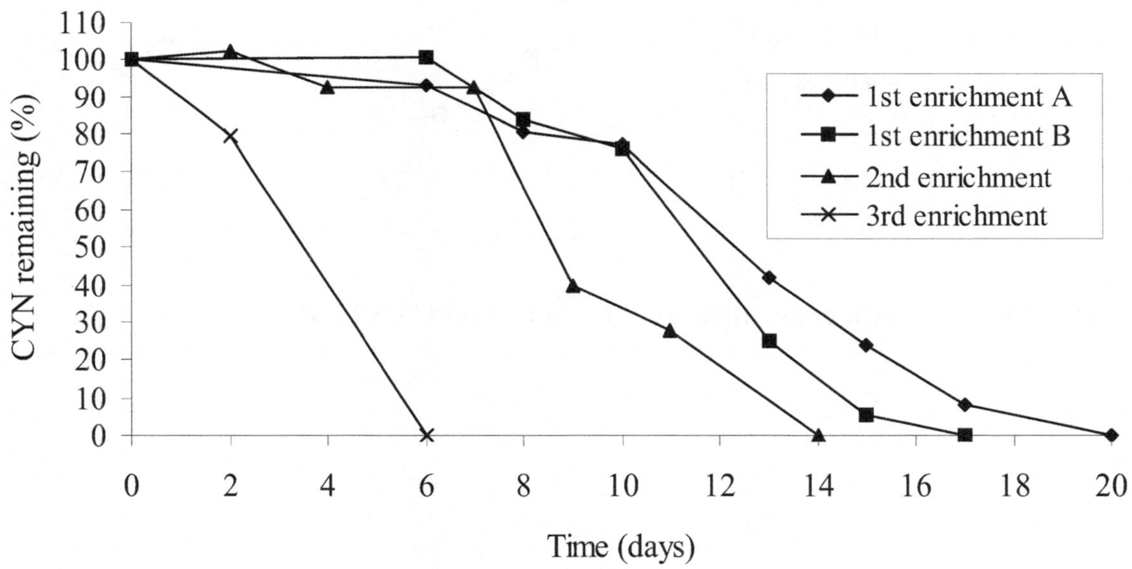

Figure 5.3 Enrichment of CYN degrading organisms from NPD using CFE-CYN as toxin source

Effect of CYN concentration on biodegradation of CYN – site 10006

At site 10006 degradation was observed in CYN concentrations examined between 3 and 76 μg L^{-1}CYN. However, complete degradation to below 1 μg L^{-1} did not occur at 3 and 6.8 μg L^{-1} CYN and 38.2% and 18.9% of CYN remained on day 40 of the experiment respectively. CYN degradation did not occur at concentrations <1.0 μg L^{-1} with 75.7% CYN remaining on day

40. The concentration of the control remained at 89.8% of the initial concentration of CYN after 40 days of the experiment.

The curve produced from plotting initial concentration of CYN versus the maximum rate of degradation shows the observed rate of reaction is proportional to the concentration of CYN (Figure 5.6). The maximum rate of degradation ranged from 0.3 to 24µg L^{-1} day $^{-1}$ in the CYN concentration range 3 to 76 µg L^{-1}CYN. The lag periods ranged from 2 days to 16 days with complete loss of toxin occurring within 16 to 30 days (Table 5.1).

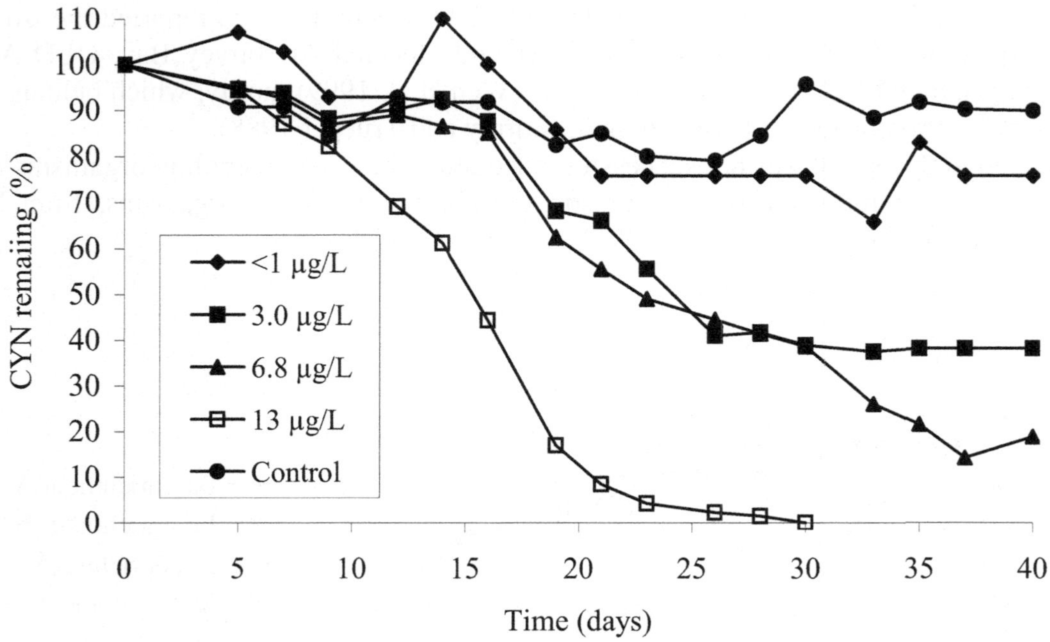

Figure 5.4 Effect of toxin concentration (low range) on CYN biodegradation.

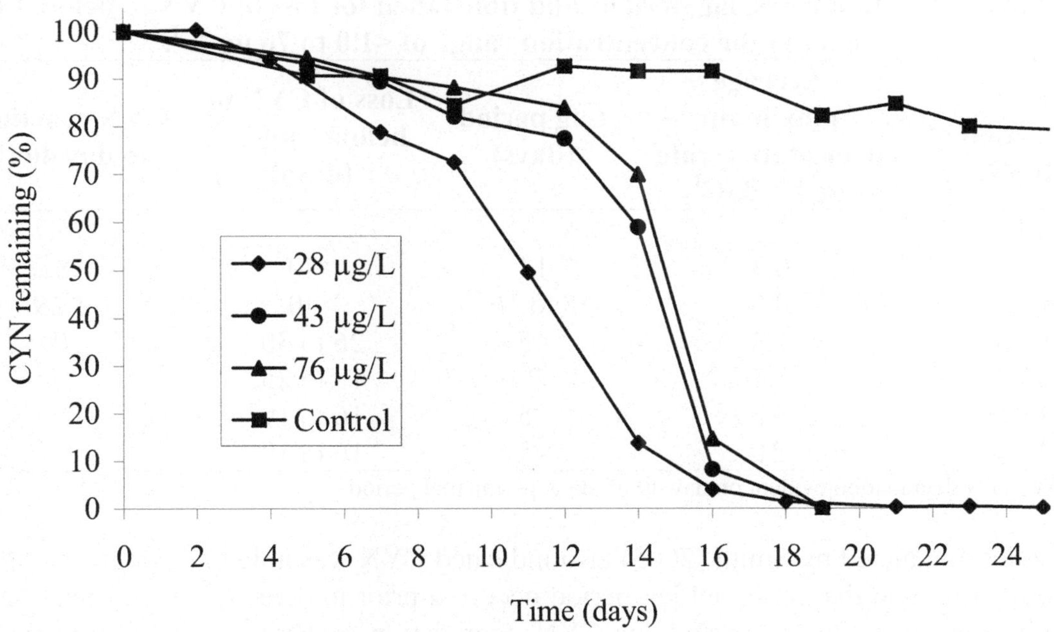

Figure 5.5 Effect of toxin concentration (high range) on CYN biodegradation.

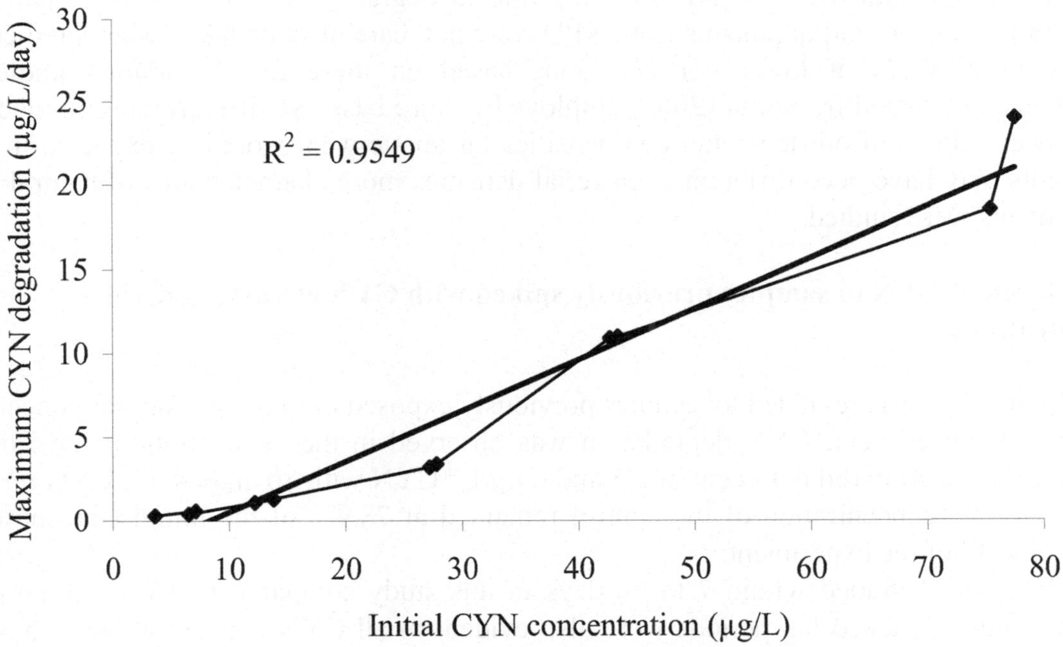

Figure 5.6. Plot of maximum rate of CYN degradation versus initial concentration of CYN

Table 5.1
Maximum degradation rates, lag periods and time taken for loss of CYN to below 1.0 µg L^{-1} toxin in the concentration range of <1.0 to 76 µg L^{-1}.

Initial concentration of CYN (µg L^{-1})	Average maximum degradation rate (µg L^{-1} day^{-1})	Lag period (days)	Loss of CYN to below 1 µg L^{-1} (days)	CYN remaining on day 40 (%)
<1.0	*	*	*	75.7
3	0.3	14	>40	38.2
6.8	0.5	5 to 16	>40	18.9
13	1.2	<5	26 to 30	0
28	3.3	2	18 to 25	0
43	11	<5	16 to 19	0
76	21	<5	16 to 19	0

*No CYN degradation evident on day 40 of the experimental period.

Studies conducted by Smith (2005) also indicated CYN was able to be degraded at a range of concentrations and that an initial lag period occurred prior to detection of degradation. Smith (2005) observed that the lag period for CYN degradation in Harvey Bay Retardation Basin (HBRB), QLD, Australia was approximately 15 days with total degradation evident by day 30 of the experiment. Smith (2005) also reported that CYN degrading organisms isolated from HBRB had the capacity to degrade CYN, to <1 µg L^{-1}, in the concentration range of 4 to 950 µg L^{-1}. Biphasic degradation was observed with an initial slow period (5 days) and then a rapid period until degradation was completed in approximately 7 to15 days.

Although organisms from HBRB were able to degrade CYN to <1µg L^{-1} at the lower concentration ranges, and organisms from NPD were not, care must be taken when predicting the degradation of CYN at lower concentrations based on these two laboratory studies. The experiments conducted by Smith (2005) employed culture based studies growing in media with high concentrations of nutrients and cell densities far above what would be expected in nature. The results may have been different if bacterial densities more characteristics of natural aquatic environments were studied.

Re-Addition of CYN to samples previously spiked with CYN at varying toxin concentrations

When CYN was re-added to samples previously exposed to toxin at a similar concentration to the initial experiment, CYN degradation was observed in the concentration range of 10 to 76µg L^{-1}. Degradation did not occur at 2.5 and 6 µg L^{-1} CYN with 80 and 64.4% CYN remaining on day 37. The concentration of the control remained at 78.9% of the initial concentration of CYN at day 37 of the experiment.

CYN was degraded within 7 to 26 days in this study compared to 16 to 30 days in the previous study. Reduced lag periods were also evident in all CYN concentrations examined in this study. The lag periods reduced from approximately 2 to 16 days in the initial experiment to <2 to 2 days duration in this study (Tables 5.1 and 5.2).

The maximum degradation rates did not increase on re-addition of toxin to the induced samples. The curve produced from plotting initial concentration of CYN versus the maximum rate of degradation shows the observed rate of reaction was again proportional to the concentration of CYN (Figure 5.9).

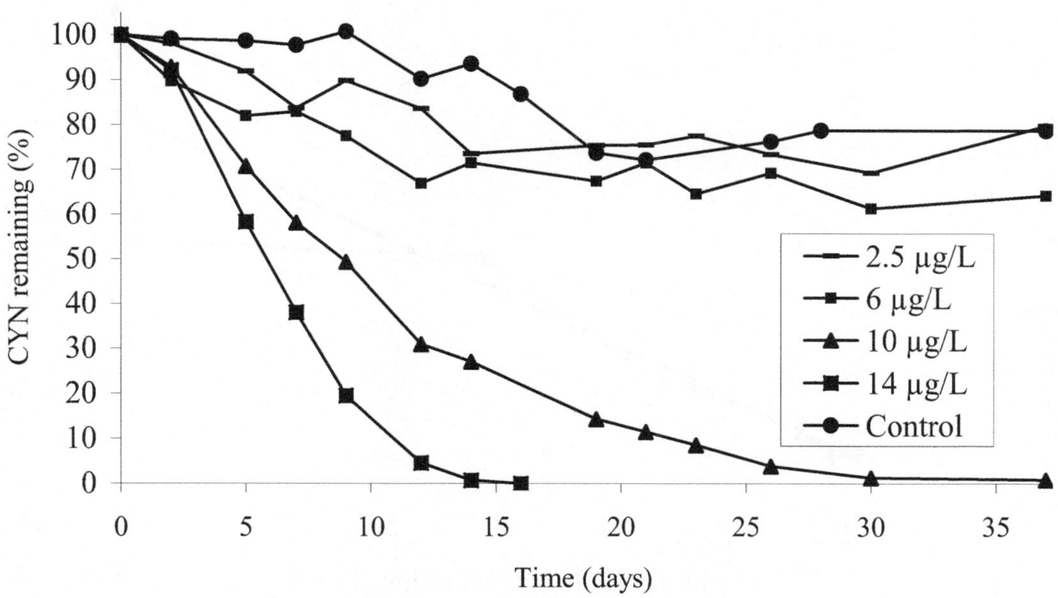

Figure 5.7 Effect of re-addition of CYN to induced samples at various toxin concentrations (low range).

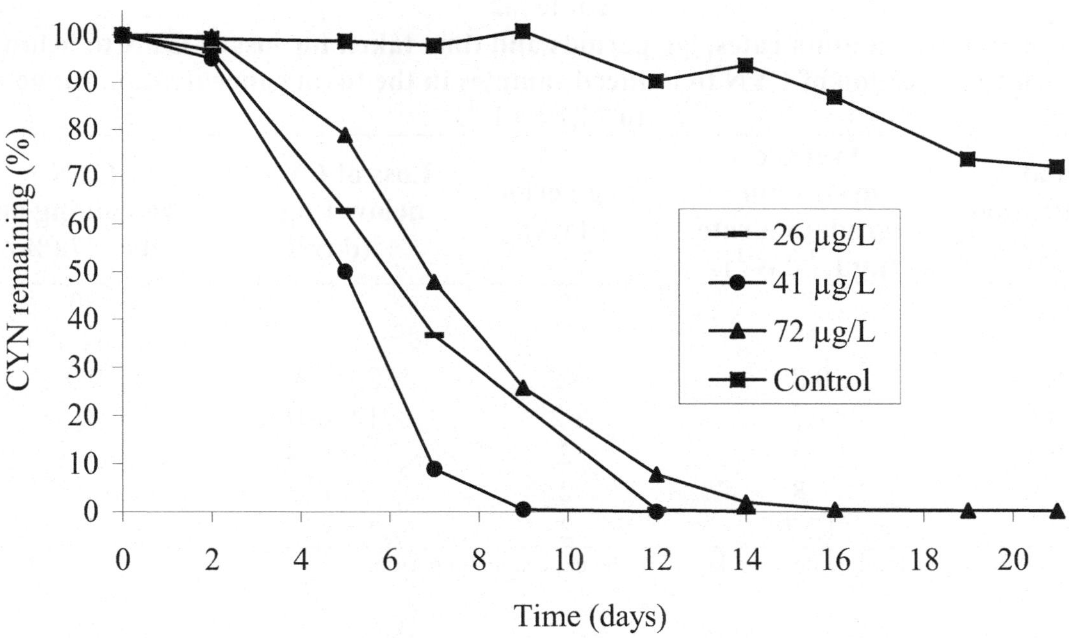

Figure 5.8 Effect of re-addition of CYN to induced samples at various toxin concentrations (high range).

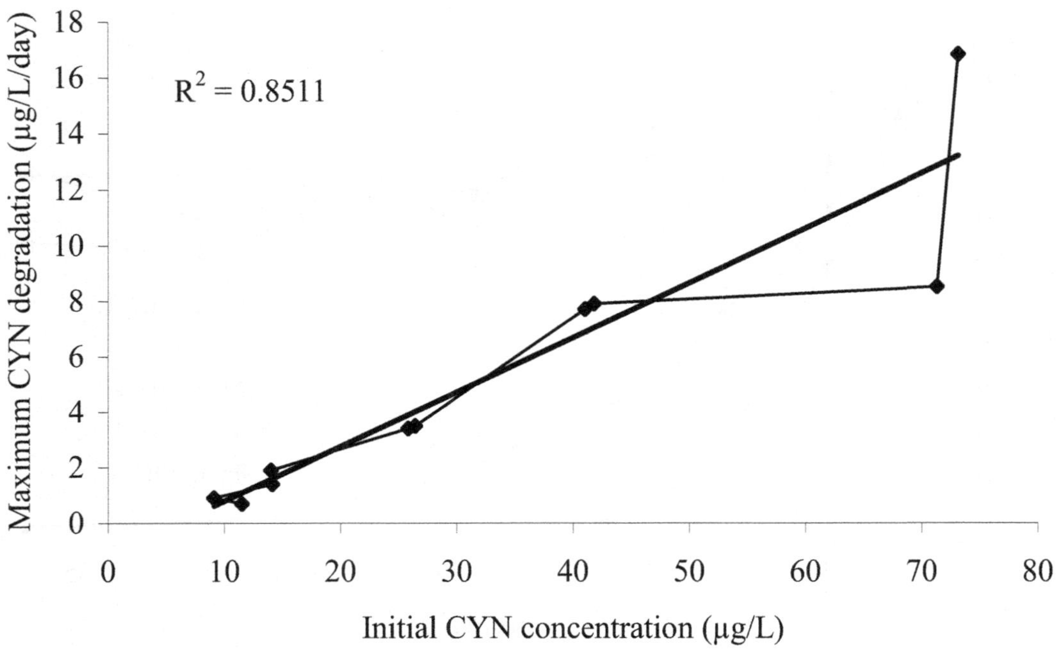

Figure 5.9 Plot of maximum rate of CYN degradation versus initial concentration of CYN on re-addition of toxin to induced samples.

Table 5.2
Maximum degradation rates, lag periods and time taken for loss of CYN to below 1.0 µg L^{-1} for re addition of CYN to induced samples in the toxin concentration range of 2.5 to 73.1 µg L^{-1}.

Initial concentration of CYN (µg L^{-1})	Average maximum degradation rate (µg L^{-1} day^{-1})	Lag period (days)	Loss of CYN to below 1 µg L^{-1} (days)	CYN remaining on day 37 (%)
2.5	*	*	*	80
6	*	*	*	64.4
10	0.8	<2	23 to 37	0.9
14	1.7	<2	12 to 16	0
26	3.5	2	7 to 14	0
41	7.8	2	7 to 12	0
72	12.7	2	9 to 21	0

*No CYN degradation evident on day 37 of the experimental period.

The reduction of the lag periods on re-addition of CYN to the induced samples was consistent with CYN studies conducted by Smith (2005). Smith (2005) reported that when CYN was re-added to an induced sample from HBRB, CYN was 62% degraded within 10 days with a reduced lag period evident. The lag period was reduced from approximately 15 days to 4 days and complete degradation occurred in < 15 days, which was less than the initial study (30 days).

This observation was also consistent with microcystin biodegradation studies conducted by Jones et al. (1994) in which an initial lag period of 14 days was evident with residual microcystin LR being present after 23 days incubation. No lag period was present upon re-addition of microcystin LR. These results are also consistent with other biodegradation studies investigating pollutants such as 4-nitrophenol (Spain, Prichard and Bourquin 1980).

The reduction of the lag period upon re-addition of CYN to the induced NPD samples may have been due to a number of factors. Previous studies suggest the most likely reason for the reduction may have been that only a small population of the organism/s responsible for the degradation was originally present. During the lag period, these organisms may have become sufficiently abundant for our analysis to detect CYN removal. This observable fact has been found by a variety of authors with different organisms (Spain, Prichard and Bourquin 1980; Ventullo and Larson 1986; Wiggins, Jones and Alexander 1987). On re-addition of CYN, the degradative organisms may have then been at a sufficient level to degrade CYN more rapidly. If the population or community of organisms was still large when the water again received toxin, it is predicted that the degradation would again occur with a reduced lag period. If the population size had decreased by the second addition, the lag time would most likely increase again.

Previous studies also suggest that the reduction of the lag period may have been due to the length of time needed for the degradative enzymes participating in the initial period of degradation to be induced (Alexander 1973; Torstensson, Stark and Goransson 1975; Stephenson, Lester and Perry 1984). The time required for enzyme induction has been shown to vary from minutes to hours (Richard 1968), although even longer periods have been reported (Button and Robertson 1985).

The reduction of the lag period upon re-addition of CYN suggested that biodegradation of CYN may commence more rapidly in water bodies that have frequent exposure to toxic blooms of organisms producing CYN.

Effect of Low Toxin Concentrations on CYN Biodegradation at an alternative site – NPD Site 10003

To gain insight into whether the limited degradation at the lower CYN concentrations was site specific, degradation of CYN, at the lower concentrations, was investigated with water from another site. At site 10003 degradation to below 1 μg L^{-1} did not occur at 1.4 and 7.6 μg L^{-1} CYN and 32% and 45% of CYN remained on day 48 of the experiment respectively. The concentration of the control sample remained at 108.8% of the initial concentration of CYN at day 48 of the experiment (Figure 5.10).

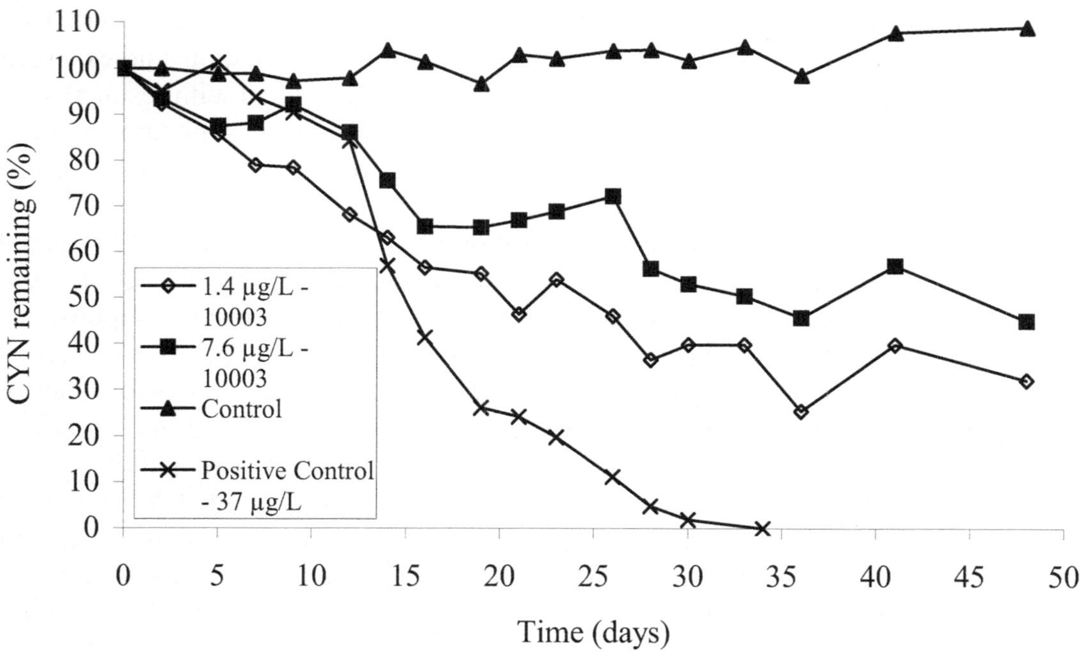

Figure 5.10 Effect of low toxin concentrations on CYN biodegradation at an alternative site – NPD site 10003.

The incomplete degradation of toxin at CYN concentrations in the range of <1 µg L^{-1} to 7.6 µg L^{-1} at sites 10006 and 10003 indicates the limited degradation is most likely not an effect of sample site variability but perhaps the existence of a threshold CYN concentration that sustains growth of CYN degrading organisms but which the CYN degrading organism cannot proliferate and hence little CYN loss occurs (Alexander 1994).

To maintain viability every organism must expend energy. When the concentration of the carbon (C) source for growth is high, the substrate is able to satisfy the needs for maintenance energy and for the processes that lead to increases in cell size, growth and multiplication. At low substrate concentrations the energy required for maintenance represents a high percent of substrate C and an even smaller percent is used for growth and multiplication. At some lower value, all the energy in the form of C is used to keep the cell alive and none is used for growth. At this concentration, although the substrate is being metabolized, the cells are not growing, and biomass is not increasing. This is known as the threshold concentration (Alexander 1994).

This observable fact has also been evident from studies investigating biodegradable synthetic compounds. These studies indicated that no biodegradation occurs in the test period below a certain concentration or that the rate of loss is less than which might be expected from the rates observed at higher concentrations (when rate is proportional to concentration) (Alexander 1994). The indigenous population of 2,4-dinitrophenol metabolizing bacteria in soil could not be maintained and would not multiply if the concentration were 0.1 mg L^{-1}, although the organisms multiplied at higher concentrations (Schmidt and Gier 1989).

The threshold is lowered if the organisms carrying out the degradation have alternative C sources available to them and these levels may be regenerated by excretion of other organism or new additions. Under these conditions the energy needs for maintenance and growth of the degrading organism maybe met by use of these other organic molecules. Organism may metabolize two or more organic substances simultaneously (Alexander 1994).

Effect of sample sites within NPD on CYN biodegradation

CYN degradation was observed in all sites examined within NPD (Figure 5.11). The lag periods ranged from <3 days to 7 days in duration with complete loss of toxin occurring within 7 to 17 days (Table 5.3). The maximum rate of degradation ranged from 2.7 to 4.6 μg L^{-1} day^{-1} at the various sites. The concentration of the control remained at 91.6% of the initial concentration of CYN at day 17 of the experimental period.

A one-way ANOVA was performed on each duplicate sample to determine if there were any significant differences between the maximum degradation rates at each site. Results showed that the *p*-value was greater than 0.05 (5% significance level) indicating there was no significant difference between degradation rates at the sites examined.

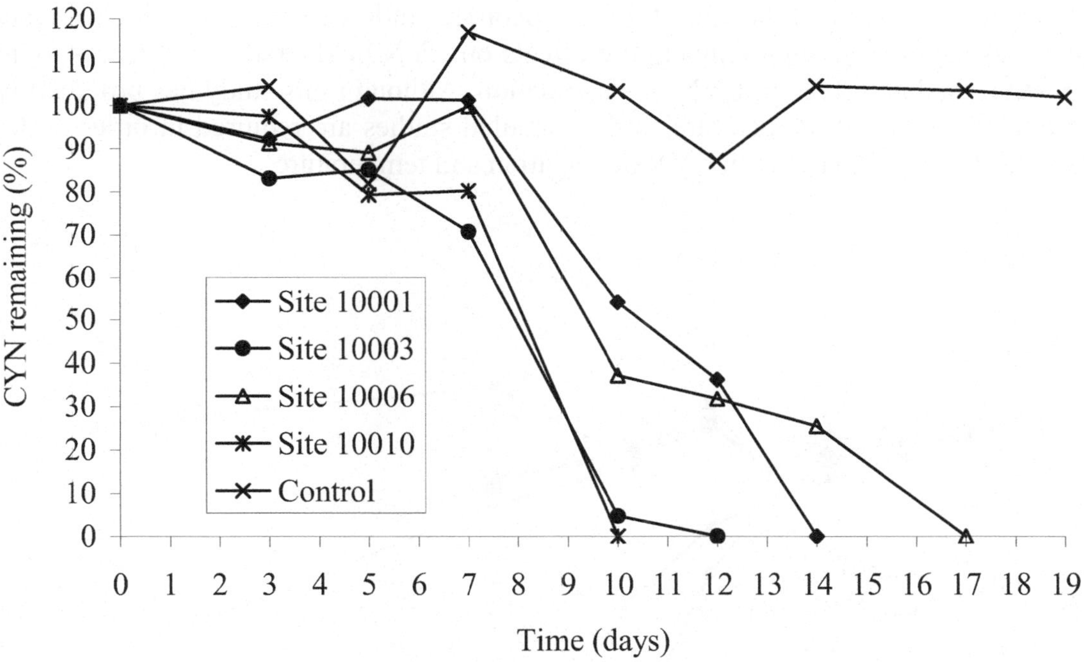

Figure 5.11 Effect of sample sites within North Pine Dam on CYN biodegradation.

Table 5.3
Maximum degradation rates, lag periods and time taken for loss of CYN to below 1.0 μg L^{-1} for at various sites within NPD.

Sample site	Average maximum CYN degradation rate (μg L^{-1} day^{-1})	Lag period (days)	Loss of CYN to below 1 μg L^{-1} (days)
10001	3.8	7	12 to 14
10003	3.6	<3	10 to 12
10006	3.7	< 3 to 7	7 to 17
10010	4.2	3	7 to 10

Observations from this study indicate the CYN degrading organisms are present and able to grow and function in the sites investigated, suggesting that if toxic *C. raciborskii* blooms occur at various sites within NPD, the degrading organisms would be able to degrade the toxin.

Effect of Temperature on CYN Biodegradaton

CYN degradation was observed at 20°C, 25°C and 35°C during the 44 day experimental period (Figure 5.12). Degradation was not completed at 20°C on day 44 of the experimental period. Results indicated 38.7% of CYN remained on day 44 of the study. The concentration of the 20°C control remained at 102.2% of the initial concentration on day 44. The average maximum rate of CYN degradation, when samples were incubated at 25°C and 35°C, was 5.1 µg L^{-1} day $^{-1}$ and 3.8 µg L^{-1} day $^{-1}$ respectively. The lag periods ranged from 0 to 3 days in duration with complete loss of toxin occurring within 17 to 24 days when samples were incubated at 25°C and 35°C (Table 5.4). The concentration of the control samples incubated at 25°C and 35 °C remained at 86.3 % and 87.8% of the initial concentration of CYN respectively on day 24 of the experimental period.

It should be noted that the aim of this temperature study was not to provide the reader with detailed information regarding temperature effects on CYN biodegradation but was to ascertain if temperature had an effect on CYN biodegradation. Although this study has provided valuable insight into the effect of temperature further detailed studies are required in order to definitely ascertain the relationship between CYN degradation and temperature.

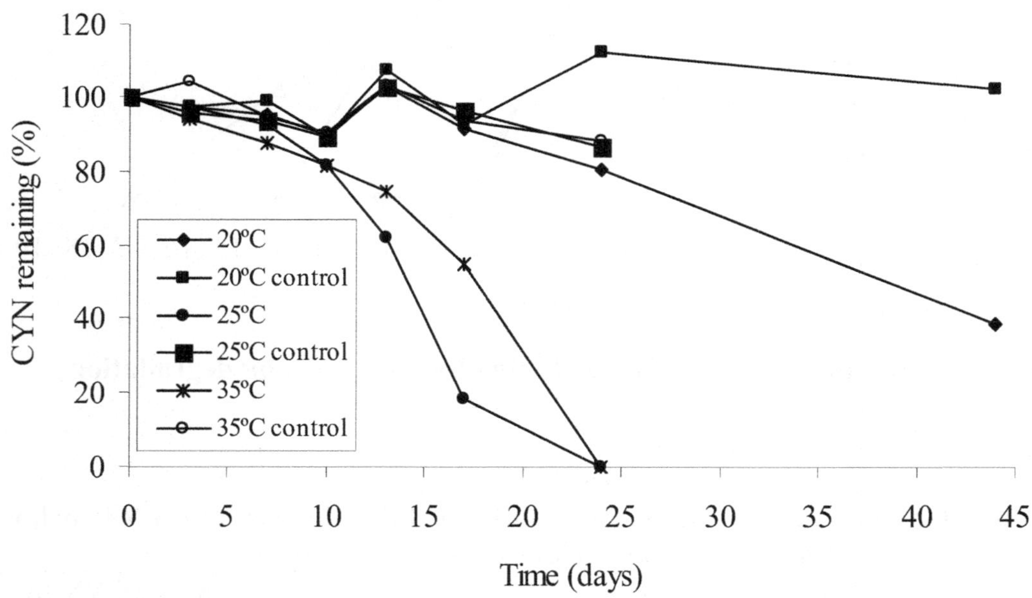

Figure 5.12 Effect of temperature on maximum CYN degradation rates.

Table 5.4
Maximum degradation rates, lag periods and time taken for loss of CYN to below 1.0 µg L^{-1} at 20°C, 25°C and 35°C

Temperatur e	Average maximum CYN degradation rate (µg L^{-1} day^{-1})	Lag period (days)	Loss of CYN to below 1 µg L^{-1} (days)
20°C	*	*	38.7% CYN remaining on day 44 of experimental period
25°C	5.1	0 to 3	17 to 24
35°C	3.8	0 to 3	17 to 24

*No CYN degradation evident on day 24 of the experimental period.

Organisms have a defined temperature range over which they are able to grow and if a temperature is too high or low for a specific organism growth will not be obtained (Brock and Madigan 1991). Organisms found in the soil or water usually have a temperature optima for growth at approximately 25°C (Brock and Madigan 1991). Low temperatures such as those encountered during winter are typically associated with low growth, low cell metabolism and little or no biodegradation. As temperatures increase microbial activity will also increase and again the biodegradation rate will be affected (Alexander 1994). Thus, the time of the year and the water temperature at which samples are taken will affect the microbial activity (Tiedje 1977; Wiggins, Jones and Alexander 1987).

Park et al. (2001) showed the degradation rates for the cyanotoxin microcystin were strongly dependant on temperature. Degradation of microcystin began within 1 day when incubated at 10 to 30°C and was degraded to below detectable concentrations after 4 days. At 5°C degradation was very slow and 30% of microcystin remained after 7 days. The highest degradation rate occurred at 30°C.

Investigations into the Inhibition of CYN Degradation when Purified CYN was used as the Toxin Source

Effect of CFE-CYN size fractions on CYN biodegradation

CYN degradation was observed when both CFE-CYN whole fraction and CFE-CYN <10K fraction was used as the toxin source at site 10006 (Figure 5.13). The concentration of the whole fraction control remained at 103.6% of the initial concentration of CYN at day 17 of the experimental period. The concentration of the <10K fraction control remained at 91.6% of the initial concentration of CYN at day 12 of the experimental period. The average maximum rate of CYN degradation when whole CFE-CYN was used as the toxin source was 3.6 µg L^{-1} day $^{-1}$ compared to 3.9 µg L^{-1} day $^{-1}$ when the <10K CFE-CYN fraction was used. The lag periods ranged from <3 days to 7 days in duration with complete loss of toxin occurring within 7 to 17 days when whole CFE-CYN was used as the toxin source. This compared to a lag period of seven days in duration with complete loss of toxin occurring between 10 and 12 days when <10K CFE-CYN was used as the toxin source (Table 5.5).

A one-way ANOVA was performed on each duplicate sample to determine if there were any significant differences between the maximum degradation rates when the two difference toxin fractions were used. Results showed that the *p*-value was greater than 0.05 (5% significance level) indicating there was no significant difference between degradation rates when whole CFE-CYN fractions were used compared to when <10K fractions were used as the toxin source. This implies that the smaller compounds within the cell free extract were utilized for fuelling degradation.

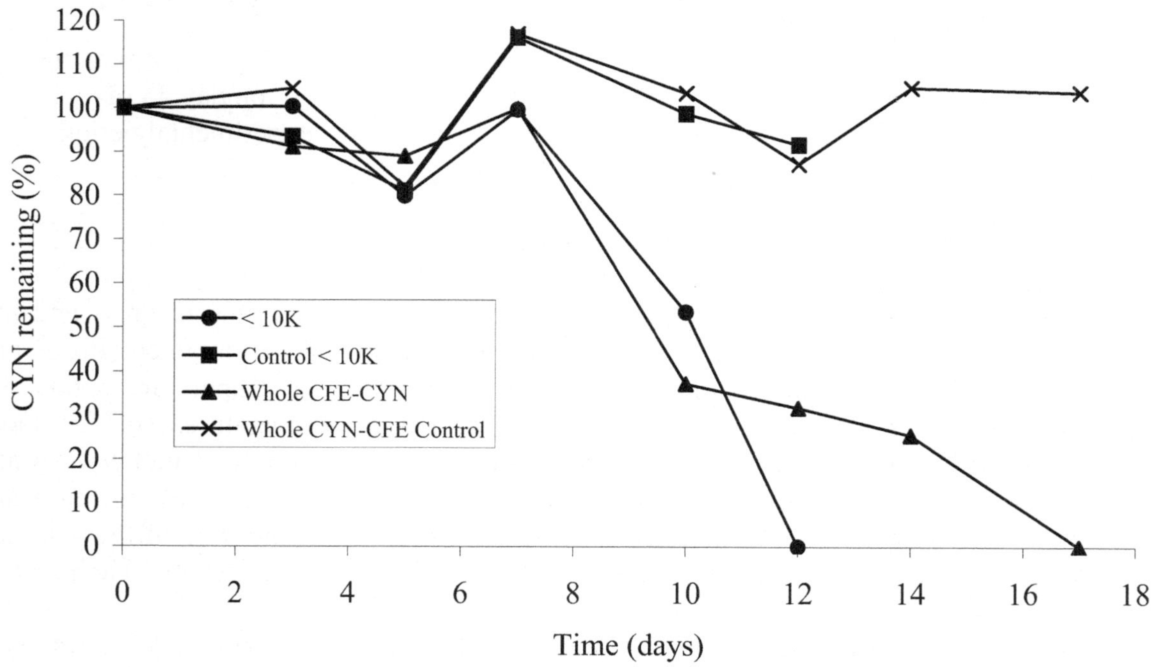

Figure 5.13 Effect of CFE-CYN size fractions on CYN biodegradation.

Table 5.5
Maximum degradation rates, lag periods and time taken for loss of CYN to below 1.0 μg L^{-1} when different size fractions of CYN were used

CYN size fraction	Average maximum CYN degradation rate (μg L^{-1} day^{-1})	Lag period (days)	Loss of CYN to below 1 μg L^{-1} (days)
Whole fraction	3.9	<3 to 7	7 to 17
<10K fraction	3.6	7	10 to12

Effect of toxin source on CYN biodegradation at various sites at NPD, QLD, Australia

CYN degradation was observed in all sample sites examined where the toxin source was CFE-CYN (Figure 5.15). The degradation periods ranged from 9 to >35 days in duration, with lag periods ranging from < 2 to 16 days in duration. The concentration of the CFE-CYN control sample remained at 133.2% of the initial concentration of CYN at day 35.

CYN degradation was not detected in sample sites examined where the toxin source was purified CYN, with one exception, site 10006 (Figure 5.14). In the sample sites where no CYN degradation was evident approximately 100% (+/-analytical error) of toxin remained at day 23 of the experimental period. Samples from site 10006 showed total loss of CYN within 19 to 23 days, with lag periods ranging from <2 to 6 days in duration. The concentration of the purified

CYN control samples remained at 122.1% of the initial concentration of CYN at day 23 of the study.

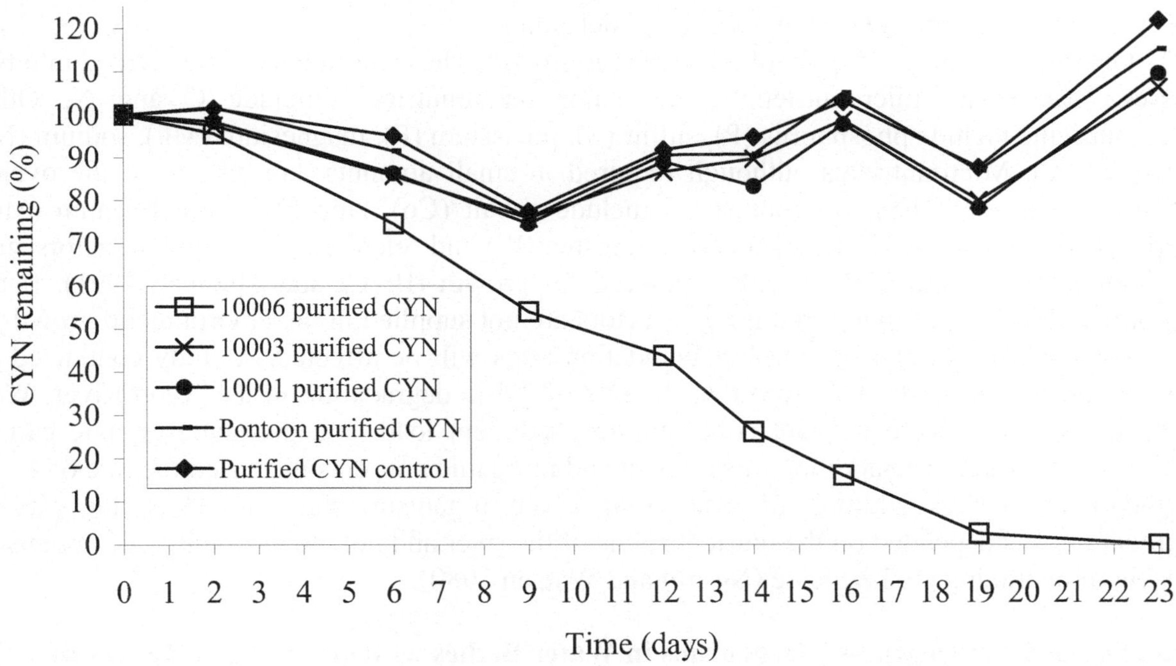

Figure 5.14 Effect of purified toxin on CYN biodegradation at various sites within NPD.

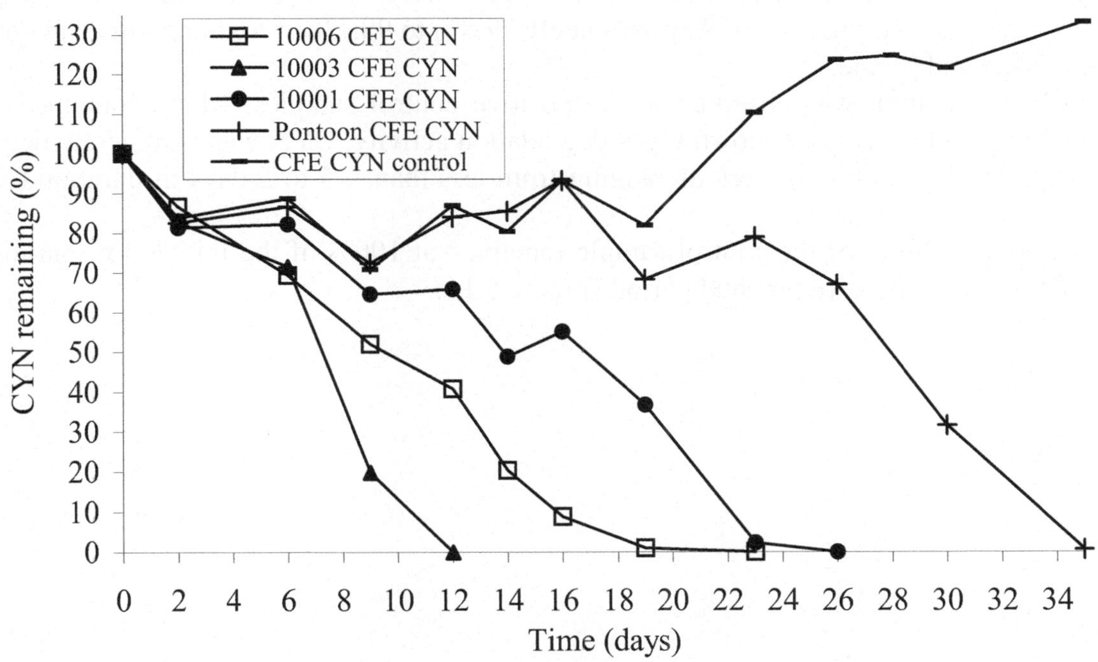

Figure 5.15 Effect of CFE- CYN as the toxin source on CYN biodegradation at various sites within NPD.

These observations suggest that at all samples sites investigated (excepting site 10006) *C. raciborskii* cells possibly provide essential growth factors required by the CYN degrading organisms for proliferation. If the cyanobacterial cells are not present, as in the case of when purified CYN was used as the toxin source, the growth of the organisms required for degradation of CYN may be retarded and toxin removal is delayed.

Organisms require a supply of nutrients for growth. These nutrients can be divided into two classes, macro and micronutrients. The major macronutrients include C and N. Other macronutrients include phosphorus (P), sulfur (S), potassium (P), magnesium (Mg), sodium (Na) and iron (Fe). Micronutrients, although required in small amounts, are critical to the overall nutrition of the organism. Micronutrients include cobalt (Co), zinc (Zn), molybdenum (Mo), copper (Cu), manganese (Mn), nickel (Ni), tungsten (W) and selenium (Se). Growth factors such as vitamins and amino acids are also required for growth (Brock and Madigan 1991). If the appropriate levels of nutrients and growth factors are not supplied in the environment, growth of the organism will be retarded and biodegradation rates will be affected. A study conducted by Nesbitt and Watson (1980) showed that the rate of 2,4-D degradation in the Avon River, West Australia, was related to nutrient concentration, sediment load, and dissolved organic carbon content of the water. The addition of sediment and inorganic nutrients increased the rate of 2,4-D degradation, whereas addition of actively involved organisms did not. Thus, the rate of degradation was depended on the nutrient status of the river and not the availability of organisms capable of degrading the herbicide (Nesbitt and Watson 1980).

Induction of CYN Degrading Organisms in Water Bodies with no Previous History of Toxic C. raciborskii Blooms

CYN degradation was not detected in water bodies without any previous history of CYN producing cyanobacteria and where *C. raciborskii* cells were not evident in the water sample at the time of the study (Figure 5.16). Approximately 77.8% to 98.1% of toxin remained at day 246 of the experimental period.

CYN degradation was observed in the positive control samples, which contained water from NPD and had previously shown CYN degradation activity. CYN was completely degraded within 15 to 20 days, with lag periods ranging from less than < 5 to 5 days in duration (Figure 5.16).

The concentration of the control sample remained at 100% of the initial concentration of CYN at days 246 of the experimental period (Figure 5.16).

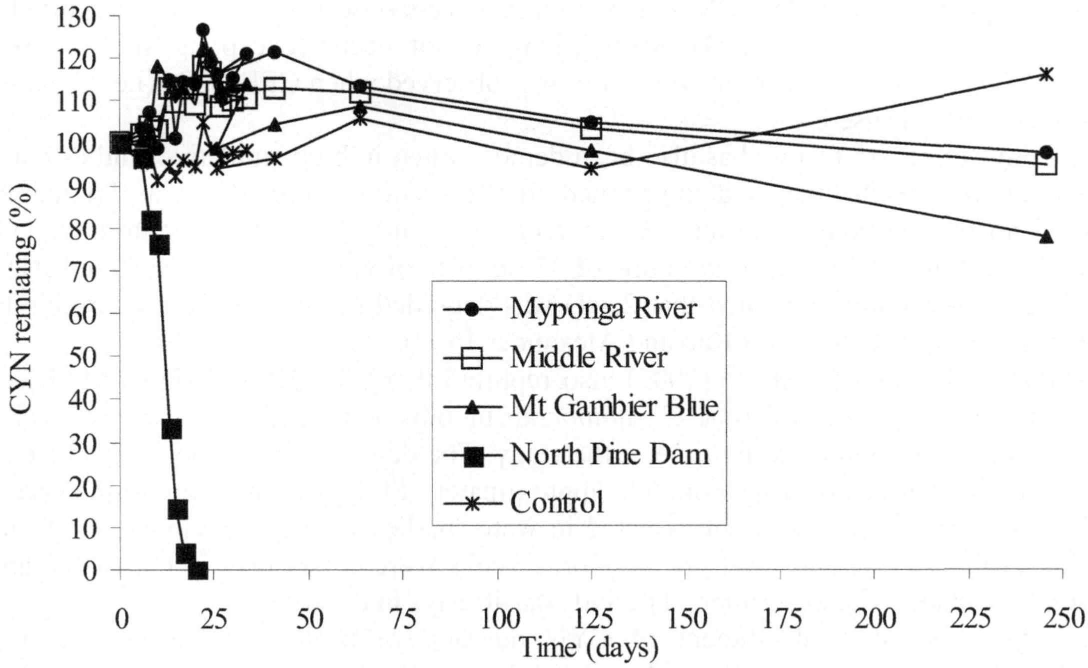

Figure 5.16 Induction of CYN degradation in water bodies with no previous recorded history of toxic *C. raciborskii* blooms.

The lack of CYN degradation in the SA water bodies may have been due to a number of factors. Firstly, the appropriate biodegradative genes in the endemic bacterial population may not have evolved. Even if the genes are present, functional expression of the gene is essential for the appropriate degradative enzymes to be produced (Daubaras and Chakrabarty 1992). If removal of CYN was to occur, the degrading microbes must not only evolve and express the degradative genes, but also establish themselves in the environment where they are to be used.

However, from an evolutionary point of view, it seems unlikely that an enzyme specialized in metabolizing one specific compound, which may occur only sporadically and in relatively low concentrations, such as CYN, would need to evolve. Literature suggested that only if the cyanotoxin was generally toxic to the bacteria would an evolutionary pressure be exerted to select for CYN degraders (Welker, Steinberg and Jones 2001). This information infers that it is more likely that the CYN degradative bacteria contain enzymes, which have normal functions that include the metabolism of other cyclic molecules. This has been reported to be the case for other cyanotoxin degradative enzymes. The enzymes required for microcystin-LR degradation were present fortuitously in the bacterium *Sphingmonas* sp. and were required for normal cellular growth and metabolism when the bacterium was grown on peptone media (Bourne et al. 1996).

If CYN were present in the water body it would be expected that these enzyme systems would also have the capacity to metabolize CYN. This did not occur in the SA water bodies indicating that perhaps CYN degrading organisms have not yet established themselves in these environments. This lack of establishment may have been due to the presence of environmental characteristics that restricted the growth of CYN degraders. Such characteristics may have included the lack of low concentrations of trace organic nutrients such as amino acids, B vitamins, fat-soluble vitamins and other organic molecules (Wiggins, Jones and Alexander 1987). If these growth factors were not present in the environment, the growth of the dependant organisms would be restricted. *C. raciborskii* cells themselves may supply these growth factors.

If the cyanobacterial cells are not present as was the case in the SA waters, then the growth of the organisms required to degrade CYN may be retarded. This observation was also reported earlier in this study (5.2.8.2) where CYN degradation did not occur (excepting site 10006) when purified CYN was used as the toxin source but was observed when CFE-CYN (i.e. *C. raciborskii* cellular material) was used.

The impact of site factors has also been demonstrated in biodegradative studies conducted by other authors. Such studies demonstrated that a specific compound was biodegraded in samples from one particular site and not another. One study found that trichloroethylene was degraded by the indigenous bacteria in one of 43 samples of water and soil examined (Nelson et al. 1986). Another study illustrated that 2,4-D was degraded in samples from eutrophic but not oligotrophic lakes (Rubin, Subba-Rao and Alexander 1982).

Studies conducted by Smith (2005) also reported that CYN degradation was observed in water bodies with a history of toxic *C. raciborskii* blooms and where *C. raciborskii* cells were detected in the water-bodies at the time of the study. The degradation periods ranged from 11 to 33 days, with lag periods ranging from 0 to approximately 11 days in duration. Smith (2005) also reported CYN degradation was not detected in water bodies with no previous history of CYN producing cyanobacteria and where *C. raciborskii* cells were not evident in the water sample at the time of the study. The experimental period was 40 days in duration.

If adaptation or establishment of CYN degradative bacteria occurred after repeated exposure to the cyanobacterium *C. raciborskii* itself or CYN, it would be necessary to determine the length of time that bacterial communities remained adapted after such exposure. Previous studies have indicated that once the organisms of interest became adapted and activity becomes marked, the active community may retain its degradative state for some time (Alexander 1994). Conversely, the effect may be short lived (Alexander 1994). There is insufficient information to permit generalizations and hence further work needs to be conducted to determine if evolution and expression of CYN degradative genes can be accomplished in water not previously exposed to toxin, such as the SA water samples, and if so how long adaptation persists after exposure.

Effect of Addition of Copper Based Compounds to Toxic C. raciborskii Blooms on the Biodegradation of CYN

No CYN degradation was detected in the samples where $CuSO_4$ was added at pH 7.2, with 82.9% remaining at day 32 of the experimental period. In the $CuSO_4$ control sample, at pH 7.2, the CYN concentration remained at 93.8% of the initial concentration of CYN at day 32.

Limited degradation of CYN was detected in the samples where $CuSO_4$ was added at pH 8.2, with 53.8% CYN remaining at day 32 of the experimental period. The $CuSO_4$ control sample, at pH 8.2, remained at 78.0% of the initial concentration of CYN at day 32. CYN degradation was observed in the positive control sample (i.e. no $CuSO_4$ added). CYN was completely degraded at day 32 with a lag period of approximately 8 to 11 days.

No CYN degradation was detected when Coptrol® was added to NPD samples at pH 7.2 and 8.2. CYN remained at 101.7% and 79.0% respectively on day 32 of the experimental period. The control samples at pH 7.2 and 8.2 remained at 92.5% and 86.1% CYN respectively on day 32.

These observation indicated that if algicides such as $CuSO_4$ or Coptrol® were applied to water bodies during toxic blooms of CYN producing organisms, at the pH levels investigated, the degradation of the toxin may be retarded, increasing the withholding period. Further confirming the importance of applying algicides at the early stages of a toxic bloom development when cell numbers and toxin levels are low (Burch, Chow and Hobson 2001).

However, the concentration of CYN used in these experiments was approaching the level previously demonstrated to be unable to be completely degraded by CYN degrading organisms in NPD (5.2.3 and 5.2.5). It is thus suggested further investigations at higher CYN concentrations and using wide variety of water bodies are conducted to confirm the effect of copper based algicides on CYN degrading organisms.

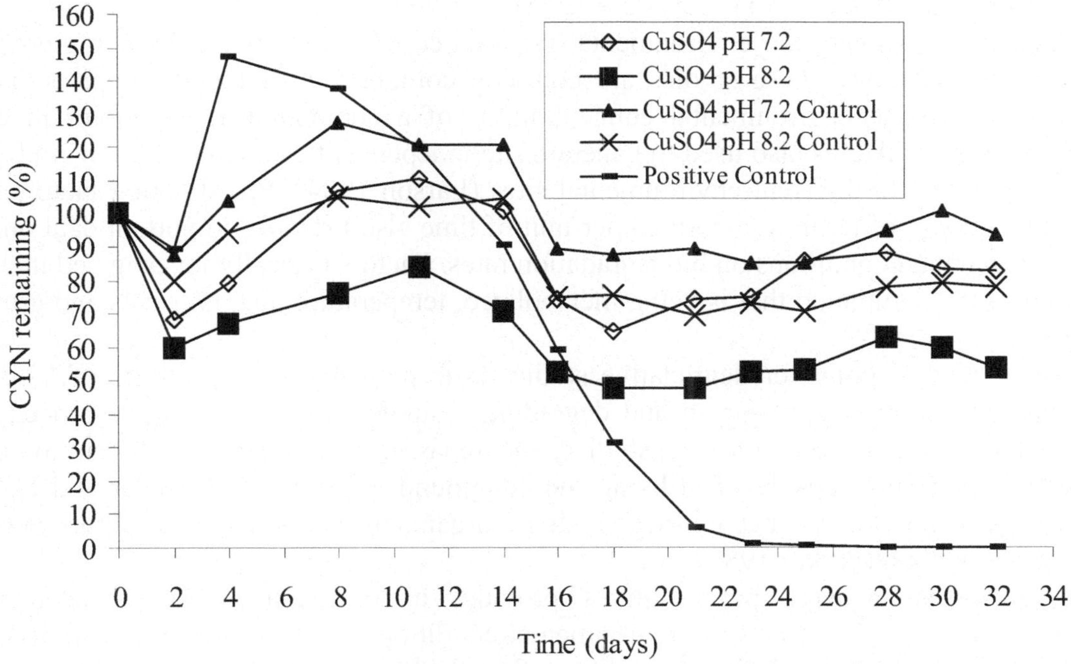

Figure 5.17 Effect of $CuSO_4$, at pH 7.2 and 8.2, on the CYN biodegradation.

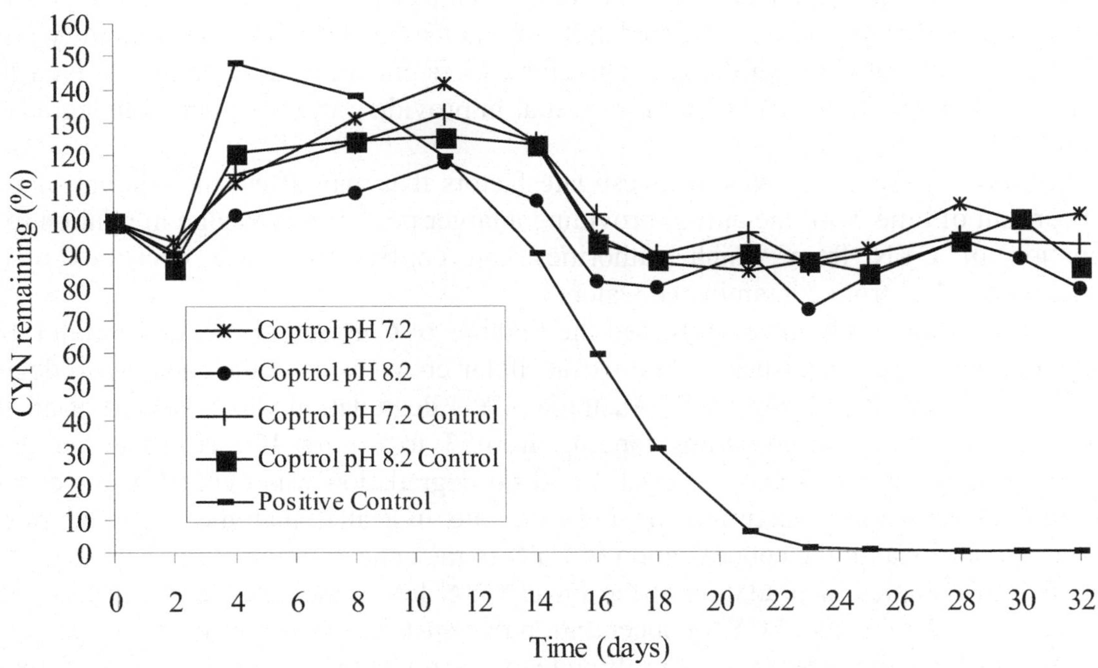

Figure 5.18 Effect of $CuSO_4$, at pH 7.2 and 8.2, on the CYN biodegradation

SUMMARY AND CONCLUSIONS

The persistence of soluble CYN in water bodies during and after bloom events has important implications for water authorities managing both recreational and drinking water supplies. The length of time it takes for soluble CYN to degrade may be of significance to public health, because of the possibility of accumulation within the water column and the adverse health effects due to exposure to the toxin (Alexander 1994).

To adequately evaluate the environmental persistence of the toxin we need to study the kinetics of biodegradation. These studies are common components of many computer based models used to estimate environmental concentrations of a compound at any point in time (Larson 1984). Kinetic data is also used to generate site independent estimates of biodegradation rates for extrapolation to different environmental sites (Larson 1984). Investigating factors that may affect the amount of toxin transformed per unit of time also provides important data on the effect of environmental conditions on biodegradation rates. Factors typically investigated include concentration of the compound that is being metabolized, temperature, pH, nutrients and growth factors.

A typical starting point for degradative studies is to consider a pure culture of a single bacterial population that is growing on and degrading a single compound. Classic procedures typically used to isolate bacteria, for such studies, encompasses preliminary enrichment to boost the number of organisms capable of utilizing the compound. (Parkes 1982, Slater and Lovatt 1984). Once the population has been enriched, single organisms can be isolated in pure culture by the agar plate method (Parkes 1982).

Although research using pure cultures provides a solid foundation for estimating biodegradation rates and the effects of environmental conditions on such rates, these studies are usually carried out under vastly different conditions than the bacterial population would normally be exposed to in nature. Hence, these studies have many compromising arbitrary features compared to the situation in the real environment. As a consequence of the differing experimental conditions between typical laboratory based culture studies and samples from the field which encompass endemic bacterial communities subjected to naturally occurring levels of nutrients, growth factors, pH, toxins and predation, extrapolations from tests on laboratory grown cultures to nature are fraught with danger. Therefore, biodegradation information derived from field samples and pure cultures combined is essential in providing more accurate biodegradation information.

The purpose of this study was to investigate factors that may affect the amount of CYN degraded per unit of time, with the aim of providing a larger pool of knowledge in relation to the biodegradation of CYN from which authorities can confidently make decisions on the biodegradation of CYN from contaminated waters.

The results of this study have confirmed the findings of Chiswell (1999) and Smith (2005) that some natural aquatic organisms or their extracellular enzymes have the capacity to degrade soluble CYN. At North Pine Dam, QLD, Australia, CYN degradation was observed at all sites investigated from initial concentrations ranging from 3 to 76 μg L^{-1}. However, complete degradation did not occur at 3 and 7.6 μg L^{-1} and no degradation was evident at lower CYN concentrations. The average maximum degradation rate indicated that the observed rate of reaction was proportional to the concentration of CYN in the concentration range 3 to 76 μg L^{-1}.

The lack of complete degradation at the lower CYN levels indicates that, perhaps within NPD the existence of a threshold CYN concentration that sustains growth of the CYN degrading organism but which the degraders cannot proliferate exists, resulting in little or no loss of toxin is present. At low CYN concentrations the organisms may utilize a more energetically favorable carbon source.

When degradation did occur it was usually delayed for a period of time, i.e. lag period. Previous studies suggested this delay is most likely due to the time taken for the degrading organisms to develop the necessary enzymes for degradation and/or the biomass of CYN degrading organisms to reach a level that loss of toxin is detected.

The outcomes of this investigation also illustrated that repeated CYN exposure reduced the lag period suggesting that biodegradation of CYN may commence more rapidly in water bodies that have had frequent exposure to toxic blooms of organisms that produce CYN. However, care must be taken when trying to predict the duration of the lag period. The prediction of a reduced lag period is based on the assumption that a higher population of CYN degrading organism is already present or that degrading enzymes have already been induced. The period may be longer or shorter depending on the environmental conditions such as temperature, aeration status and the availability of other resources (Ludzack and Kinkead 1956; Atlas and Batha 1972; Wiggins, Jones and Alexander 1987; Alexander 1994).

Further experiments conducted during this study suggested that successful removal of CYN appeared to be dependant on the presence of *C. raciborskii* cellular material and/or a previous history of toxic *C. raciborskii* blooms. It is apparent that the loss of toxin via biological mechanisms is unlikely to occur if these conditions are not met.

The water temperature also affects the average maximum rate of degradation of CYN and the time taken for total loss of toxin. For CYN degrading organisms located in NPD 25ºC appears to be the optimum temperature for degradation when compared to 20ºC and 35ºC. Further studies are required to provide more detailed information on the effect of temperature on CYN degradation.

Toxic CYN producing blooms of cyanobacteria not only have important implications for water authorities because of the possibility of adverse health effects due to exposure to the organisms and /or the toxin but the bloom also may have significant effects on the water treatment operations and production of tastes and odours. As a result many water authorities will apply copper-based algicides to the water body with the aim of controlling such events. $CuSO_4$ has generally been the algicide of choice although and chelated copper algicides have been developed as a result of the reduced effectiveness of $CuSO_4$ in hard alkaline waters (Burch, Chow and Hobson 2001; Burch, Hobson and Chow 2002).

Results of this study have shown that the addition of these chemical algicides to a toxic bloom of *C. raciborskii* may have an unfavorable effect on the biodegradation of CYN. Results indicate that if $CuSO_4$ or chelated copper algicides are applied to a bloom when toxins levels are in the range of approximately 10 μg L^{-1} CYN the loss of soluble toxin will be retarded. However, the concentration of CYN used in these experiments was approaching the level previously demonstrated to be unable to be completely degraded by CYN degrading organisms in NPD. It is thus suggested further investigations at higher CYN concentrations and using wide variety of water bodies are conducted to confirm the effect of copper based algicides on CYN degrading organisms.

CHAPTER 6: DEVELOPMENT OF A MODEL FOR CYANOBACTERIAL GROWTH, TOXIN PRODUCTION AND BIODEGRADATION

INTRODUCTION

Numerical hydrodynamic and ecological models facilitate the transfer of knowledge from field and laboratory experimentation to reservoir management by providing a predictive tool for management strategies. Control strategies for the management of cyanobacteria populations in reservoirs can be simulated before implementation to gain an understanding of whether the strategy is likely to have the desired outcome.

The major processes leading to cyanobacterial growth are hydrodynamics and biogeochemistry. The one-dimensional hydrodynamic model DYRESM (Dynamic Reservoir Simulation Model) characterises reservoir stratification behaviour based upon solar radiation, wind mixing and inflows. The ecological model CAEDYM (Computational Aquatic Ecosystem Dynamics Model) can be coupled with DYRESM to simulate the aquatic chemistry of the reservoir based upon chemical inflow concentrations, sediment release and uptake by phytoplankton. Phytoplankton growth is simulated in response to the prevailing hydrodynamics and chemistry. The primary group of concern is the cyanobacteria which have specific characteristics coded into the model to describe their nutrient uptake for field populations and growth. To simulate toxin, taste and odour production and degradation in field settings, a new metabolite subroutine was developed and added to the reservoir hydrodynamics and ecological model DYRESM-CAEDYM.

DYRESM-CAEDYM MODEL DESCRIPTIONS

The hydrodynamic numerical model DYRESM, developed by Imberger et al. (1978) is based on fundamental energy, mass, and momentum equations describing the one-dimensional structure detected in small to medium lakes and reservoirs. The model divides the water body into horizontal layers. The layers are Lagrangian, which are able to expand and contract, thus accommodating volumetric change due to inflow, outflow, and evaporation. The thickness of the layers also changes when vertical movement accommodates the bathymetry of the water body. Consequently no numerical diffusion is considered (Fischer et al. 1979). DYRESM is used to predict the variation of water temperature and salinity with depth and time (Antenucci and Imerito 2000). The assumption of one dimensionality is used because fluctuations in lateral directions are small compared with vertical variations (Hamilton and Schadlow 1997). The density stratification occurring in reservoirs and lakes inhibits vertical motion whilst horizontal density variations defuse expeditiously by lateral and longitudinal advection, which occurs at a faster rate than vertical convection. An additional function of DYRESM is to provide a one-dimensional platform for the CAEDYM ecological model.

CAEDYM is an autonomous ecological model that can be linked to a suite of hydrodynamic models, e.g. DYRESM and ELCOM, but can be run independently to assess specific ecological processes (Herzfeld and Hamilton 2000). CAEDYM is a general ecosystem simulation model with the capability to simulate the biogeochemical processes of phytoplankton, zooplankton, fish, macrophytes, macroinvertebrates, dissolved oxygen, biological oxygen demand, nutrient dynamics, tracer/colour, pH, iron, manganese and aluminium.

One dimensionality

Density stratification in lakes and reservoirs can inhibit vertical movement whilst horizontal density variations are rapidly dampened by horizontal advection and convection. Small to medium lakes and reservoirs can experience this phenomena and subsequently one-dimensional modelling can be used to predict the physical structure. It is necessary to validate the one-dimensionality of a lake or reservoir and this is achieved with the use of the Lake Number L_N (Imberger and Patterson 1990). The Lake Number is categorised by the stratification stability and the magnitude of the wind. Figure 6.1 shows a typical density profile in a stratified lake or reservoir. The stratification in Figure 6.1 is subjected to a wind field that provides a surface friction velocity u^*.

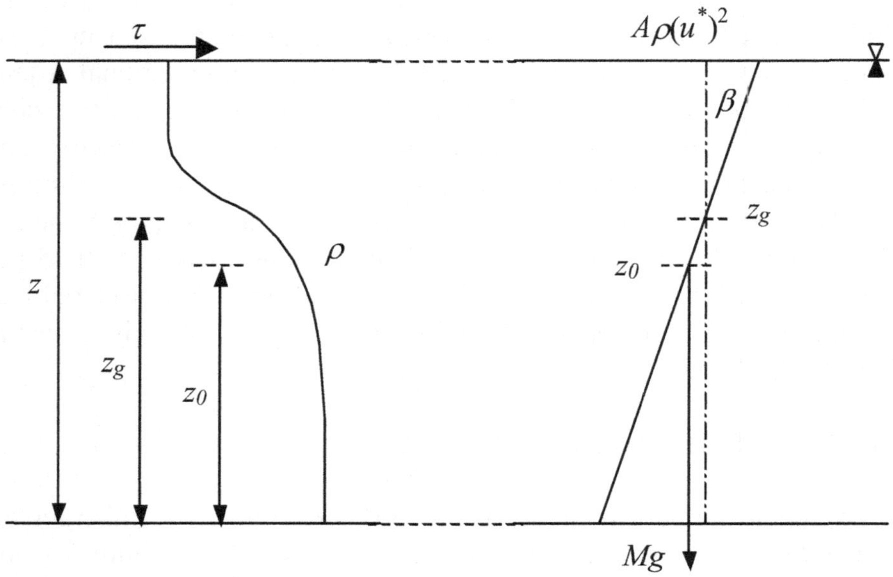

Figure 6.1 Lake Number variables, z is the height of the water column, z_0 is the centre of gravity of the water mass with a density stratification $\rho(z)$ at height z. M is the total mass of water and β is the angle subtended to the vertical by the line segment connecting the centre of mass to the centre of volume (reproduced from Antenucci and Imerito 2000).

As a wind stress τ acts on the surface layer a net force results that acts to overturn the density structure of the water column. Considering the moment around the water bodies centre of volume z_g at equilibrium, gives:

$$\left(z - z_g\right) \int_{A(z)} \rho_0 \left(u^*\right)^2 dA = \left(z_g - z_0\right) Mg\beta \qquad (6.1)$$

The Lake Number:

$$L_N = \frac{(z_g - z_0)Mg\beta}{\int\limits_{A(z)} \rho_0 (u^*)^2 \, dA (z - z_g)} \tag{6.2}$$

and z_0 is defined by:

$$z_0 = \frac{\int_0^z \rho(z) z A(z) dz}{\int_0^z \rho(z) A(z) dz} \tag{6.3}$$

The thermocline at height z_T, corresponding to the centre of the metalimnion, reaches the surface when maximum deflection occurs and β is defined by:

$$\beta = \frac{z - z_T}{\sqrt{A(z)}} \tag{6.4}$$

Substituting Equation 6.4 into 6.3 gives:

$$L_N = \frac{(z_g - z_0)Mg\left(1 - \frac{z_T}{z}\right)}{A^{1/2}\left(1 - \frac{z_g}{z}\right)\int_{A(z)} \rho_0 (u^*)^2 \, dA} \tag{6.5}$$

The wind stress is assumed to be constant over the entire water surface; therefore Equation 6.5 can be reduced to:

$$L_N = \frac{(z_g - z_0)Mg\left(1 - \frac{z_T}{z}\right)}{A^{3/2}\left(1 - \frac{z_g}{z}\right)\rho_0 (u^*)^2} \tag{6.6}$$

The restoring force is greater than the disturbing force for $L_N >> 1$ and the deflection of z_0 is minimal. Subsequently the density structure of the water body is approximately horizontal and the assumption of one-dimensionality is valid. In real terms a Lake Number > 1 implies severe stratification dominating the forces imposed by the surface wind stress. The ramifications of severe stratification is minimal turbulent mixing in the metalimnion and/or the hypolimnion. The validity of the one-dimensional assumption must be extended to other disturbing factors such as inflow and the effects of the Earths rotation. Consider the disturbance force due to inflows; an inflow lake Number L_{Ni} criterion is developed:

$$L_{Ni} = \frac{(z_g - z_0)Mg\left(1 - \frac{z_T}{z}\right)}{A^{1/2}uQ\left(\frac{z_g - z_i}{z}\right)} \tag{6.7}$$

where u = the inflow speed
z_i = the depth of the inflow

Similarly, when L_{Ni} >>1, one-dimensionality is valid, and the stratified conditions dampen the horizontal instability induced by the inflows. The Earths rotation can induce horizontal movement in a water body is defined by the ratio (Patterson, Hamblin and Imberger 1984; Antenucci and Imerito 2000);

$$R = \frac{R_i}{B} \tag{6.8}$$

where B = the maximum width of the lake or Reservoir
R_i = the internal Rossby radius of deformation defined as,

$$R_i = \frac{\sqrt{g'h}}{f} \tag{6.9}$$

where h = the surface layer depth
g' = the effective gravity defined in Equation 6.4

Subsequently for $R > 1$ the water bodies one-dimensionality is valid. Prior to using DYRESM, the one-dimensional assumption checks should be performed, as these are not undertaken in the numerical model itself (Antenucci and Imerito 2000).

The numerical model DYRESM

DYRESM is based on a Langrangian layer system where the slab volumes are maintained between prescribed limits via slab partitioning and amalgamation, (Figure 6.2), as opposed to a fixed grid approach, e.g. PROTECH (Elliot, Irish and Reynolds 2002), which would be defined as a Eulerian Scheme.

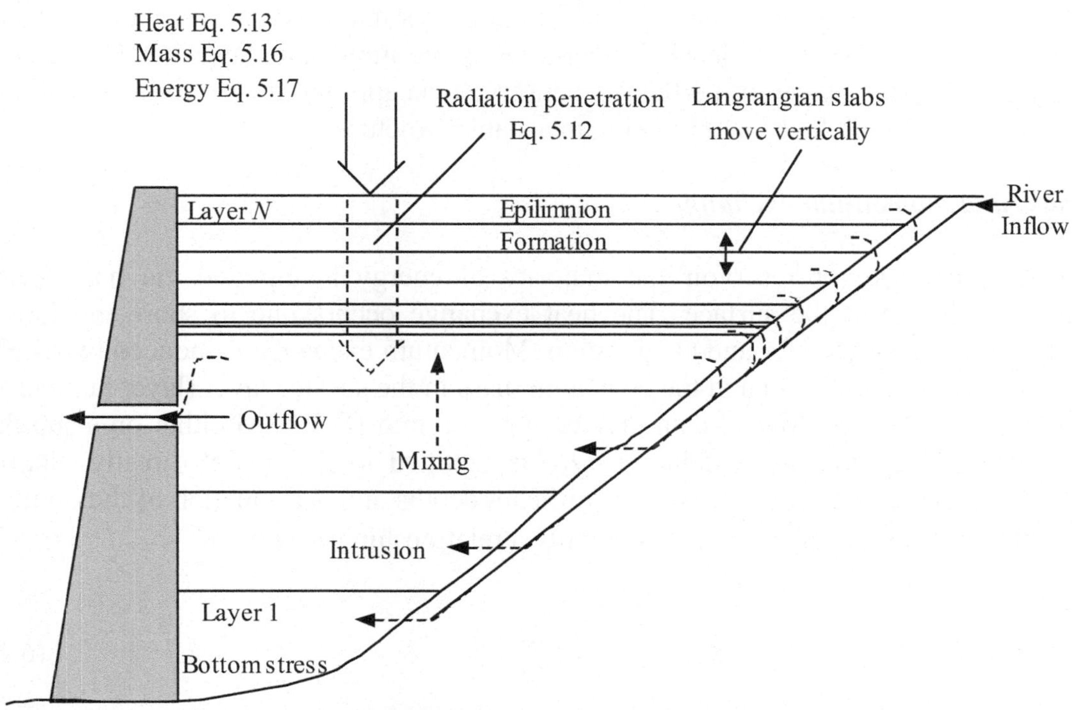

Figure 6.2 Schematic diagram of DYRESM based on Fischer et al. (1979).

The layer structure accommodates the seasonal behaviour found lakes and reservoirs where conditions can rapidly change from isothermal to stratified. During stratified conditions a mixed layer exists at the top of the lake or reservoir (epilimnion) where the biogeochemical properties are relatively constant so that only one layer is required. Whereas, across the metalimnion large vertical gradients exist so that many layers are required. Conversely isothermal conditions require minimal layers as the water body is well mixed. The layer volume, area and thickness relationship calculates the volume and area as a function of layer thickness or the thickness and area as a function of layer volume. The layers are numbered from the bottom up, as shown in Figure 6.2. Intuitively, the smaller the number of layers prescribed to a simulation run, the less the actual run-time. However to achieve adequate resolution and avoid an excessive number of layers developing, the upper and lower thickness limits are set.

Layer density

As the water bodies temperature decreases the density usually decreases, except under special circumstances where the water temperature is less than 4 °C. On the contrary, as the salinity increases the density increases. The water layer density (kg m^{-3}), given its temperature (°C), salinity (psu), pressure (bar) is calculated using the UNESCO (1981) equation of state for the density of salt water (Antenucci and Imerito 2000):

$$\rho(T,S,P) = \frac{\rho(T,S,0)}{\left(1 - P/K(T,S,P)\right)} \tag{6.10}$$

To ensure that the layer structure is stable the individual layer densities are checked from the top down. If the upper layer density is higher than the immediate layer below, the two laws are amalgamated and the layer properties are conserved via appropriate governing equations. The process is repeated for all layers culminating in a stable structure.

Heat, mass and momentum exchange

In a typical lake or reservoir the majority of energy is inputted via heat, mass and momentum exchange at the surface. The heat exchange occurs due to short and long wave radiation, sensible heat transfer and evaporation. Momentum exchange is induced by wind stress and mass transfer is due to rain and the change in mass in the surface layer (layer number N) due to latent heat flux. Meteorological data may be inputted into DYRESM either on a sub-daily or daily time-step. The short-wave radiation Q_{SW} is inputted to the model directly. Short-wave radiation will penetrate the water body to significant depths and is a function of the on the water bodies light attenuation coefficient η. Commonly a relationship given by:

$$\frac{Q_Z}{Q_{SW}} = e^{-\eta(H-z)} \tag{6.11}$$

where Q_Z is the radiation remaining at depth z (Fischer et al. 1979).

In DYRESM 55% of the incoming solar radiation is assumed to be non-penetrative, with the residual 45% (Gates 1966) being dispersed throughout the water column layers according to:

$$\Delta Q_{SW_j} = Q_{SW_j}\left(1 - e^{-\eta(z_j - z_{j-1})}\right) \tag{6.12}$$

The penetrative short-wave radiation has a wavelength < 700 nm and is known as Photo-synthetically Active Radiation (PAR). The following heat fluxes are all absorbed in the water surface layer, i.e. non penetrative. The net long-wave radiation can be inputted either as incoming long-wave, net long wave (defined as Q_{in}-Q_{out}) or cloud cover data may be used. The heat flux due to conduction, known as the sensible heat transfer Q_{SH} is calculated within the model by:

$$Q_{SH} = C_S \rho_A C_P U (T_A - T_S) \Delta t \tag{6.13}$$

where $C_S \approx 1.3$ E-3, the sensible heat transfer coefficient
$C_P = 1003$ Jkg^{-1} K^{-1}, the specific heat of air at constant pressure
$\rho_A = 1.2256$ kg m^{-3}, the density of air
U = the wind speed at the water surface in m s^{-1}
T_A = the air temperature
T_S = water surface temperature
Δt = the time-step

The loss of heat due to evaporation Q_{LH} is given by:

$$Q_{LH} = min\left(0, \frac{0.622}{P} C_L \rho_A L_E U (e_a - e_s(T_S))\Delta t\right) \quad (6.14)$$

where C_L = 1.3 E-3, the latent heat transfer coefficient
L_E = 2.453 E-6 Jkg^{-1}, the latent heat of evaporation of water
P = the atmospheric pressure in milli-bars
e_a = the air vapour pressure
e_s = the saturation vapour pressure at the water surface

No condensation effects occur in the numerical model so that $Q_{LH} \leq 0$. The total non-penetrative energy Q_{NP} received in the surface layer is given by:

$$Q_{NP} = Q_{LW} + Q_{SH} + Q_{LH} \quad (6.15)$$

where Q_{LW} is the long wave energy flux

There are three methods of calculating Q_{LW} within DYRESM depending on the method of data input, e.g. incident long wave radiation, percentage cloud cover or net long wave radiation. The change in mass across the surface layer is given by:

$$\Delta M_i = \underbrace{\frac{-Q_{LH} A_N}{L_w}}_{latent\ heat\ mass\ flux} + \underbrace{\frac{\rho_N A_N R_h \Delta t}{N_d}}_{rain\ mass\ flux} \quad (6.16)$$

where L_w = the latent heat of evaporation of water
R_h = the daily total rainfall
N_d = the number of seconds in a day

The properties of the rain are assumed to be the same as in the surface layer. Wind stress along the water surface causes momentum exchange into the water column and is included in DYRESM. The wind energy available to mix is defined as E_{stir} and is available only to the surface layer and given by (Imberger and Patterson 1981):

$$E_{stir} = \eta_S \rho_i A_i (u^*)^2 \Delta t \quad (6.17)$$

where η_s = the wind stirring efficiency
ρ_N = the surface layer density
A_N = the area of he surface layer
u^* = the shear velocity calculated by:

$$u^* = \left(\frac{C_D \rho_a}{\rho_N} \right)^{1/2} u_a \qquad (6.18)$$

where u_a = the wind velocity at a reference height of 10 m

Mixing mechanisms

Mixing within the layer structure is undertaken with three specific mechanisms, wind mixing, convective overturn and shear transfer between layers. The available kinetic and potential energies of the individual layers, starting at the surface layer, are used to determine if adjacent layers will mix. If adequate energy is available, the layers will mix with any excess energy used to determine if the subsequent layers are able to mix. The amalgamation of layers ceases when there is insufficient energy left, which is carried over to the next time step. The epilimnion height (h_E) and property values are those of the surface layer. The hypolimnion height (h_H) and properties are that of the layer immediately beneath the epilimnion.

The magnitude of the bottom stress is calculated to allow for re-suspension of sediment within CAEDYM. The stress is calculated using the water velocities in the epilimnion and the hypolimnion. The epilimnion velocity (u_E) is equal to the surface layer velocity calculated from:

$$u_N = \frac{(u^*)^2}{\Delta z_N} \Delta t \qquad (6.19)$$

Subsequently the hypolimnion velocity is calculated by:

$$u_H = \begin{cases} \dfrac{u_E h_E}{h_H} & \forall\ h_E < h_H \\ u_E & \forall\ h_E \geq h_H \end{cases} \qquad (6.20)$$

Using the drag coefficient c_D the bottom stress's for the epilimnion and hypolimnion can be calculated from:

$$\tau_E = C_D \rho_E u_E^2 \qquad (6.21)$$

$$\tau_H = C_D \rho_H u_H^2 \qquad (6.22)$$

The mechanisms of hypolimnetic mixing are separated into boundary and internal mixing. To simulate the action of the benthic boundary layer, aliquots from each hypolimnetic layer are mixed into the lowest epilimnion layer, the thermocline layer (Antenucci and Imerito 2000). Starting at layer 1, internal mixing is achieved by removing a percentage of each layer and mixing it with the layer above.

The inflow to the water body is modelled by inserting the volume into the layer of equal density taking into account the associated entrainment as the inflow passes layers of greater density. The layer structure is maintained within the limits set by the user and amalgamation of layers occurs accordingly. Outflow will occur with withdrawal and overflow events. The water is removed from the corresponding layer to the withdrawal height; if the volume of this layer is less than the withdrawal volume, then water is extracted from layers above the withdrawal height.

Destratification algorithm

DYRESM incorporates a bubble plume algorithm used to simulate an aerator employed for destratification. The plumes generated by the aerator diffuser are modelled with buoyant plume equations assuming that the plumes are non-interacting and axisymmetric. The initial volumetric flow rate of entrained water is calculated with:

$$Q_P = \alpha \frac{6\pi}{5} b_1 L_R B^{1/3} z_1^{5/3} \tag{6.23}$$

where α = the entrainment coefficient (within the range 0.04 – 0.14)
b_1 = a constant (= 4.7, Fischer et al. 1979)
L_R = the ratio of plume radius to plume length (= 0.1)

The initial buoyancy flux is calculated with:

$$B = g \underbrace{Q_{Air} \left(\frac{P_{Air}}{P_{Diffuser}} \right)^{0.71}}_{Corrected\ free\ flow\ rate} \tag{6.24}$$

Note that the free airflow rate is corrected for adiabatic compression at the diffuser height (assuming the air is an ideal gas). To calculate subsequent entrainment the air flow rate is adjusted in each layer to account for the adiabatic expansion of the air bubbles and a combined buoyancy flux of the entrained water and air bubbles is calculated by:

$$B_i = gQ_i - g\left(\frac{\rho_i - \rho_P}{\rho_i} \right) Q_P \tag{6.25}$$

where ρ_i = the density of the current layer.

Subsequently the entrained volume in layer i is calculated by:

$$Q_{Pi} = \alpha \frac{6\pi}{5} b_1 L_R B^{1/3} \Delta z^{5/3} + Q_{Pi-1} \tag{6.26}$$

The insertion of the entrained water into a neutrally buoyant layer occurs when Equation 6.24 becomes negative. For a comprehensive description of DYRESM refer to the Science Manual by Antenucci and Imerito (2000).

The ecological model CAEDYM

The model is designed to offer substantial flexibility for the user with the choice of state variables. An extensive description of CAEDYM is contained in the Science Manual authored by Herzfeld and Hamilton (2000).

Phytoplankton

The phytoplankton groups that can be modelled with CAEDYM include any combination of freshwater phytoplankton including the cyanophytes, chlorophytes diatoms and dinoflagellates. The biomass of each phytoplankton group is represented by its total chlorophyll *a* concentration. The generalised maximum growth rate at 20 °C is a function of *PAR*, phosphorus, nitrogen, silica, carbon and temperature that is represented by:

$$r' = r_{20°C} \times min\{f(I), f(N), f(P), f(Si), f(C)\} \times f(T) \tag{6.27}$$

where $f(T)$ = a temperature function described by Equation 6.30
$f(C)$ is used for explicitly modelling the internal carbon
$f(Si)$, $f(P)$, $f(N)$ and $f(I)$ represent limitation by silica, phosphorus, nitrogen and *PAR* respectively.

Light limitation is modelled with the Webb model, which ignores photo-inhibition, and is given by:

$$f(I) = 1 - e^{\left(-I/I_k\right)} \tag{6.28}$$

where I = the incoming irradiance
I_k = a parameter for the initial slope of the associated *PI* curve

Michaelis-Menten kinetics is used to simulate nutrient limitation of phytoplankton growth. For phosphorus limitation the Michaelis-Menten term is:

$$f(P) = \frac{PO_4}{PO_4 + K_P} \tag{6.29}$$

where K_P = the half saturation constant for phosphorus on growth rate

The equation for the bio-available dissolved carbon takes the same form as Equation 6.29 with K_P replaced with the half saturation constant for carbon K_C. The half saturation constant for nitrogen (K_N) is set to zero for nitrogen fixing cyanobacteria.

Figure 6.3 shows $f(P)$ as a function of $[PO_4]$.

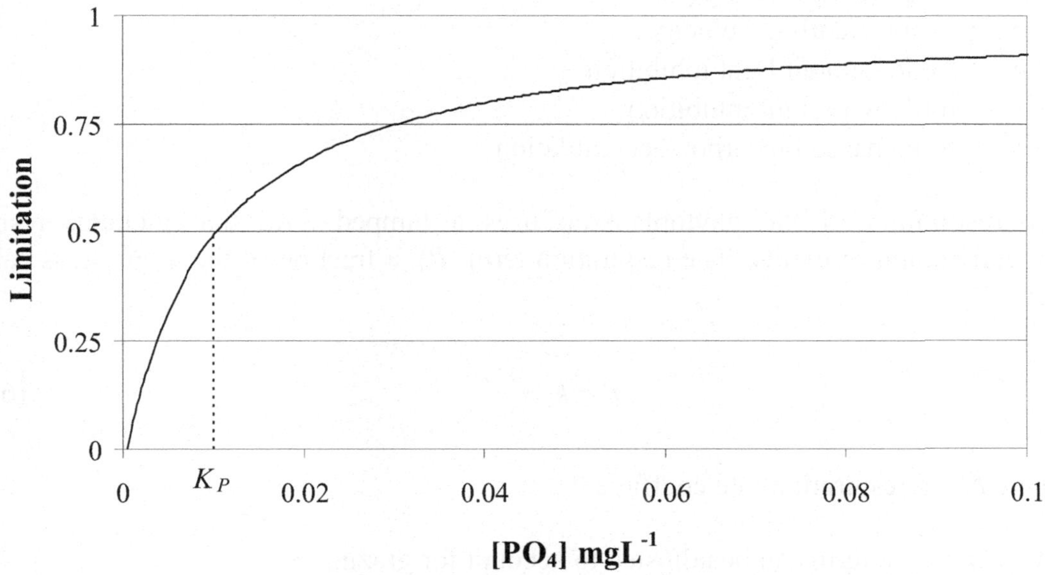

Figure 6.3 Michaelis-Menten nutrient limitation for dissolved inorganic phosphorus.

Figure 6.3 shows that the limitation rate is proportional to the concentration of phosphorus, and as nutrient concentration increases the limitation order decreases from one to zero.

Maximum growth for phytoplankton occurs at an optimum temperature T_{opt}. For $T < T_s$, the standard temperature, temperature limitation to growth is calculated by:

$$f(T) = \vartheta^{T-20} \tag{6.30}$$

where ϑ = a non-dimensional temperature multiplier

When the temperature increases above T_{opt}, $f(T)$ reduces to zero where the temperature is defined as T_{max}. The temperature limitation equation for $T > T_s$ has the form:

$$f(T) = \vartheta^{T-20} - \vartheta^{k(T-a)} + b \tag{6.31}$$

where k, a and b are unknown which are solved using the boundary conditions:

$$\begin{aligned} @T = T_s, \quad & f(T) = 1 \\ @T = T_{opt}, \quad & \partial f(T) / \partial T = 0 \\ @T = T_{max}, \quad & f(T) = 0 \end{aligned} \tag{6.32}$$

There are several options for vertical migration and settling of phytoplankton in CAEDYM, namely:

- Stokes settling velocity
- Constant settling velocity
- Motile without light inhibition
- Motile with light inhibition
- Motile based on carbon accumulation

The respiration of the phytoplankton uses a lumped term that includes respiration, excretion and natural mortality. The respiration term (*R*), a fraction of the biomass, is calculated by:

$$R = k_r \vartheta^{T-20} \tag{6.32}$$

where k_r = a respiration rate coefficient

The respiration term can be adjusted to account for grazing.

Dissolved oxygen

The modelled reactions of dissolved oxygen in the water column are associated with the following processes:

- Sediment chemistry
- Air/water surface exchange
- Phytoplankton photosynthesis and respiration
- Nitrification

There are two options to model the utilisation of oxygen at the sediment surface, either with a static model or a dynamic model. The oxygen flux at the sediment surface is related to dynamic oxygen budgets (see Herzfeld and Hamilton 2000). In the static model, sediment oxygen flux is a function of the adjacent ambient water temperature and dissolved oxygen concentration.

The flux of oxygen across the air-water boundary is modelled using the method presented in Wanninkhof (1992) where:

$$F = k(C_a - C_w) \tag{6.33}$$

where F = the oxygen flux (positive into water column)
k = the oxygen transfer coefficient ,
C_a = the concentration of oxygen in the air boundary layer, a function of temperature and salinity
C_w = the oxygen concentration in the water surface boundary layer

Input files

The input files required to run the model include:

1. a file which contains configuration information required for the simulation
2. a description of the lake morphometry, elevation, latitude, lake bathymetry at the inflow and outflow height.
3. an initial profile of temperature and salinity
4. meteorological data at the appropriate time step including short wave radiation, long wave radiation, air temperature, vapour pressure, wind speed and rainfall.
5. information on the stream inflow volume, temperature and salinity
6. information on the withdrawal volume
7. a parameter file with constants for processes described by the model

Modelling toxin production and degradation

Species of concern

Toxins,taste and odour compounds for a number of species will eventually be considered for the model of metabolite production and degradation. However, for the first composition and these model coding details, we will consider *Microcystis*. It will be necessary to link the remaining species to the algorithms described in 'processes'.

Processes to be considered

To model toxin kinetics the following processes need to be considered

1. production
2. cell lysis
3. adsorption
4. degradation

Conceptual model

Phytoplankton lysis causes toxin release. This toxin can be degraded/removed from the water body in two ways; microbial degradation which is influenced by temperature; and adsorption, which is influenced by convection within the water body (Figure 6.4) For the purpose of this model, we assume that the water body is well mixed at all times.

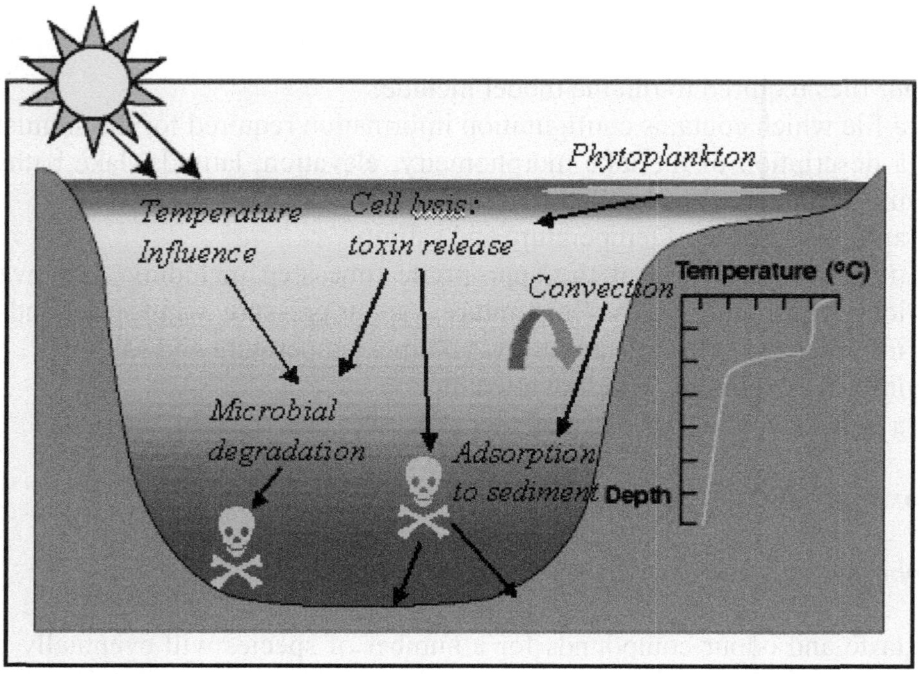

Figure 6.4 Conceptual model of the processes being considered.

Production

At present, the exact purpose of toxin metabolites in cyanobacteria is unknown. Various theories have been suggested over past research documents, and are discussed below. There has been previous suggestion that toxins such as microcystin have a secondary function (Carmichael 1992), and more recently, some have also suggested that it may have a primary function. Shi, Carmichael and Miller (1995) proposed that microcystin is not stored in *Microcystis aeruginosa* cells, but could be involved in specific cellular activity, (ie. a primary function) such as the regulation of protein phosphorylation.

Despite the lack of definite knowledge on the purpose of microcystin synthesis, their chemical structure and properties are well documented. Toxins of cyanobacteria tend to be highly soluble in water, due to their chemical structure. Microcystins are cyclic peptides, where most strains, including –LR, support hydrophilic interactions (Lyck 2001). In *Microcystis aeruginosa*, microcystin usually remains contained within the cells and is released only upon cell lysis (Lam et al. 1995, Park et al. 2001, Lahti et al. 1997, Chorus and Bartram 1999). Since most microcystins are hydrophilic, artificial removal from water bodies is further complicated.

In cyanobacterial blooms, microcystin-LR is reported to be the most commonly occurring toxin within sample analyses (Sivonen 1996). It is also one of the more toxic microcystin strains (Lyck 2001).

There are many environmental factors which influence the absolute production and strain composition of microcystin. Nutrient availability is a particularly important factor. In a study by Long, Jones and Orr (2001) it was found that under nitrogen limited conditions, *Microcystis aeruginosa* displayed an apparent linear relationship between the specific growth rate and cell quota of microcystin. This study was based on lab cultures.

This increases linearly with growth of cyanobacteria. Of the produced toxin, percentage α is particle bound and percentage β is dissolved such that α + β = 100%. For microcystin, set α = 100%; β = 0%.

The microcystin cell quota produced is determined as in Equation 6.35. This will be used in 'cell lysis' to calculate the microcystin released upon lysis of the cells. A positive linear relationship exists between microcystin cell quota and specific growth rate (Figure 6.5).

$$Q_{mcyst} = \mu(Q_{mcyst\ max} - Q_{mcyst\ min})/\mu_{max} \tag{6.35}$$

where μ = specific growth rate, day^{-1}
Q = cell quota, fmols $cell^{-1}$

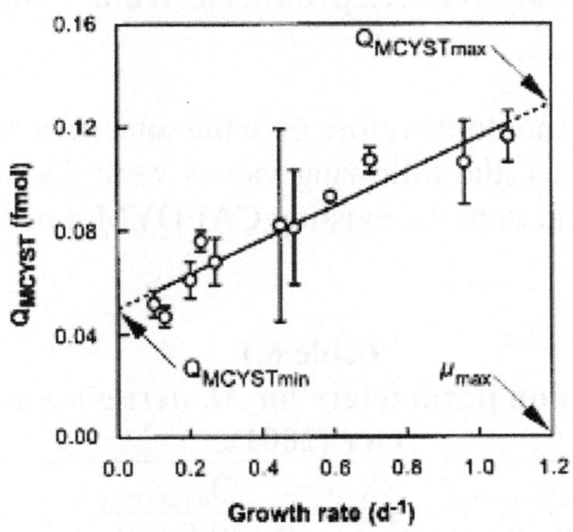

Figure 6.5 Cellular microcystin quota of *M. aeruginosa* grown in N-limited chemostats. (Reproduced from Long, Jones and Orr 2001).

Net microcystin production rate correlates positively with growth rate to form the hyperbolic function, as in Figure6.6.

$$R = \mu(\mu.Q_{mcyst\ max} - \mu.Q_{mcyst\ min})/\mu_{max} + \mu.Q_{mcyst\ min} \tag{6.36}$$

where R = net microcystin production rate
μ = specific growth rate
Q = cell quota, fmols $cell^{-1}$ (i.e. R=0 at 0 growth)

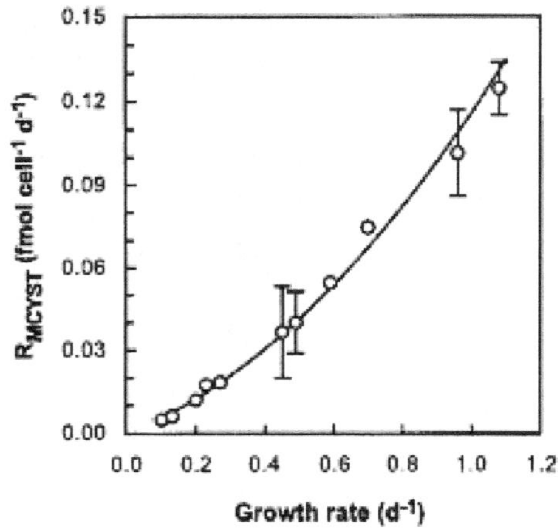

Figure 6.6 Net microcystin production rate as a function of specific growth rate of *M. aeruginosa* in N-limited chemostats. (Reproduced from Long, Jones and Orr 2001).

The cumulative sum of R should therefore give the total microcystin produced.

Under N-limited chemostats, the following values were derived for Q and R (Table 6.1). Use whichever remains consistent with the existing CAEDYM model units.

Table 6.1
Derived Microcystin production parameters for *M. aeruginosa*. Source: Long, Jones and Orr (2001).

Measurement	$Q_{mcyst\ min}$	$Q_{mcyst\ max}$	$R_{mcyst\ max}$
Cell quota basis	0.050 fmol cell^{-1}	0.129 fmol cell^{-1}	0.155 fmol cell^{-1} day^{-1}
Dry wt basis	0.71 mg g^{-1} $_{(dry\ wt)}$	7.6 mg g^{-1} $_{(dry\ wt)}$	9.12 mg g^{-1} $_{(dry\ wt)}$ day^{-1}

Cell lysis

This is linked to death rate (in existing CAEDYM module). We assume that upon death, all cells lyse, and that upon lysis, 100% of microcystin becomes dissolved. i.e. for lysed cells, set α = 0%; β = 100%. Hence the total microcystin released is:

$$Q_{mcyst\ dissolved} = C_{lysed} * Q_{mcyst} \quad (6.37)$$

Where C_{lysed} = number of cells lysed or dry wt of lysed cells (with respect to the units used for 'production')

Adsorption

The Freundlich isotherm is a measure of adsorption rate of a particular substrate onto an adsorbent. It can be applied to various situations where adsorption processes occur, including herbicides (Cox et al. 1997, Fushiwaki and Urano 2001), taste and odour compounds MIB and geosmin (Cook, Newcombe and Sztanjnbok 1998), and cyanobacterial toxins such as microcystins (Miller et al. 2001).

Consequently, aside from the study by Miller et al. (2001), for the purpose of modelling adsorption degradation of microcystin, studies concerning adsorption by Cox et al. (1997) on polar herbicides, and Cook, Newcombe and Sztanjnbok (1998) on MIB and geosmin have been considered to determine general adsorption properties and their consequent trends with regard to the eventual model.

The amount of toxin adsorbed is given by the Freundlich isotherm:

$$S = k_f . C_{eq}^{\ n} \tag{6.38}$$

Where S = toxin adsorbed into sediment (mg kg^{-1}) at time t

C_{eq} = [toxin] in solution at equilibrium (mg L^{-1})

k_f = Freundlich adsorption coefficient (L kg^{-1})

n = Freundlich (linearity) exponent

(Cox et al. 1997; Cook, Newcombe and Sztanjnbok 1998; Miller et al. 2001)

Equation 6.38 is empirical; hence values for C_{eq}, k_f and n must be experimentally determined. For initial purposes, set n = 1.

In general, linear adsorption isotherms were observed by Miller et al. (2001) for the toxin solution (mL):sediment (g) v/w ratio, with the exception of the case 25:2 where k_f =3.8, n =0.673. It is apparent that an increasing ratio gives a lower k_f value (Miller et al. 2001). This poses a question for further investigation; does a smaller ratio lead to a trend of non-linear isotherms?

A linear relationship is expected between a log-log plot of toxin adsorbed per unit of sediment vs toxin concentration at equilibrium. The toxin concentration at equilibrium varies with the sediment dose and sediment composition (Fukiwaki and Urano 2001). From the graph below (Figure 6.7), we can deduce that if less toxin is adsorbed, then the concentration of toxin at the corresponding equilibria point will increase. Since the relationship in the graph below is positive and linear, a lower sediment dose correlates to lower adsorption rate per unit of sediment, (ie. less is adsorbed per unit of sediment per unit of time) (Cook, Newcombe and Sztanjnbok 1998). Miller et al. (2001) also note that adsorption is heavily dependent on hydrophobic and hydrophilic interactions of the toxins.

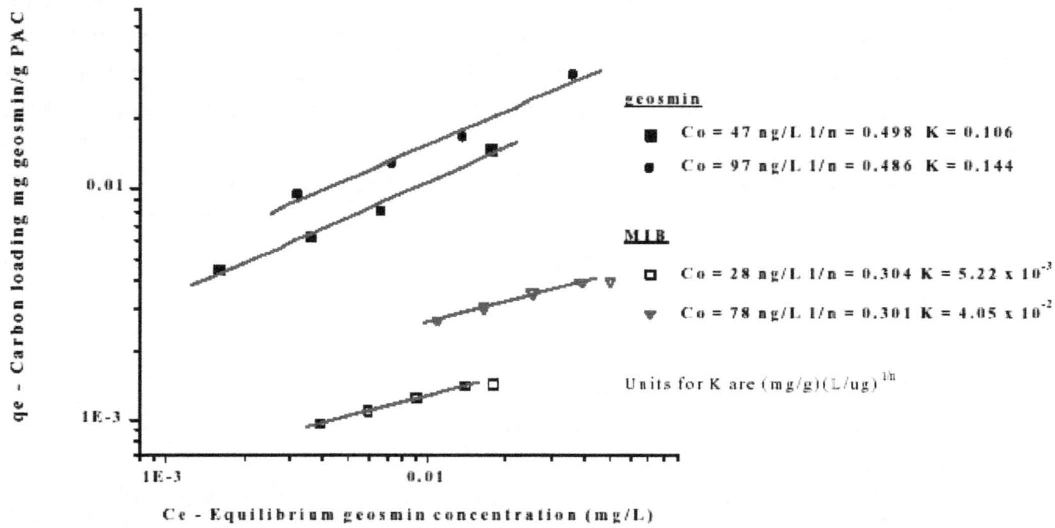

Figure 6.7 Sample of expected relationship. (Reproduced from Cook, Newcombe and Sztanjnbok 1998).

Chorus and Bartram (1999) suggested that in general, a maximum of 20% of the total microcystin concentration is adsorbed by soils and lake sediment in reservoir systems, where soils retain adsorbed toxin more effectively than sediment columns. It is also to be noted that after adsorption, bacteria in the sediment then act to degrade the toxin. When cells rather than toxin is adsorbed however, it is likely that the bacterium within the soils will degrade the cells and are thus likely to cause toxin release on a secondary level (Chorus and Bartram 1999). Within the model, it is to be assumed that all bacterial degradation is due to bacteria in the water column.

Degradation (bacterial, UV)

It has been suggested that microcystin degradation as a result of ultra-violet and visible light occurs only in the presence of a catalyst, such as titanium dioxide, TiO_2 (Feitz et al. 1999, Lawton et al. 1999). Highly oxidising conditions, for example, under high levels of ozone, O_3 or chlorine, Cl have also resulted in degradation of microcystins (Keijola et al. 1988, cited by Lawton et al. 1999).

In contrast, it has been suggested that intense UV radiation alone can degrade microcystins (Benoufella et al. 1994, cited by Lyck 2001). Further investigation must be done to confirm and otherwise determine the relationship between oxidation of microcystins and light intensity. However, UV degradation is not being considered here

Bacterial degradation can potentially account for over 90% of total toxin degradation, depending on the water body considered, initial concentration of toxin and water temperature, (Jones et al. 1994, Lahti et al. 1997, cited by Chorus and Bartram 1999). Lahti et al. (1998) note that in the field situation, microcystins may persist for several weeks after a cyanobacterial bloom.

In studies concerning bacterial degradation of microcystins, there are some authors who report a lag period of around 3 days before such degradation begins (Takenaka and Watanabe 1997; Lahti et al. 1998; Christoffersen, Lyck and Winding 2002), while others detected degradation begin immediately (Park et al. 2001; Christoffersen, Lyck and Winding 2002).

Rapala et al. (1994) suggests that a lag may be associated with the first bloom occurring in the water body. This is explained by the presence and store of inocula in the water, where prior to a bloom, there is little inocula present, so therefore the population of inocula must grow in order to effectively degrade the toxin. Soon after a bloom, this inocula is likely to remain for a period, and hence if the water sample has previously experienced cyanobacterial toxin release, the inocula will still be present. Hence, degradation is immediately evident (Rapala et al. 1994; Park et al. 2001; Christoffersen, Lyck and Winding 2002).

Bacterial degradation is linked to population growth and thus toxin increase as an exponential expression, (Equation 6.5). See Figure 6.8.

$$Toxin(t)=T_o e^{-kt} \tag{6.39}$$

where k is a function of temperature

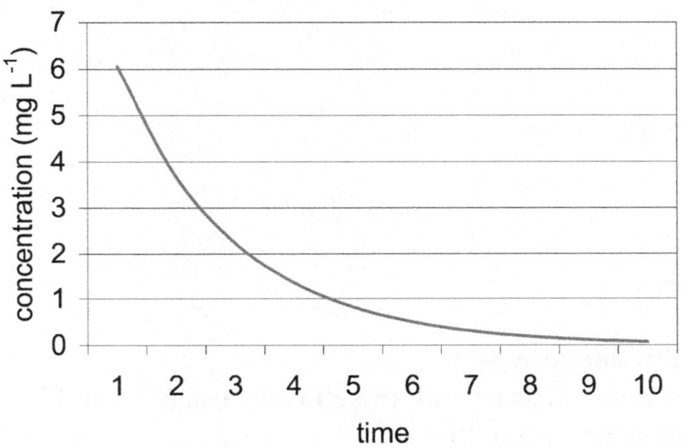

Figure 6.8 Predicted shape of curve for degradation, see Equation 6.5.

We can deduce a general formula to describe the relationship between the degradation rate and temperature. The degradation rate, k is a linear function, and therefore must be mathematically defined k in Equation 6.39 as in Equation 6.40. (see Figure 6.9).

$$k = m.T \tag{6.40}$$

where T = temperature
m = constant

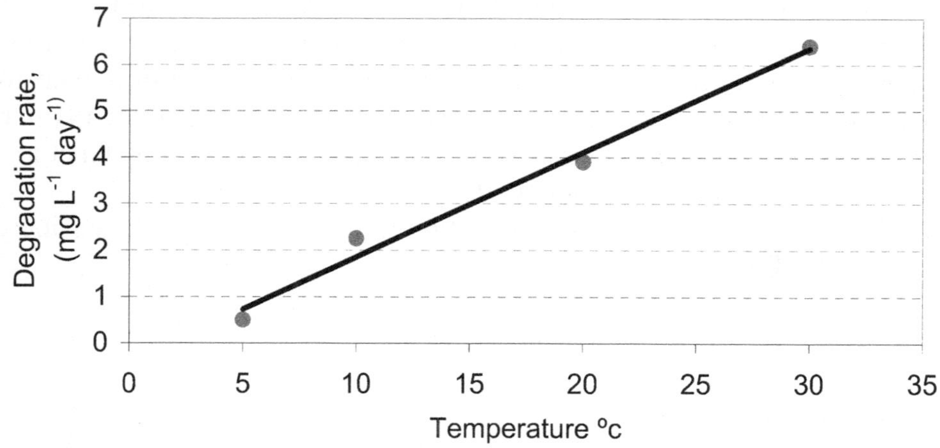

Figure 6.9 Degradation rate of microcystin-RR as a function of temperature, (reproduced from Park et al. 2001). This provides the predicted shape of curve for Equation 6.5. m value is yet to be defined.

Hence Equation 6.41:

$$Toxin(t)=T_oe^{-m.T.t} \tag{6.41}$$

Summary of algorithms

To calculate **production**, use 6-36.
To calculate toxin released as a result of **cell lysis**, use 6-35; 6-37.
To calculate **adsorption**, use 6-38.
To calculate **degradation**, use 6-41.

DISCUSSION

The production of the metabolite sub-routine describing the production and degradation of toxin or taste and odour compounds is an important addition to the reservoir and lake model. By using the model to predict the risk from cyanobacteria in any given reservoir the operators are better equipped to manage the hazards in the both the reservoir and the water treatment plant.

There is a data requirement for model use but the expense associated with this is considered modest compared with the improved operation, health and aesthetic outcomes that are realised with better system understanding and management.

The application of the model in a predictive capacity allows operators the ability to test management scenarios before implementation. Candidate scenarios could include, but are not limited to:

Nutrient reduction

What reduction in catchment derived nutrient load is required to reduce the cyanobacterial population to a concentration that will ensure the toxin or taste and odour thresholds are not reached?

Artificial destratification

What is the impact of artificial destratification on cyanobacterial growth and toxin production or geosmin production?

Treatment plant design

What is the likelihood and likely concentration of cyanobacterial toxins in the source water and what treatment options should be incorporated into plant design to remove these compounds and produce aesthetically acceptable water?

Recreational waters

What is the probability that recreational waters will experience cyanobacterial blooms?

CHAPTER 7:APPLICATION OF THE CYANOBACTERIAL TOXIN MODEL AT NORTH PINE DAM, QUEENSLAND AUSTRALIA

INTRODUCTION

North Pine Dam was selected as the location to test the toxin production and biodegradation model. North Pine Dam is a major water supply reservoir to the city of Brisbane, Australia and has in recent years experienced severe issues with blooms of toxic cyanobacteria including *Cylindrospermopsis raciborskii*. *Cylindrospermopsis raciborskii* has typically been classified as characteristic of semi-tropical water bodies (Neilan et al. 2003), however, recently its presence has also been identified at higher latitudes (Briand, Leboulanger and Humbert 2004). Similarly, it has been increasingly prevalent in water bodies both in Australia and countries around the world. Located at semi-tropical latitudes of southern Queensland, North Pine Dam has experienced increasing problems associated with this cyanobacterium.

The first recorded occurrence of *Cylindrospermopsis raciborskii* in North Pine Dam was in the late 1970's (Harris and Baxter 1996). The toxin produced is *cylindrospermopsin*, an alkaloid cytotoxin (McGregor and Fabbro 2000), while it also produces compounds that taint the taste and odour of potable water and affect water quality.

The aim of this research was to apply the toxin production and degradation model to North Pine Dam to test the ability of the model to predict the timing and magnitude of not only *Cylindrospermopsis* but also the toxin cylindrospermopsin.

METHODS

Study Site

North Pine Dam is located at 27 ° 15'S, 152 ° 55'E in south-eastern Queensland, 30 km north of Brisbane. Otherwise known as Lake Samsonvale, it is a major water supply reservoir to the City of Brisbane and also used as a recreational lake. The reservoir was impounded during December 1973, and first reached full supply during November of 1976, (King and Everson 1978). There are three main rivers that feed into the dam; North Pine River; Kobble Creek; and Terror's Creek. The site used for field data inputs, calibration and validation was at the dam wall. This particular site was the deepest in the reservoir and is monitored routinely for a suite of water quality parameters including phytoplankton.

Occurrences of cyanobacteria were first observed prior to 1976, particularly *Microcystis* species (King and Everson 1978, Harris and Baxter 1996). King and Everson (1978) also observed cyanobacteria and suggested that algal blooms had the potential become a nuisance in the dam. In retrospect, this has become evident in more recent years. *Cylindrospermopsis raciborskii*, another toxic cyanobacteria is also prevalent in North Pine Dam, the first observed and recorded occurrence was in 1978 (Harris and Baxter 1996). Between 1976 and 1978, prior to the dam first reaching its full storage capacity, elevated concentrations of chlorophyll *a* were observed, particularly during the wet season from November to April. There has also been suggestion of possible inoculation of the dam from the river inflows, (King and Everson 1978).

It has been identified that cyanobacteria dominate the phytoplankton community in North Pine Dam. The three dominant cyanobacteria species, *Aphanocapsa* spp., *Planktolyngbya* spp. and *Cylindrospermopsis raciborskii* (in order of dominance) comprise some 75 to 85% of the total phytoplankton community. *Achnanthidium* has been identified as the major diatom genus, while *Ankistrodesmus* is the major chlorophyte genus. Determination of the phytoplankton species assemblage and species dominance was based on cell count data. The species of *Aphanocapsa*, *Planktolyngbya*, *Achnanthidium* and *Ankistrodesmus* are yet to be identified.

At present, published information on characteristic growth parameters of these dominant species is limited. Consequently, on occasions where data has been unavailable, values adopted have been based on functionally similar phytoplankton, for example, *Planktolyngbya* and the average values for several cyanobacteria species. This information has been obtained from tables compiled by Reynolds (1984).

Growth characteristics

Growth rates of algal species are extremely variable, and also highly dependent on the historical and prevailing conditions of a system. In a review by Dokulil and Teubner (2000) the growth rates of several cyanobacteria species in lake ecosystems were summarised. *C. raciborskii* had a minimum and maximum recorded growth rate of 0.002 d^{-1} and 0.138 d^{-1} respectively, with a mean of 0.042 d^{-1} in field observations. Under laboratory conditions, it was found that an Australian strain recorded a maximum growth rate of 0.7 day^{-1}, (Briand, Leboulanger and Humbert 2004) *Planktolyngbya subtilis* had minimum, maximum and mean growth rates respectively of 0.010 d^{-1}, 0.081 d^{-1} and 0.032 d^{-1} based on change in biomass, (Dokulil and Teubner 2000).

The optimum temperature for growth of *C. raciborskii* is relatively high, between approximately 24 to 30 ºC (Istvánovics et al. 2000, Briand, Leboulanger and Humbert 2004). Dokulil and Teubner (2000) found in reviewing previously published data that the median temperature preferred by *C. raciborskii* was 22 ºC, with the maximum temperature preference was approximately 28 ºC. Istvánovics et al. (2000) also noted that based on previous data records, temperature was not the only condition that triggered blooms.

As a heterocystic cyanobacteria, *C. raciborskii* is able to utilise both atmospheric and dissolved forms of nitrogen. It has been recorded that higher dissolved concentrations of nitrogen have toxic effects on heterocystic species such as *C. raciborskii*, (Sprõber et al. 2003) but this is probably an extreme condition as cells grow happily in media with 1mM nitrate.

Istvánovics et al. (2000) investigated the effect of P supply on the growth of *C. raciborskii*. Based on data obtained at various growth rates under batch and continuous experiments, and under different P limiting conditions, Istvánovics et al. (2000) found that the Michaelis-Mentin model was a reasonable representation of P uptake characteristics. This is also the model adopted by CAEDYM. For a specific growth rate of 0.56 per day, the cellular quota of P was 6.1 mg (mg $C)^{-1}$, and the half saturation constant was 5.7 ± 1.3 (μg P) L^{-1}. It was also noted that the high affinity of *C. raciborskii* to P, and its ability to store internal P is an important factor contributing to its competitive ability.

Under field observations a PAR intensity of 30 – 60 μmol m^{-2} s^{-1} has been recorded for the optimal growth of *C. raciborskii*, (Dokulil and Mayer 1996). It is well known that dominance of cyanobacteria in phytoplankton community assemblages is often linked with eutrophic systems (Dokulik and Teubner 2000). Distinct correlations were also recorded between increased biomass of various cyanobacteria species and secchi depths, (Dokulil and Teubner 2000) which can be interpreted as a surrogate for the euphotic depth of the system. In particular, *C. raciborskii* appears to be able to sustain positive net increase in biomass at a large range of water clarity with secchi depths, ranging from approximately 1m to 16m, at temperatures over 15 ºC. Below 15ºC, there were no reported observations of growth at greater secchi depths in the literature reviewed by Dokulil and Teubner (2000)

In Queensland dams *C. raciborskii* has been observed to be seasonally dominant throughout the later and earlier months of the year, particularly between Oct and May, (Harris and Baxter 1996, McGregor and Fabbro 2000). Characteristically, the reservoirs exhibit thermal stratification of the water column during these months. A strong association between the seasonal variation in abundance of *C. raciborskii* and *Planktolyngbya* has also been observed in many Queensland reservoirs (McGregor and Fabbro 2000).

C. raciborskii blooms have frequently been correlated to warm water temperatures strongly stratified water columns, high surface irradiances (PAR) of approximately 2200 μmol m^{-2} s^{-1} and basic water column pH, are deemed 'ideal' conditions, (Fabbro and Duivenvroorden 1996, McGregor and Fabbro 2000). The time at which bloom peaks occur appear to correlate to latitude at which particular water bodies are located in Queensland. Peaks generally occurred between December and February.

Simulation inputs

Of the 5 species of interest, *C. raciborskii*, *Aphanocapsa*, *Planktolyngbya*, *Achnanthidium* and *Ankistrodesmus*, only 3 have been simulated. These species were simulated on the basis of available data. These three species included two cyanobacteria species, and one diatom, *Planktolyngbya*, *C. raciborskii* and *Achnanthidium* respectively.

Data for the period 2001-02 was collected from a number of sources to meet the model input requirements. Meteorological data was obtained from South East Queensland Water (SEQ Water) and the Australian Bureau of Meteorology. Water quality data, including nutrient inflows for the simulation setup, and algal cell counts for calibration and validation were also obtained form SEQ Water.

In preliminary simulations, a number of methods were tested to find an appropriate method to interpret nutrient data for input into the model. Linear regression between available data points was attempted, however the long periods between some data points resulted in some unusual outputs with poorly defined regression curves. Effectively, there was insufficient data across larger time scales with which to correlate nutrient concentrations with flow magnitude (methods as described by Linden et al. 2004) to predict likely nutrient inflows given daily inflow volume readings.

Some raw data sets were provided in intervals greater than that required. In these instances, linear regression was used to obtain complete data sets in the required temporal subdivisions.

The cylindrospermopsin simulation was performed in Dyresm-CAEDYM, and is based on the assumption that cylindrospermopsin is released into the water column only when cells lyse on mortality. An average toxin content of 0.17 μg (μg Chl a^{-1}) was assumed based on a study by Saker, Neilan and Griffiths (1999). Degradation of toxin was modelled by assuming a first order relationship between temperature and the degradation rate. From experimental evidence (chapter 5) a degradation rate constant of 0.204 has been assumed, based on a maximum observed degradation rate of 5.1 μg L^{-1} day-1 at 25°C

RESULTS

Simulation output: Hydrodynamics

Accurate representation of the physical reservoir environment is necessary as this forms the habitat template to which the biogeochemical and growth model is coupled. Temperature data are presented for the period 2001-2002 (Figure 7.1). Well mixed conditions are evident in winter June-August) with varying degrees of stratification in the other months. The DYRESM outputs were compared with temperature data collected from a thermistor string permanently deployed in the reservoir. The hydrodynamic output from DYRESM is accurate over each complete simulation period. Overall, the percentage error at any one time period was less than 0.25 %. A four week period in the middle of the simulation from 27 August to 24 September 2001 was chosen for a validation comparison between modelled and observed results (Figure 7.2). In this four week snapshot the relative error was consistently less than 0.15%. This General trend was exhibited across the entire simulation period (Figure 7.3).

The good representation of the hydrodynamics by DYRESM provided confidence that this model was adequately simulating the physical habitat and could be coupled to the model describing cyanobacterial growth.

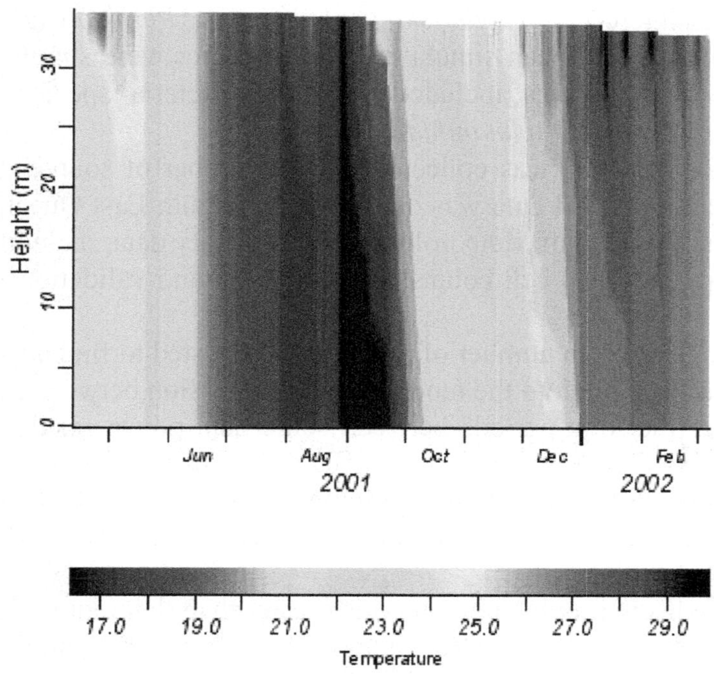

Figure 7.1 Simulated temperature profile for North Pine Dam from April 2001-March 2002.

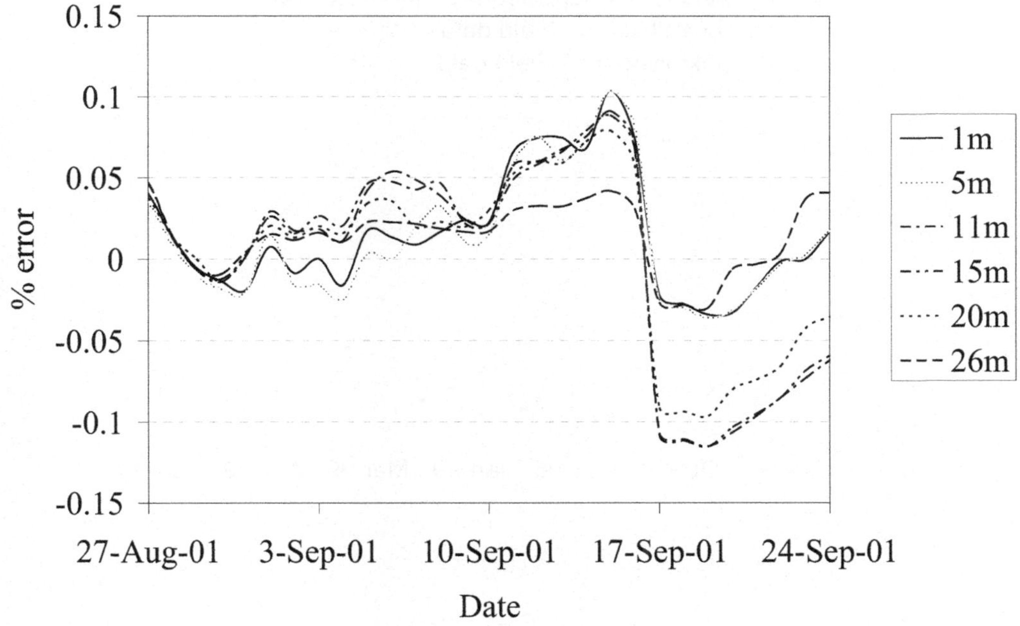

Figure 7.2 Relative error between field and DYCD temperature data, snapshot over 4 weeks.

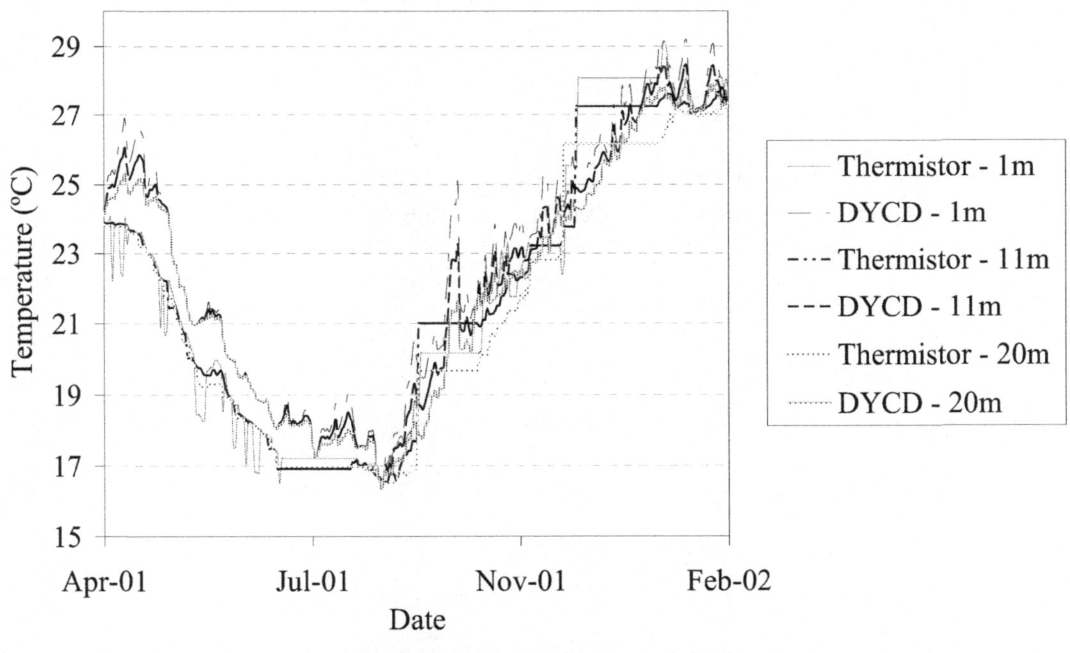

Figure 7.3 Comparison of temperature data from thermistor (field) recordings and DYCD output for selected discrete depths.

The cyanobacteria contributing to the majority of the phytoplankton community were *Cylindospermopsis*, *Achnanthidium*, and *Planktolyngya*. There is strong seasonal variation in their growth (Figure 7.4) with highest concentrations typically occurring in late spring through to early autumn. Phytoplankton growth for the 2001-2002 period was simulated using the ecological model CAEDYM.

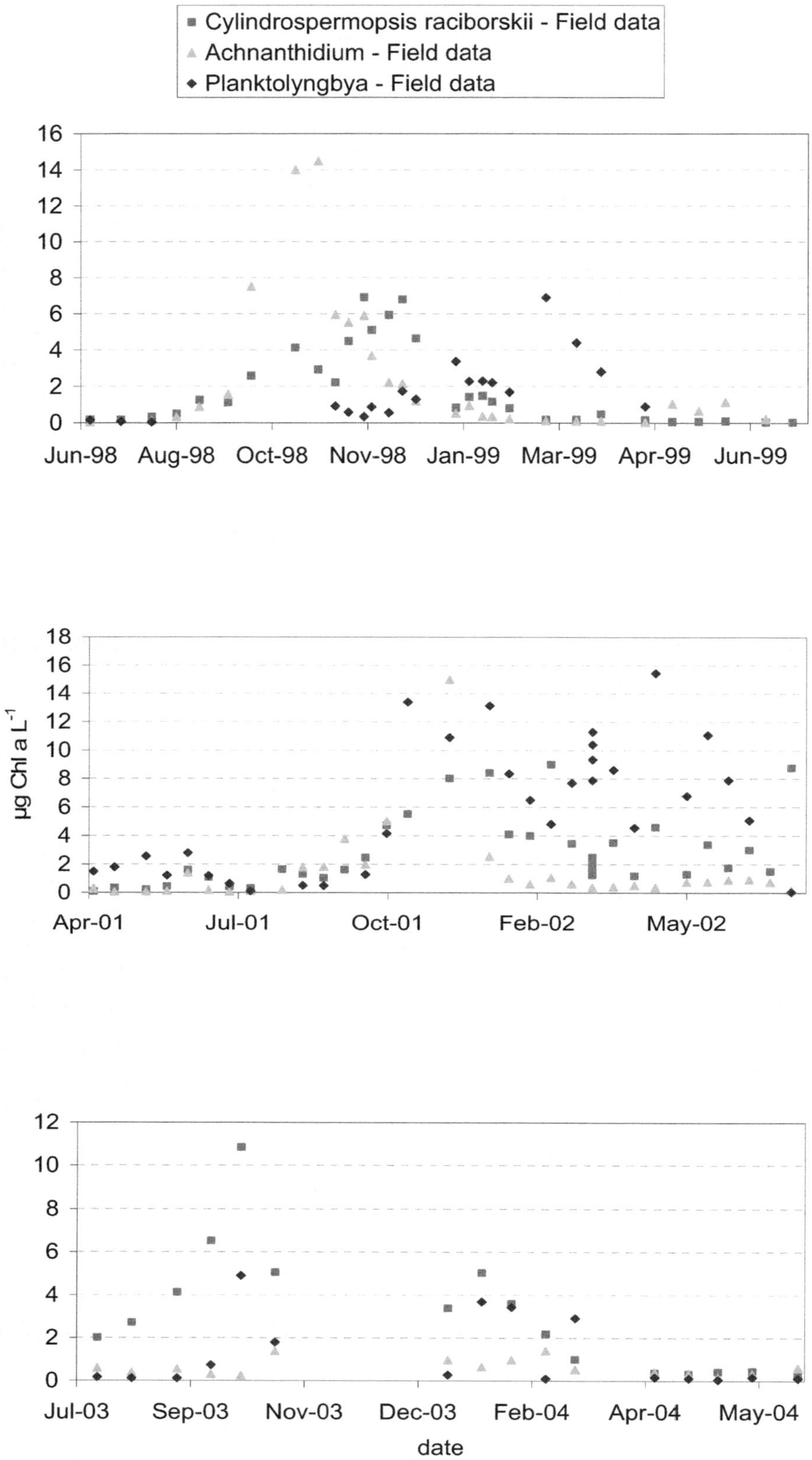

Figure 7.4 Phytoplankton population (*C. raciborskii*, Planktolyngbya and Achnanthidium) in, Top: 1998-99, Middle: 2001-02, Bottom: 2003-04 periods.

The timing and magnitude of blooms was similar for the field and the simulated data sets. For *Cylindrospermopsis*, comparison between field data and simulation outputs were strongly correlated; R^2 value of 0.83 and show a good visualfit (Figure 7.5). The good fit a preliminary indication of appropriate parameter selection for the DYCD simulation, and further supported by estimated total biomass across the season, at less than 19% error for both cyanobacterium groups. Predictions of *Achnanthidium* and *Planktolyngbya* also compared well with concentrations recorded in the reservoir (Figure 7.6-7.7) Predictions of *Achnanthidium* concentration were comparatively less accurate (Table 7.1), however, for a there is a ten week period of unrecorded field data which corresponded with a significant portion of the *Achnanthidium* bloom, and would tend to skew the modelled-observed comparison.

Planktolyngbya gives the least accurate fit in terms of shape and time of rise and fall in Chl *a* concentrations. In particular, the latter months of the simulation, where we should expect that Chl *a* concentrations will remain high, indicating that the bloom persists for a number of months after the first estimated peak (Figure 7.8). However, *Planktolyngbya* has the least error when comparing integrated concentrations over the simulation period of 14%.

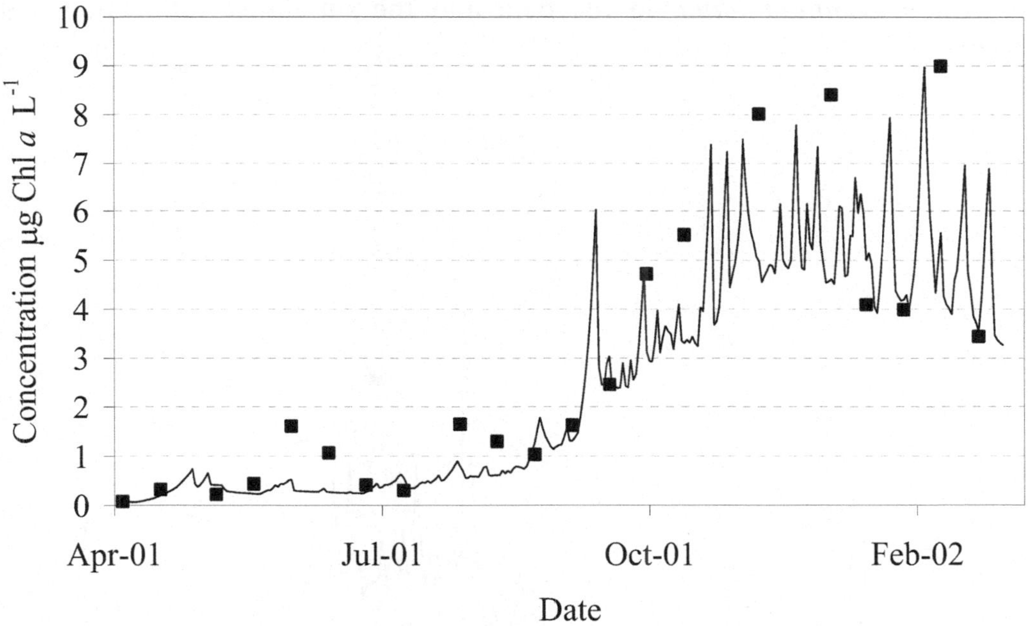

Figure 7.5 Plot of *Cylindrospermopsis raciborskii* in the field and the simulated calibration equivalent concentrations.

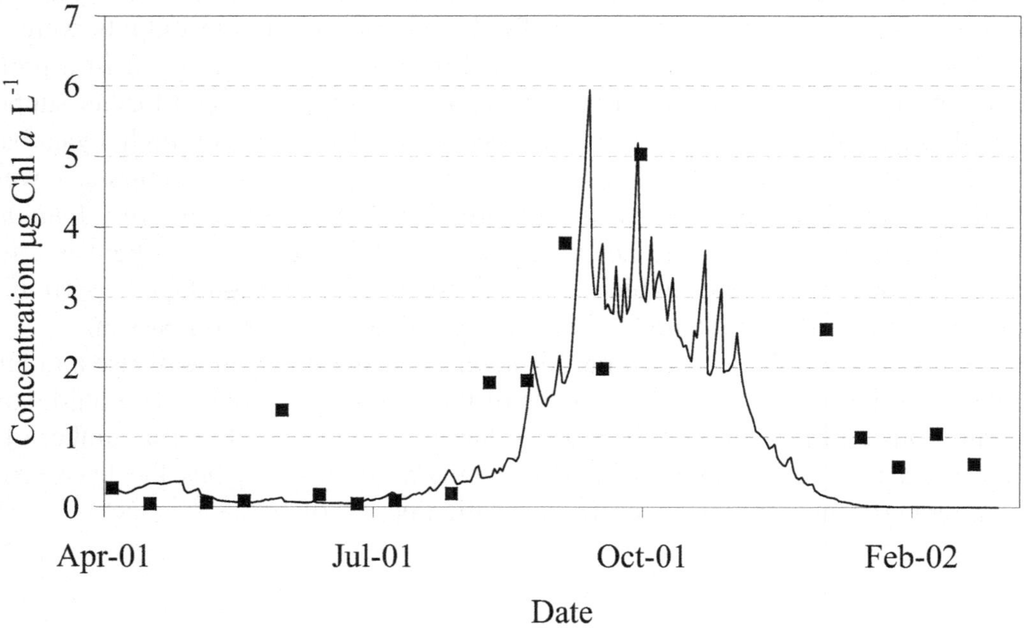

Figure 7.6 Plot of *Achnanthidium* in the field and the simulated calibration equivalent concentrations.

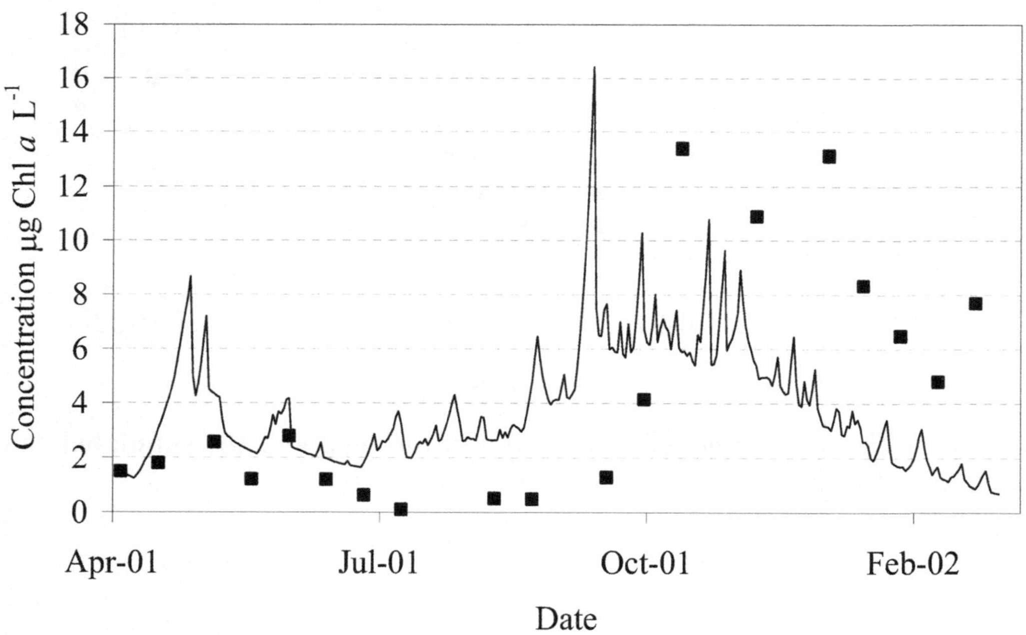

Figure 7.7 Plot of *Planktolyngbya* in the field and the simulated calibration equivalent concentrations.

Table 7.1
Comparison of the integral of phytoplankton species in the calibration run, 2001-2002.

Species	Integrated field samples	Integrated DYCD output	% error
Cylindrospermopsis raciborskii	1012.8	821.2	19
Achnanthidium	515.0	274.1	47
*Planktolyngbya*1	423.6	1222.5	14

The good fit between predicted *Cylindrospermopsis* growth and the observed concentrations meant that the Cadeym model could be extended to include toxin production and degradation. Toxin concentration was measured in North Pine dam between Spetember 2001 and February 2002 for another project and Dr Maree Smith made these results available for this study. The peak Cylindrospermopsin toxin concentration did not correspond with the peak *Cylindrospermopsis* concentration (Figure 7.8), which highlights that this feature may be difficult to simulate for this period.

The toxin model is closely linked to growth of the cyanobacteria and because cyanobacteria biomass and toxin do not correlate well there is a similar misrepresentation of toxin concentration by the model (Figure 7.9). The Dyresm-Cadeym output does not fully capture the true observed toxin behaviour in North Pine dam. A peak in cylindrospermopsin concentration is observed approximately at the peak of the bloom (late Jan 02), while the field data suggests the actual peak is a few months earlier (late Oct 01). The minimum concentration as estimated by DYCD is also above the minimum observed in the field.

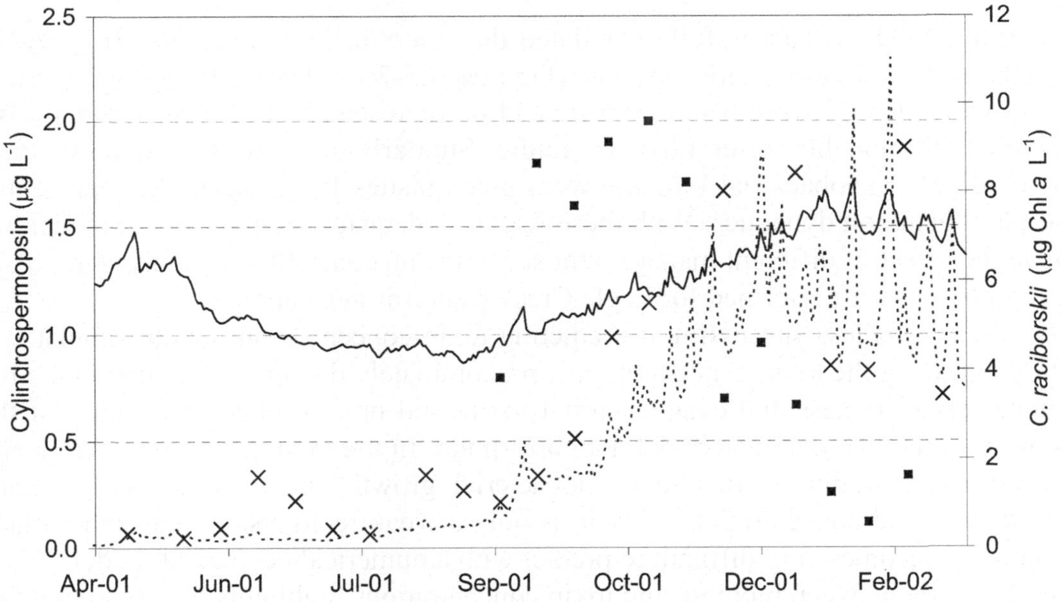

Figure 7.8 Prediction and observation of Cylindospermopsis and cylindrospermopsin in North Pine Dam. - Dotted line = *C. raciborskii*; 'x' = Field *C. raciborskii* concentration; Continuous line = DYCD output of cylindrospermopsin in water column; Points = cylindrospermopsin concentration as measured in the field.

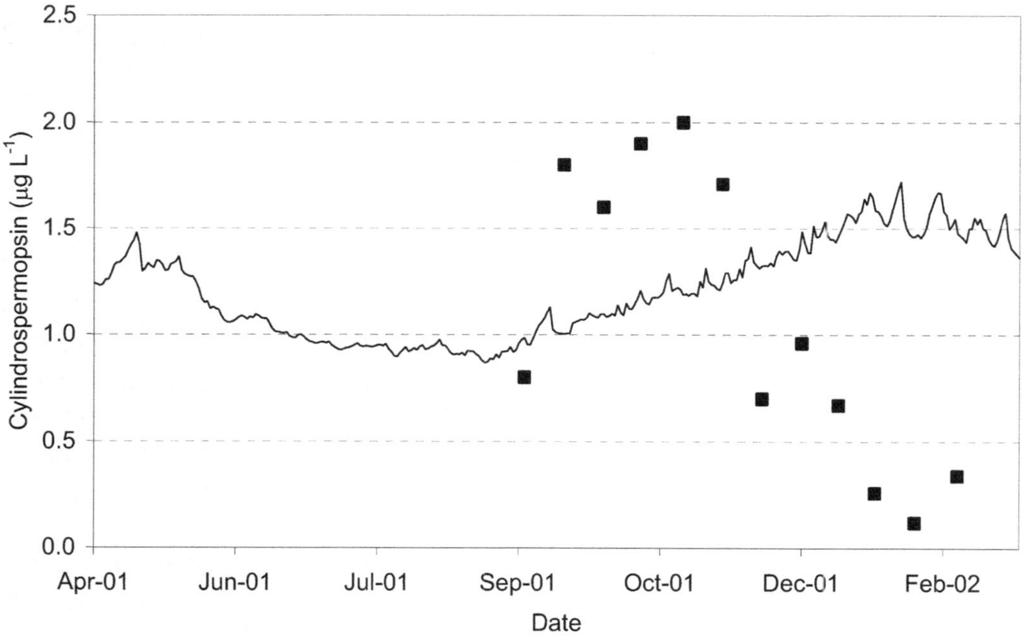

Figure 7.9 Observed (squares) and simulated (line) cylinspermopsin concentration in North Pine Dam.

DISCUSSION

Dyresm-CAEDYM successfully simulated the reservoir hydrodynamics (Figures 7.1-7.3) and growth of the phytoplankton species (Figures 7.5-7.7). The criteria for determining a satisfactory validation of hydrodyanmics were a close representation of temperature measured in the reservoir with modelled reservoir temperature. Similarly, accurate prediction of the timing and magnitude of cyanobacterial biomass were prerequisites for good model representation of reservoir phytoplankton dynamics. With these features adequately simulated the model was used to simulate the effect of different management scenarios on controlling cyanobacteria. A number of these simulations are performed for Eagle Creek reservoir in Chapter 8.

The model was less successful at predicting toxin concentration for the simulated period. The failure to predict the toxin concentration is predominately due to the fact that in the reservoir the cyanobacterial biomass and cyanobacterial toxins did not correlate and timing of the peak concentration of the two parameters did not correspond. In the toxin production and degradation model toxin concentration is tied to cyanobacterial growth and so relies on a relationship between biomass and concentration. If there is only a weak relationship and high variability in real situations it becomes very difficult to predict with a numerical ecological model.

The disparity between biomass and toxin concentrations highlights the complexity of toxin dynamics and the difficulty water quality managers' face in predicting, controlling and treating cyanobacteria. The cause of this difference could be that under different nutrient or light resource conditions the rate of toxin production changes. In the model, the effect of light and nutrient limitation is linked to the rate of toxin production however, it is directly related to the influence that resource limitation has on the growth of cells. There may be other factors that stimulate or retard the production of toxin other than conditions needed for cell growth. The difference between the observed biomass and toxin concentrations could also be because different strains of the same cyanobacterial species are in competition with each other. Chorus and Bartram (1999) noted that strains may differ in toxicity by more than three orders of magnitude. Two strains of the same cyanobacteria are not likely to be easily distinguishable and therefore biomass measures would represent the combined populations of both species. If for example, a bloom contained two strains, toxic and non-toxic, then the amount of toxin being produced relative to the biomass would increase or decrease as the proportion of each species changed. The population dynamics of two strains would be difficult to simulate with a computer model without detailed knowledge of the competitive advantage toxin production might confer.

New genetic techniques to identify the genes responsible for toxin production will enable further insight into toxin production and improve predictions of toxicity. The ability to measure the population of toxic cyanobacteria genetically could allow the dynamics between different strains to be investigated and may lead to a better understanding of the conditions favourable to toxic strains. These techniques could also be used to investigate whether environmental factors impact on toxin production and what biochemical mechanisms, if any, control toxin synthesis.

Cyanobacteria may produce several types of toxins simultaneously. The most notable is *Microcystis aeruginosa* that can produce a number of variants of microcystin. However, usually one variant is dominant in any given strain (Chorus and Bartram 1999) and so the model, which simulates one toxin per species, is applicable for most cases.

CHAPTER 8: CAEDYM MODEL OF EAGLE CREEK RESERVOIR, INDIANA, USA

INTRODUCTION

Eagle Creek Reservoir is a shallow dam located in Indiana USA used for flood mitigation and also to supply the local area with drinking water. The latitude of the reservoir (39.9° north) means that it has a climate similar to Mediterranean countries with warm summers that can promote the growth of nuisance cyanobacteria. Severe taste and odour episodes have occurred in the reservoir as the result cyanobacterial blooms. Due to stringent regulations on taste and odour compounds in the water produced by the local treatment plant, several management scenarios are under consideration to mitigate the impact of nuisance cyanobacteria. One scenario is the installation of a bubble plume aerator to artificially mix the reservoir. The action of a mixer can reduce the growth of cyanobacteria by mixing the cells below the depth that light can penetrate. Another scenario is to lower the input of nutrients into the reservoir from the surrounding catchment.

Hydrodynamic modelling of the Eagle Creek Reservoir was previously undertaken using DYRESM to test designs for a possible artificial destratification system. These studies suggested that because of its shallow depth and large inflows of cold water, the Eagle Creek Reservoir was not well suited to artificial mixing by a bubble plume aerator.

The ecological model CAEDYM can be configured to run in parallel with the hydrodynamic model used previously in order to model algal populations and other biological process in the reservoir. In this study, CAEDYM was used together with the hydrodynamic model to simulate the impact of artificial mixing and nutrient input reduction on the population of alga in the reservoir.

EAGLE CREEK DAM INPUT FILES

The input files required to run the DYRESM and CAEDYM models include:

1. a file which contains configuration information required for the simulation
2. a description of the lake morphometry, elevation, latitude, lake bathymetry at the inflow and outflow height.
3. an initial profile of temperature, salinity and nutrients
4. meteorological data at the appropriate time step including short wave radiation, long wave radiation, air temperature, vapour pressure, wind speed and rainfall.
5. information on the stream inflow volume, temperature, salinity and nutrient concentrations
6. information on the withdrawal volume
7. a parameter file with constants for processes described by the model

All of the morphometry, inflow and meteorological data used for the hydrodynamic modelling (i.e. DYRESM only) were used unchanged. Some of the actual files, for example the initial profile and inflow files, were edited to include nutrient data.

Eagle Creek Nutrient Input

There are three inputs into the eagle creek reservoir, for the purposes of hydrodynamic modelling (DYRESM) only one input was used and the combined flow of the three catchments was combined into one. In a one dimensional model this approximation was not expected to affect the results of the simulation. This approximation was continued when preparing input files for the CAEDYM model. Of the three inputs 80-90% of the total flow into the reservoir comes from the main inflow. The nutrient data used to prepare the model inputs was from the CEES sample location ECP-3 which was located close to the reservoir on the main inflow.

Phosphorus

Total P model – Eagle Creek. The model inputs relating to phosphorus are of key importance in determining the algal dynamics in the model simulations. The total amount of P must be partitioned into several soluble and particulate fractions.

There have been studies indicating that the total P inputs are related to flow (Linden et al. 2004). Given the importance of the input values of P and the low number of actual measurements throughout the simulation period, interpolation between samples could mean high levels of P observed during high flow events would be applied when the flow was low. This could have a detrimental effect on the results produced by the model particularly in terms of algal biomass and speciation/succession. The data was investigated to see whether a flow based model (Equation 8.1) could be applied to the eagle creek inflow.

$$[TP] = a + b\ln(Q) \tag{8.1}$$

The trend in total phosphorus described by the model of Linden (2004) (see Figure 8.1) is such that the concentration is low at low flows and increases with flow (Q). The rate of increase is initially very steep and decreases with increasing flow until the model flattens out to a maximum value for very high flows. A graph of the available data appeared, at a first glance, to fit the overall trend of this model and a least squares fit of the equation was attempted. The curve generated using this process was not suitable as a model because at low flows the curve produced values less than zero. Attempts to adjust the parameters to resolve this resulted in a curve that did not describe the data very well.

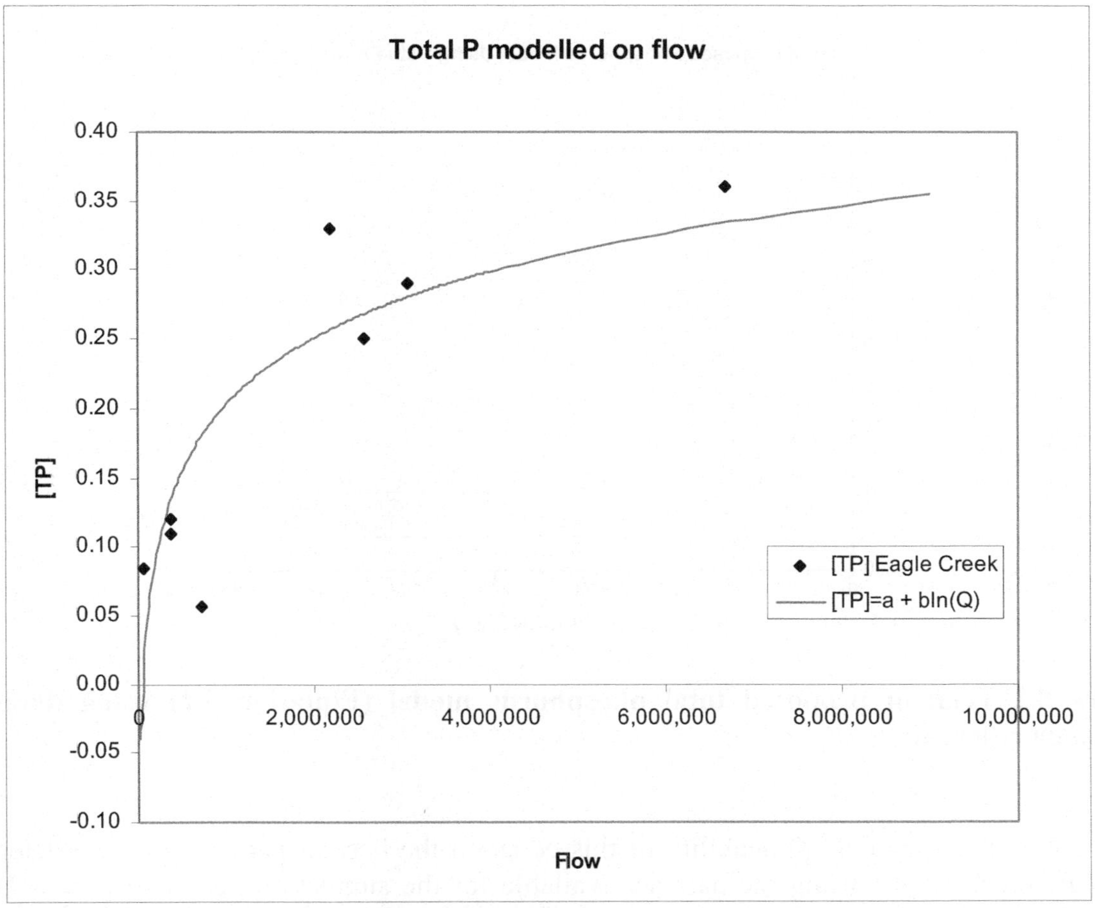

Figure 8.1 Total Phosphorus model based on Equation 8.1.

An alternative model equation was investigated (Equation 8.2). Due to the lack of data available for Eagle Creek, the proposed curve was trialled using a more complete data set provided by Linden (private communication) to test whether the shape of this curve adequately describes the behaviour of a natural system. The parameters were estimated using a least squares fit of the data set. The curve defined by the best fit parameters (Figure 8.2) described the data at least as well as the model of Linden (2004) ($R^2 = 0.56$ *c.f.* 0.55).

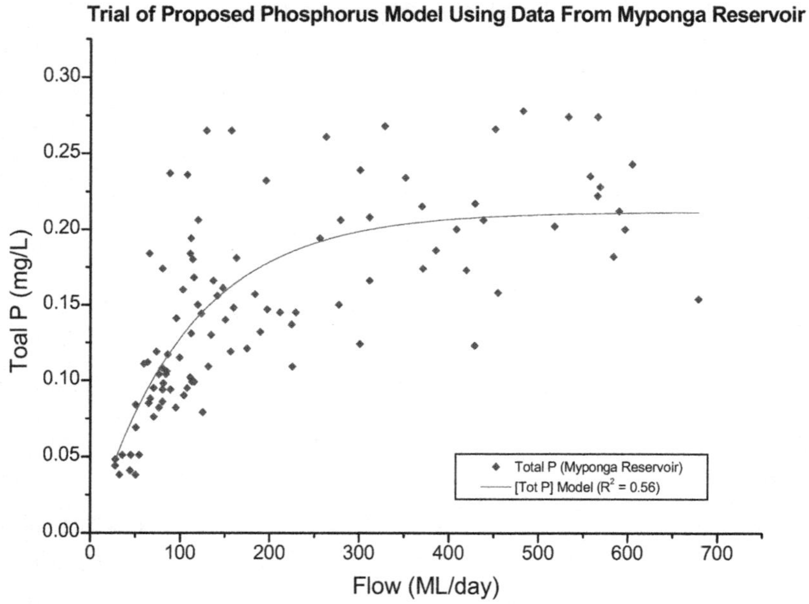

Figure 8.2 Trial of proposed total phosphorus model (Equation 8.2) using data from Myponga reservoir.

Given the apparent applicability of this equation the best fit parameters were determined for eagle creek inflow using the data set available for the simulation period (Figure 8.3). This model appeared to describe the data set adequately. One useful aspect of this model is that the P_{min} & P_{max} parameters can be adjusted easily to model scenarios where the amount of phosphorus input is altered by management strategies. Changing these parameters does not affect the overall curve shape which is defined by K_p.

$$[TP] = P_{\min} + \left(1 - P_{\max}.e^{-K_p.Q}\right) \tag{8.2}$$

where P_{min} = the total phosphorus concentration when the flow is zero

P_{max} = the maximum amount that the total P will increase at high flows i.e. P_{min}+P_{max} = Maximum P at high flows.

K_p = a rate constant that determines how the total P increases in relation to the flow

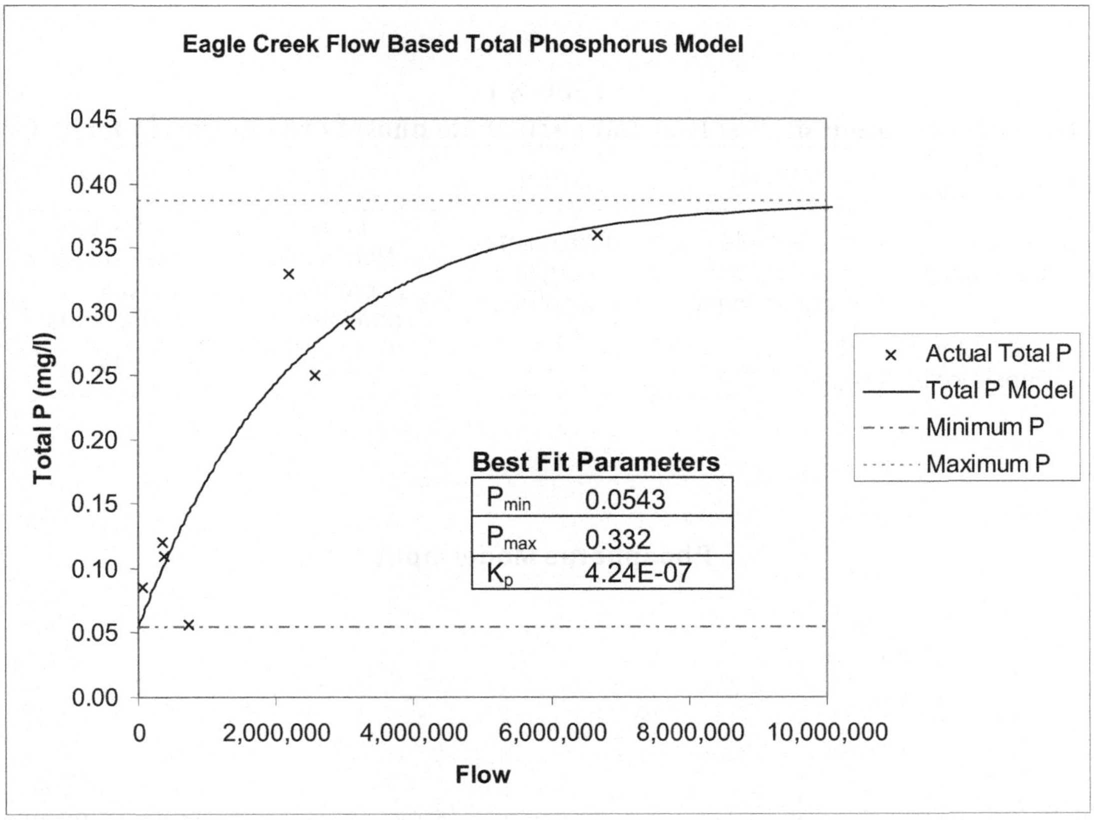

Figure 8.3 Flow based total phosphorus model used to calculated Eagle Creek inflow loadings

Partitioning of Soluble and Particulate P Species. The CAEDYM model requires the Phosphorus input to be partitioned into several soluble and particulate fractions. The CAEDYM nutrient inputs were estimated by calculating total P using the above model and then dividing this among the different fractions using the proportions shown in Table 8.1; the rational for these values are described below.

Samples from Eagle Creek inflow were analysed for levels of soluble reactive phosphorus however six of the eight samples were below the detection limit. The CAEDYM input PO4, which represents soluble inorganic P, was estimated using the average ratio of SRP:TotP observed in the two samples above the detection limit (40%). The remaining 60% of total P was assigned to the particulate fractions of phosphorus.

Suspended sediment samples (8) from the eagle creek inflow were analysed for different fractions of P these results were averaged and used to calculate proportions of species required for model input. The "reactive" and "acid hydrolysable" phosphorus fractions made up 24% of particulate P and were thought to be representative of adsorbed PO4 and particulate orthophosphate species. These two fractions were collectively assigned to the CAEDYM parameter for particulate inorganic P (PIP). The organic P fraction was on average 76% of total particulate P and was divided equally between the labile and refractory organic phosphorus parameters required by the model.

The loadings of soluble and particulate phosphorus estimated by applying the partitioning factors to the total phosphorus model are shown in Figure 8.4.

Table 8.1
Relative concentrations of dissolved and particulate phosphorus species in Eagle Creek

CAEDYM Parameter	[PO4]	[PIP]	[POPL]	[POPR]
Description	Soluble inorganic phosphorus	Particulate inorganic phosphorus	Labile Particulate organic phosphorus	Refractory Particulate organic phosphorus
% Total Phosphorus	40%	14%	23%	23%

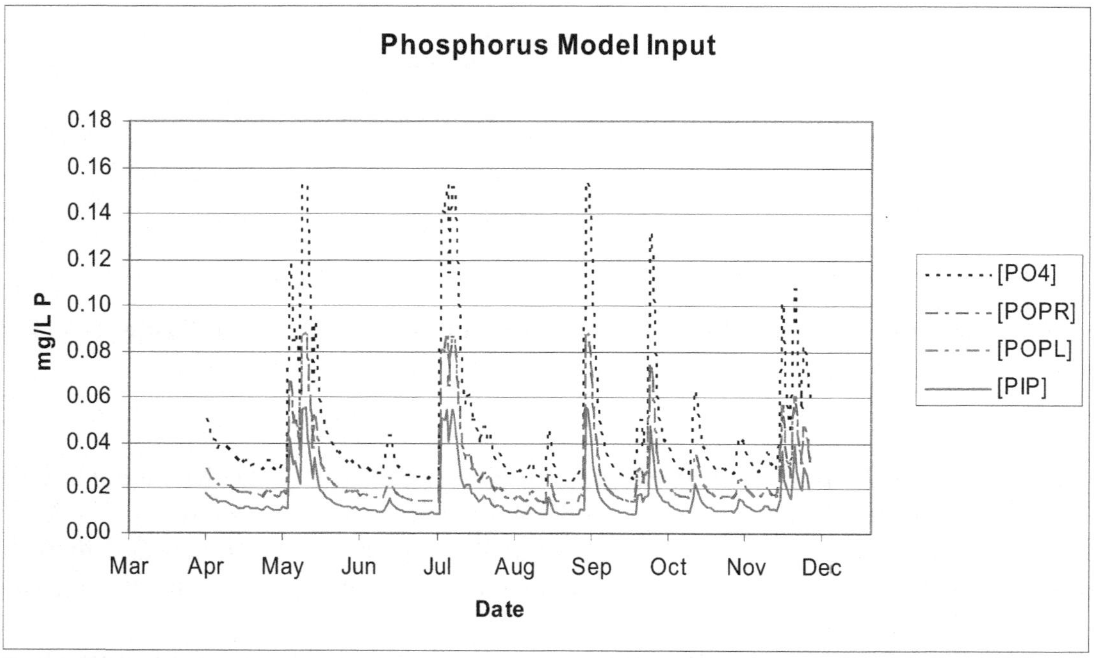

Figure 8.4 Eagle Creek inflow soluble and particulate phosphorus loadings.

Organic Carbon

Samples from the Eagle Creek inflow were analysed for dissolved organic carbon (DOC) and total organic carbon (TOC). The nutrient parameter DOCL (labile DOC) was estimated by linearly interpolating between the available DOC data. The nutrient parameter POCL (labile particulate organic carbon) was estimated by interpolating between the available TOC data and subtracting the corresponding DOC value (see Figure 8.5).

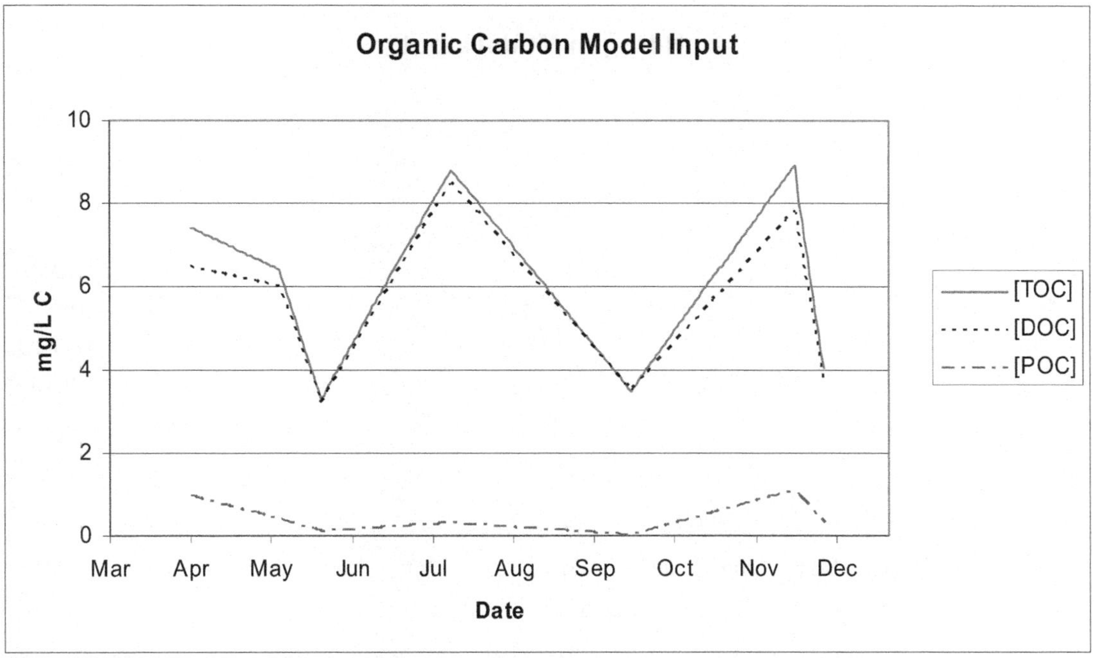

Figure 8.5 Organic loading input for the Eagle Creek inflow.

Nitrogen

Samples from the Eagle Creek inflow were analysed for nitrate, nitrite, ammonia and total Kjeldahl nitrogen. Similar to phosphorus above, the amount and timing of nitrogen inputs can have an effect on the biomass and speciation of algae in simulations. The data available for the simulation period was examined for any direct relationships to flow however none were apparent and therefore the model input was estimated by interpolating between the available data. Given the high levels of nitrogen in the inflowing water this approximation was not expected to have a large impact on the overall result of the model.

All of the samples were below the detection limit for nitrite; the nutrient parameter NO2 was assumed to be zero for the whole simulation period. The interpolated ammonia/ammonium concentration data were assigned to the nutrient parameter NH4. The measured concentrations of nitrate were interpolated and assigned to the nutrient parameter NO3. The total organic nitrogen was determined by subtracting the measured ammonia concentration from the total Kjeldahl nitrogen and then these values were then divided among dissolved and particulate fractions as follows. In a study of many of the worlds rivers, Meybeck (1982) suggests that the nitrogen to carbon ratio in particulate organic mater is approximately 10%. The particulate nitrogen concentration (represented by PONL) was estimated by multiplying the interpolated particulate organic carbon by 10%. The remainder of organic nitrogen was assigned to the labile dissolved organic nitrogen parameter DONL. The calculated model input values for the influent nitrogen species are presented in Figure 8.6.

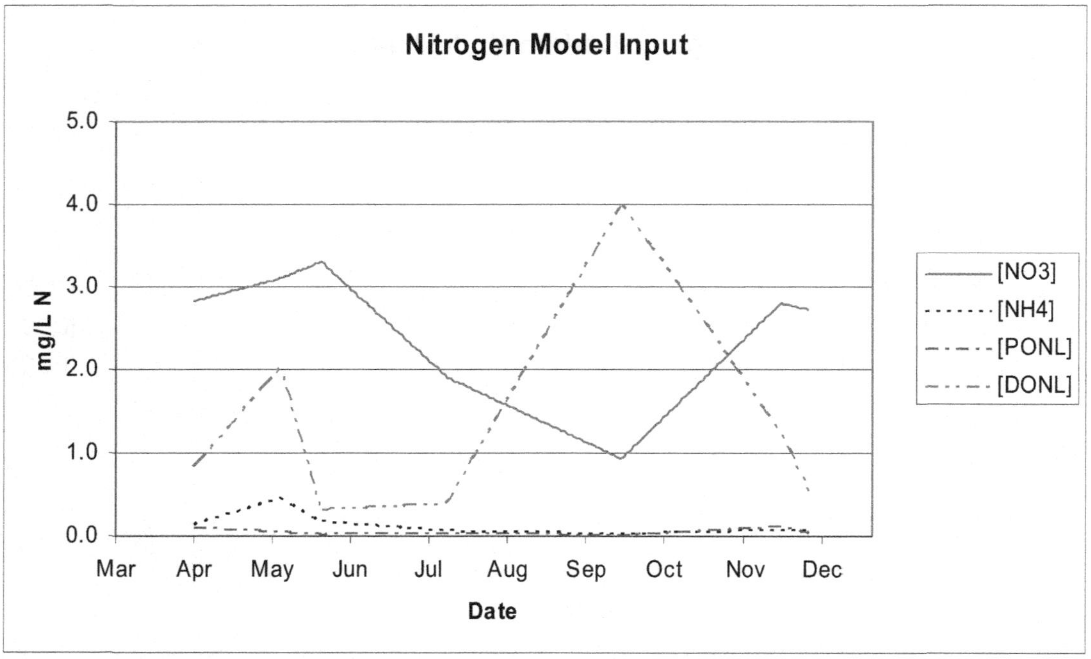

Figure 8.6 Soluble and particulate nitrogen loadings for Eagle Creek inflow.

Silicon

Dissolved silica is an important model parameter for diatoms. Samples for the eagle creek inflow were analysed for total and soluble silica. The soluble silica concentrations were interpolated and assigned to the model parameter SIO2 as shown in Figure 8.7.

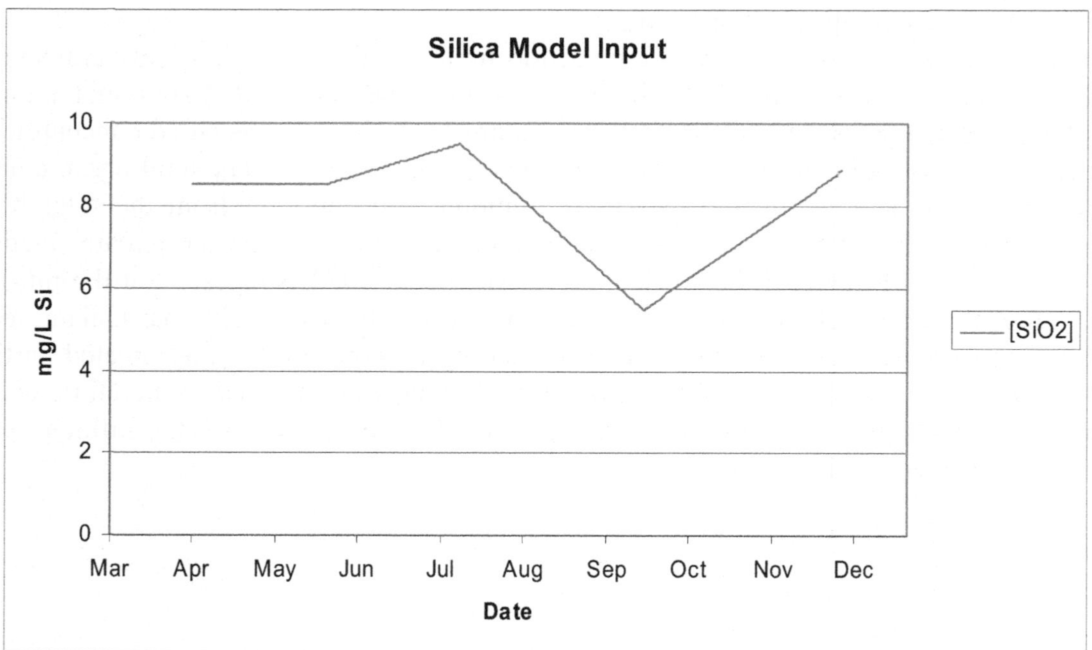

Figure 8.7 Model input for influent soluble silica.

Dissolved Oxygen

Dissolved oxygen concentrations were measured *insitu* using a multi-parameter probe however the organisation that measured these values had reservations about the accuracy of these results. The measured DO concentrations were examined in relation to the saturated oxygen concentration determined using the temperature of the inflow and Equation 8.3. This examination (see Figure 8.8) revealed that the measured values were typically 85% of the saturated DO concentration at the prevailing water temperature. This amount of saturation appeared to be a reasonable since rivers are typically well saturated (Wetzel 2001) and therefore the model input DO was estimated as 85% of the saturated DO concentration.

$$[DO]_{sat} = \frac{468}{31.6 + T} \tag{8.3}$$

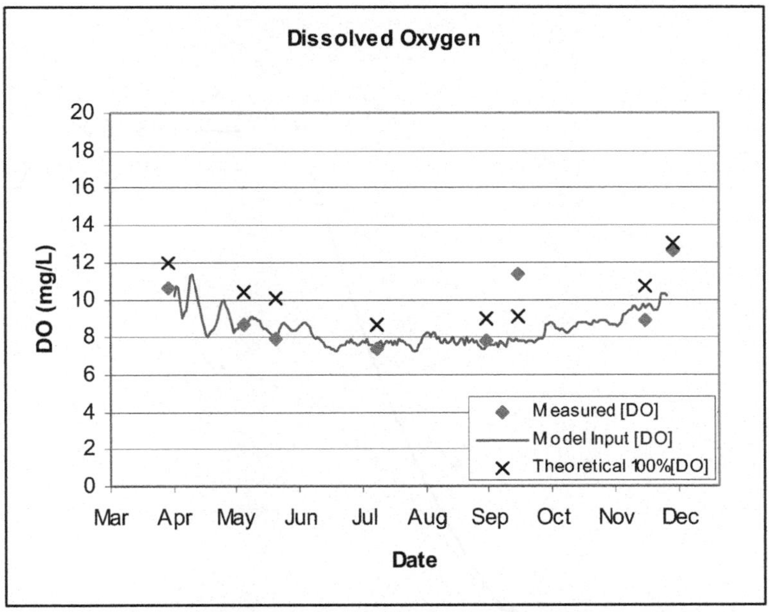

Figure 8.8 Comparison of measured DO concentration in Eagle Creek inflow with theoretical 100% saturation and model input based on 85% saturation.

Suspended Solids

Samples from the Eagle Creek inflows were analysed for suspended solids. The data from all three locations exhibited a strong relationship to flow. Studies have suggested that the total suspended solid loading can be explained by a simple power model as shown in Equation 8.4 (Thomas and Maybeck 1992, Moog and Whiting 2002).

$$[TSS] = a.Q^b \tag{8.4}$$

where a =constant

b = contstant

Q = flow

Parameters for this equation were estimated for eagle creek using a least squares optimisation. While the best fit curve appeared to describe the data quite well (see Figure 8.9) this model resulted in very large amounts of suspended solids at high flows. At the highest flow during the period (18019028m^3/day) the model resulted in suspended solids concentration of 35103mg/L. The modelled amount of solids in the reservoir was far in excess of what was observed in reality. An alternative model (Equation 8.5) was trialled that had a similar shape for low to medium flows but levelled off to a maximum at high flows (see Figure 8.9).

$$[TSS] = \frac{a}{1 + e^{-(Q-b)/c}} \qquad (8.5)$$

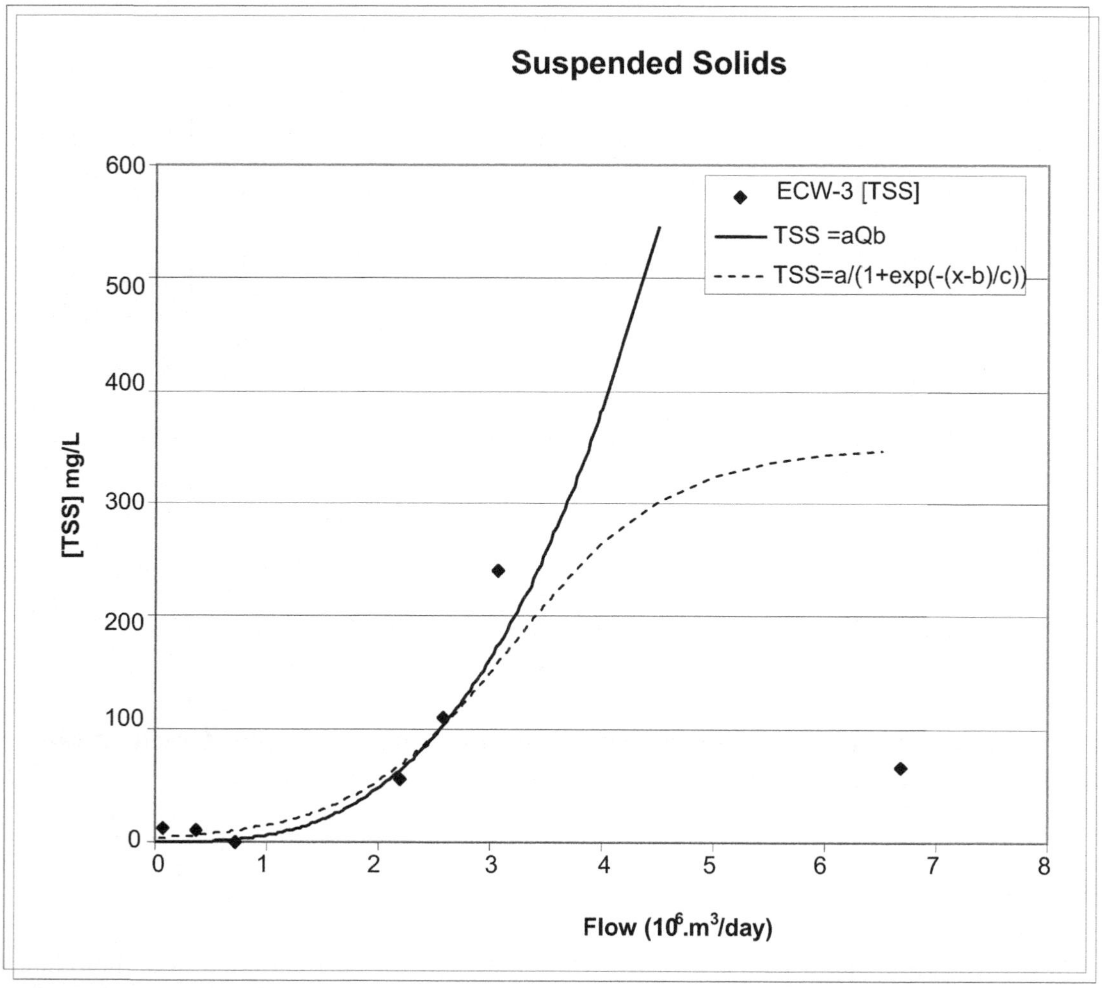

Figure 8.9 Comparison of two flow based models with total suspended solids field data from Eagle Creek inflow.

Parameterisation of Eagle Creek

There are a numerous reservoir and biological processes described by CAEDYM such as particulate settling, sediment redox and algal growth. All of these processes are modelled using equations that require parameters to describe the behaviour of the system under study. The values of the input parameters have a large impact on the results of the model.

Parameters for all of the processes being described by the CAEDYM model are entered into a single input file. The model software downloaded from the Centre for Water Research web site has an example CAEDYM parameters file that contains values for all of the necessary parameters. This file was used as the initial basis for the CAEDYM simulation and then many of the parameters in this file were changed to reflect the local conditions and dominant algal species in the Eagle Creek Reservoir. The values were sourced from journal papers, previous modelling studies and by comparison of field data with modelled values (calibration).

Algae

Four species of algae were simulated using CAEDYM

- *Anabaena* and *Microcystis* represent cyanobacteria with and without heterocysts. Modelling of a*nabaena* and the production of geosmin is of particular interest for investigating the effect of management techniques on drinking water quality.
- *Aulacoseira* represented the dominant species of diatom observed in the reservoir throughout the simulation period
- Green alga were represented by a single "group" in CAEDYM; the natural population did not have one dominant species as was the case for the diatoms so one generalised group was created.

Anabaena. The parameters used to describe Anabaena were based on previous modelling studies (Lewis, Brookes and Lambert 2004) of the strain of *Anabaena circinalis* found in the Myponga Reservoir, South Australia. The parameters are displayed in Table 8.2. A value of zero for the half saturation of nitrogen effectively means that the algae are never limited by low nitrogen levels in the reservoir. Using this value is one way of modelling alga that can fix nitrogen. A positive value for settling velocity translates to floatation and is useful for describing algae that can control their buoyancy.

Table 8.2
***Anabaena* model parameters**

Parameter	Value
Minimum biomass threshold	0.010 (mg/L)
Average C:Chla ratio	42.000 (mg C)/(mg Chla)
Growth rate	0.750 /day
Respiration rate	0.074 /day
Fraction of respiration relative to total loss	0.7
Temperature multiplier for respiration	1.08
Temperature multiplier for growth	1.08
Standard temperature	18.000 deg C
Optimum temperature	33.000 deg C
Maximum temperature	39.000 deg C
Freshwater salinity limitation	
Optimum salinity	3
Salinity limitation at S	36.0 : 1.000
Non-photoinhibited light limitation	
Photosynthesis-irradiance curve parameter	62.5 uE/m^2/s
Half saturation constant for phosphorus	0.005 mg P/L
Half saturation constant for nitrogen	0.000 mg N/L
Specific attenuation coefficient	0.02200 ug Chla/L/m
Constant settling velocity	
Settling velocity	0.167E-04 m/s
Critical shear stress for resuspension	0.020 N/m^2
Resuspension rate	0.230E-03 mg/m^2/sec
Half saturation constant for resuspension	1.000 mg/m^2
Sediment survival time	1.000 days

Microcystis. The parameters used to describe *Microcystis* (Table 8.3) were based on previous modelling studies (Robson and Hamilton 2004) of the strain of *Microcystis aeruginosa* found in the Swan River estuary, Western Australia.

Table 8.3
***Microcystis* model parameters**

Parameter	Value
Minimum biomass threshold	0.010 (mg/L)
Average C:Chla ratio	40.000 (mg C)/(mg Chla)
Growth rate	1.200 /day
Respiration rate	0.080 /day
Fraction of respiration relative to total loss	0.7
Temperature multiplier for respiration	1.08
Temperature multiplier for growth	1.08
Standard temperature	20.000 deg C
Optimum temperature	30.000 deg C
Maximum temperature	35.050 deg C
Freshwater salinity limitation	
Optimum salinity	4
Salinity limitation at S	25.0 : 1.000
Non-photoinhibited light limitation	
Photosynthesis-irradiance curve parameter	500.0 uE/m^2/s
Half saturation constant for phosphorus	0.006 mg P/L
Half saturation constant for nitrogen	0.030 mg N/L
Specific attenuation coefficient	0.02000 ug Chla/L/m
Constant settling velocity	
Settling velocity	0.167E-04 m/s
Critical shear stress for resuspension	0.010 N/m^2
Resuspension rate	0.230E-03 mg/m^2/sec
Half saturation constant for resuspension	1.000 mg/m^2
Sediment survival time	2.000 days

Green Algae. During the simulation period a wide range of green algae grew in the eagle creek reservoir. The "default" parameters for green algae were considered suitable to describe the overall trends in the chlorophyta population and were not changed. The values used are shown in Table 8.4.

Table 8.4
Chlorophyta model parameters.

Parameter	Value
Minimum biomass threshold	0.010 (mg/L)
Average C:Chla ratio	40.000 (mg C)/(mg Chla)
Growth rate	1.300 /day
Respiration rate	0.050 /day
Fraction of respiration relative to total loss	0.7
Temperature multiplier for respiration	1.08
Temperature multiplier for growth	1.06
Standard temperature	20.000 deg C
Optimum temperature	26.000 deg C
Maximum temperature	35.050 deg C
Freshwater salinity limitation	
Optimum salinity	14
Salinity limitation at S	36.0 : 1.000
Non-photoinhibited light limitation	
Photosynthesis-irradiance curve parameter	100.0 uE/m^2/s
Half saturation constant for phosphorus	0.085 mg P/L
Half saturation constant for nitrogen	0.110 mg N/L
Specific attenuation coefficient	0.02000 ug Chla/L/m
Constant settling velocity	
Settling velocity	-.230E-06 m/s
Critical shear stress for resuspension	0.010 N/m^2
Resuspension rate	0.230E-03 mg/m^2/sec
Half saturation constant for resuspension	1.000 mg/m^2
Sediment survival time	2.000 days

Diatom. Diatoms, in particular *Aulacoseira*, were frequently the dominant algal phyla observed in the eagle creek reservoir during the simulation period. Using the original parameter values found in the CAEDYM input file, the model appeared to adequately describe the biomass of diatoms during the spring period (see Figure 8.10). During summer however, the modelled population of diatoms fell while the field data showed an increase in numbers. The modelled decline in the population of diatoms appeared to coincide with an increase in the modelled surface water temperature. The parameters for the temperature dependant growth limitation function for the diatom group was examined with a view to altering the diatom growth characteristics to better describe the population native to the Eagle Creek Reservoir.

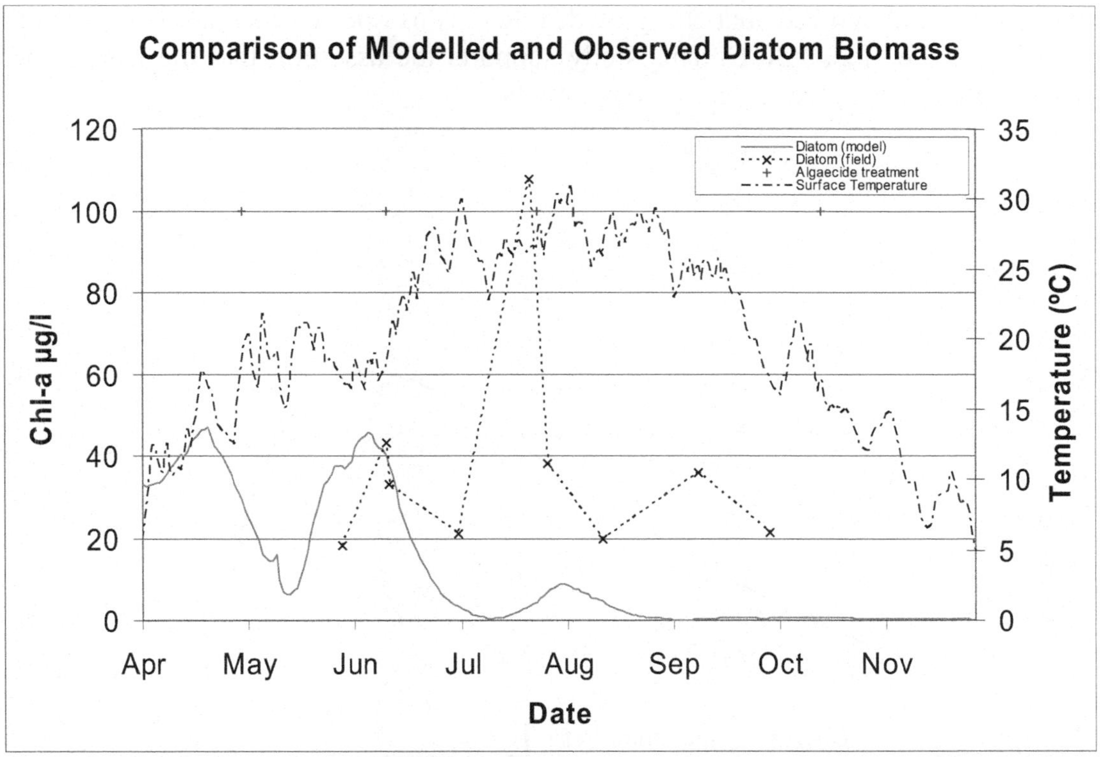

Figure 8.10 Comparison of CAEDYM simulated (averaged to 5m depth; T_s:20, T_{opt}:20, T_{max}:32) and observed diatom biomass. Field Biomass was estimated using cell counts.

The temperature dependant growth limitation function is shown in Equation 8.6 where ϑ is a non-dimensional temperature multiplier and k, a and b are constants. Three temperature values; standard (T_s), optimum (T_{opt}), and maximum (T_{max}); are input into the model and these are used to define the shape of this curve by solving for constants that satisfy the following conditions.

1. The value of the function is one at T_s.
2. The function has a maximum at T_{opt}.
3. The value of the function is zero at T_{max}.

$$f(T) = \vartheta^{T-20} - \vartheta^{k(T-a)} + b \tag{8.6}$$

The values used in the initial simulation were:

- $T_s = 20$°C
- $T_{opt} = 20$°C
- $T_{max} = 32$°C

The growth curve defined by these values is shown in Figure 8.11. During the summer period of the simulation the average surface water temperature was 26°C. It is clear that the diatoms modelled using the above parameters would be growth limited at this temperature. Given that the species of *Aulacoseira* native to the Eagle Creek reservoir had very strong growth during this warm period it seems likely that it is well adapted to these conditions. The effect of temperature on the growth of *Aulacoseira granulata* var. *angustissima* has been measured (Coles and Jones 2000) and showed that the growth rate of this species increased with temperature between 15-30ºC i.e. T_{max} >30ºC. This study supports the assertion that the native species of *Aulacoseira* could be adapted to warm temperatures.

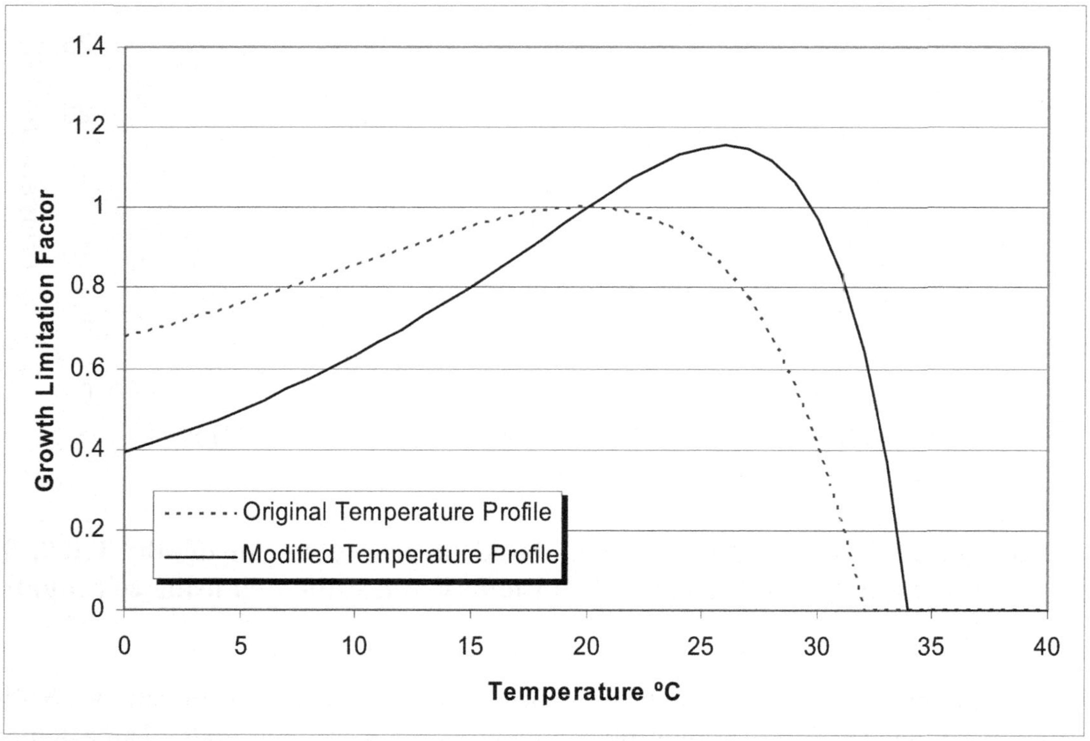

Figure 8.11 Comparison of temperature limitation curves for original (T_s:20, T_{opt}:20, T_{max}:32) and modified (T_s:20, T_{opt}:26, T_{max}:34) parameters.

The temperature parameters were changed to reflect a more warm-water tolerant species of diatom. The values used to define the temperature-growth curve shown in Figre 8.11 were:

- $T_s = 20$ºC
- $T_{opt} = 26$ºC
- $T_{max} = 34$ºC

The results of a simulation run using these values are shown in Figure 8.12. In this simulation the diatom population does not decrease during the summer months and was the dominant phyla throughout the simulation.

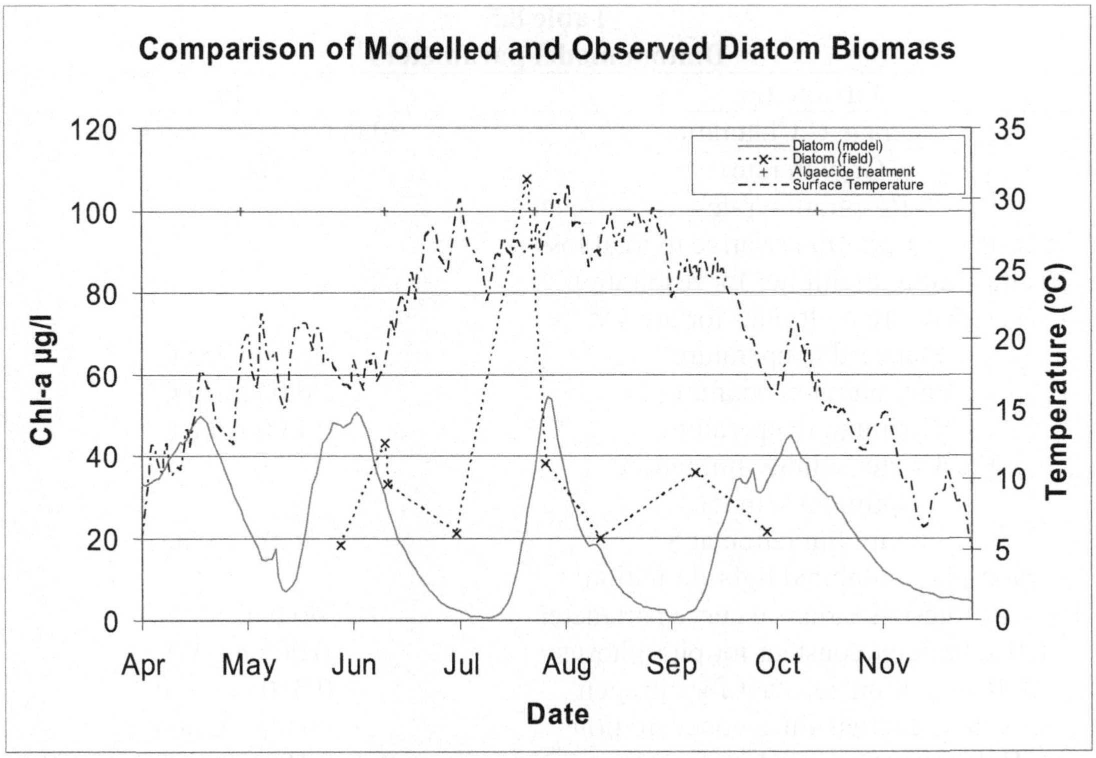

Figure 8.12 Comparison of CAEDYM simulated (averaged to 5m depth; Ts:20, Topt:26, Tmax:34) and observed diatom biomass. Field Biomass was estimated using cell counts.

The values of the parameters used to describe the diatom population are shown in Table 8.5.

Table 8.5
Diatom model parameters

Parameter	Value
Average C:Chla ratio	40.000 (mg C)/(mg Chla)
Growth rate	1.200 /day
Respiration rate	0.090 /day
Fraction of respiration relative to total loss	0.7
Temperature multiplier for respiration	1.07
Temperature multiplier for growth	1.06
Standard temperature	20.000 deg C
Optimum temperature	26.000 deg C
Maximum temperature	34.000 deg C
Freshwater salinity limitation	
Optimum salinity	1
Salinity limitation at S	36.0 : 1.000
Non-photoinhibited light limitation	
Photosynthesis-irradiance curve parameter	60.0 uE/m^2/s
Half saturation constant for phosphorus	0.008 mg P/L
Half saturation constant for nitrogen	0.010 mg N/L
Constant internal silica concentration	20.000 (units)
Half saturation constant for silica	0.100 mg Si/L
Specific attenuation coefficient	0.02000 μg Chla/L/m
Constant settling velocity	
Settling velocity	-0.120E-05 m/s
Critical shear stress for resuspension	0.010 N/m^2
Resuspension rate	0.230E-03 mg/m^2/sec
Half saturation constant for resuspension	1.000 mg/m^2
Sediment survival time	2.000 days

Sediment oxygen demand

During periods of thermal stratification, oxidative processes in the hypolimnion can consume oxygen. During persistent stratification the concentration of oxygen in the hypolimnion can be depleted to such an extent that reductive processes can occur leading to release of metal ions or nutrients from the sediment.

CAEDYM models a number of oxygen processes such as surface exchange and phytoplankton photosynthesis but of particular importance to hypolimnetic oxygen concentrations is the sediment oxygen demand. Oxygen uptake by the sediment is modelled by a static model (see Equation 8.6 defined by two parameters

1. the sediment exchange rate (rSOs)
2. the half saturation constant for sediment oxygen demand (K_{SOS})

$$f_{O_2}^{DSF} = rSOs \cdot f(T) \cdot \frac{DO}{Ksos + DO} \tag{8.6}$$

Parameters for this equation were found by calibrating the model against field data. Field data was collected by lowering a DO probe from a boat and collecting data at one metre intervals. Additional data was collected for the latter half of the simulation period using three DO probes suspended from a buoy. The positions of the two sampling locations used for the calibration are shown in Figure 8.13 relative to the depth profile of the reservoir. The buoy was located to one side of the reservoir in ~7.3 m of water and had probes located at 1m, 4m & 6.5m. While the lowest sensor attached to the buoy was quite close to the sediment, for the purposes on calibration this field data was compared with simulation data at 6.5 m depth (~4.5m from sediment).

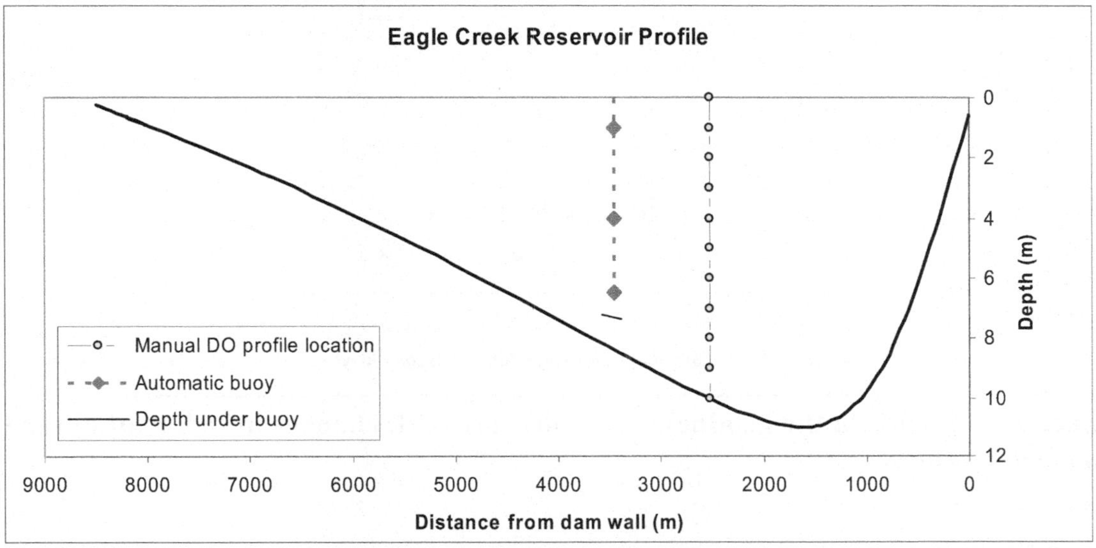

Figure 8.13 Position and depth of sampling locations relative to the depth profile of Eagle Creek Reservoir.

Different values for the sediment exchange rate and half saturation constant were trialled and the model results compared with the field data. The square of the residual was calculated for each data point and the cumulative result was plotted (see Figure 8.14) in order to find parameters that minimise the error between field and simulated values.

Opposing trends in the goodness of fit measure were observed for different levels of the water column. Near the sediment the greater the rate of sediment exchange the better the apparent fit of the model results. In contrast, the surface showed the opposite trend where, as the sediment exchange increased so did the sum of squares. Examination of the results derived from the manual profile data showed that this trend reversed at ~5m depth. At this depth the trend had a minimum at a rate of 3.0g/m^2/day and a half saturation constant of 3.5 mg/L. The results derived from the buoy data showed this trend reversed between 1m and 4.5 m.

Over the summer period many of the manual DO profiles had an inflection point (i.e. metalimnion) at approximately five metres depth (for example, see Figure 8.14). Using the parameters (above) that produced the best fit at this depth, the results of the simulation appeared to satisfactorily describe the depth profile of dissolved oxygen during this period. These values also lie near the overall minimum observed for the manual profiles and approximately at the point where the surface & bottom trends derived from the buoy data cross (this is not a minimum in Figure 8.15 because the downward trend of 4.5m & 6m outweigh the upward trend observed for 1m).

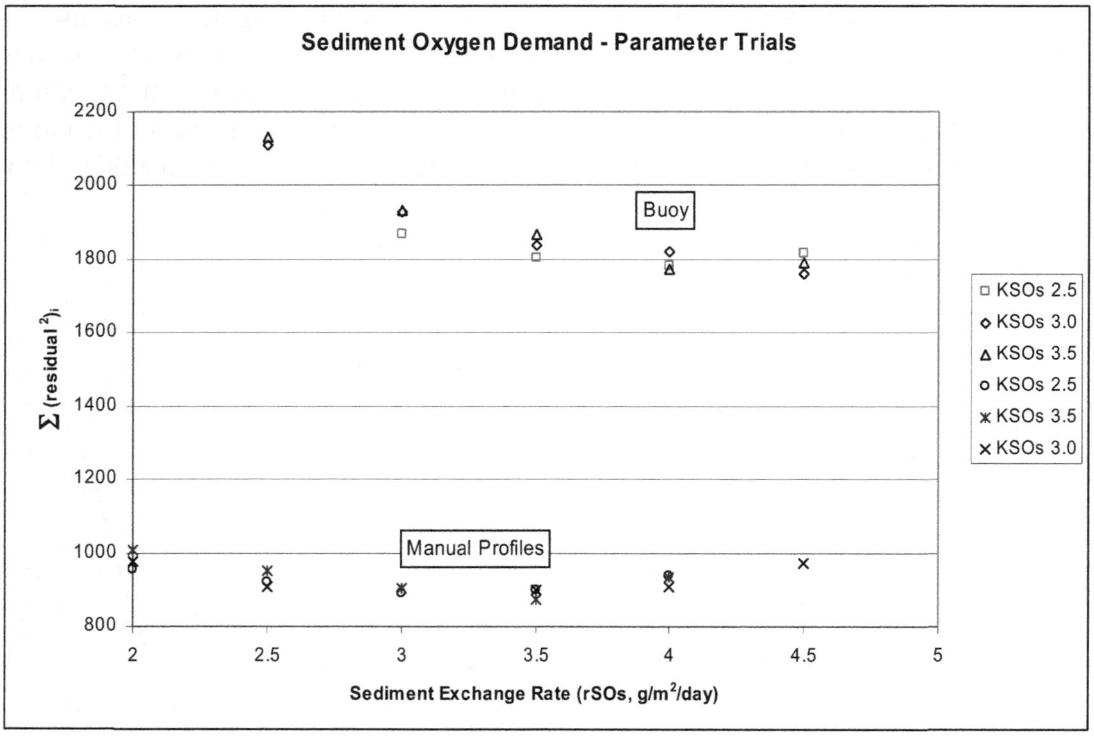

Figure 8.14 Trends in the goodness of fit measure with changes in sediment oxygen demand parameter values.

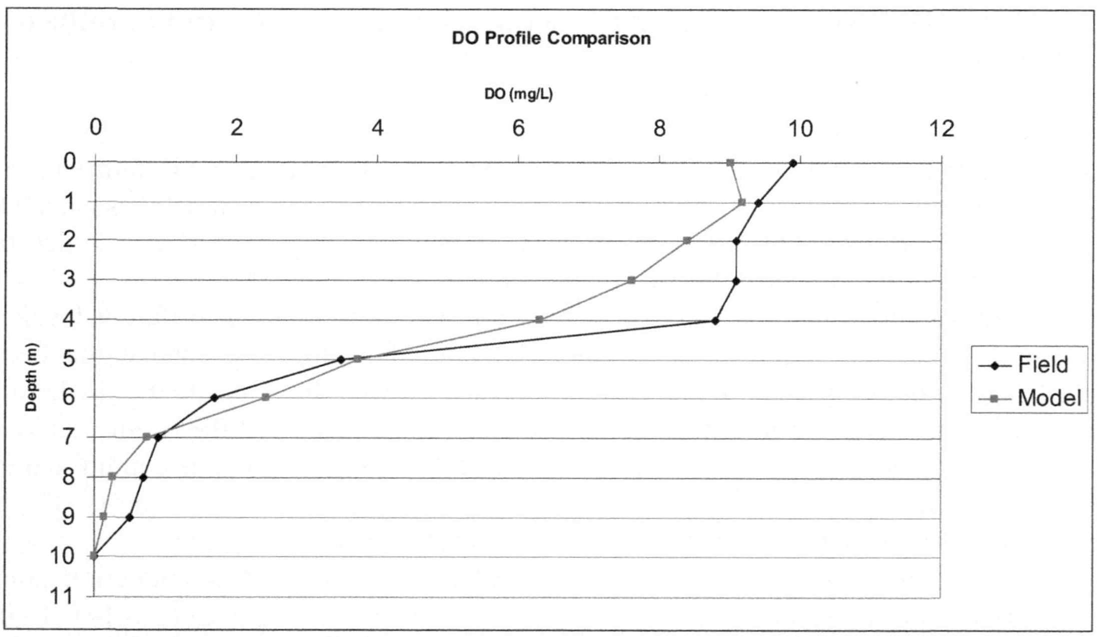

Figure 8.15 Comparison of dissolved oxygen profiles from field data and the results of a model simulation (rSOs: 3.0, K_{SOS}:3.5mg/L).

Secondary Metabolite Model

While large amounts of algal cells in water supplies can cause problems with water treatment (such as filter clogging) and water quality (eg. increased BRP) these cells can often produce secondary metabolites that are equally problematic because they cause taste and odour problems or are toxic. A model for predicting the concentration of secondary metabolites in the water column has recently been developed and added to CAEDYM.

The internal metabolite concentration is linked to the growth rate for the particular algal species and therefore metabolite production is effected by available nutrients and light. The relationship between the two is determined by a linear equation:

$$IX = IX_{min} + \frac{\mu_g}{\mu_{max}}\left(IX_{max} - IX_{min}\right) \quad (8.7)$$

where μ_g/μ_{max} = a measure of current growth rate relative to the maximum growth rate
IX_{min} = the internal metabolite concentration when the growth rate is zero
IX_{max} = the concentration during maximum production (Long, Jones and Orr 2001)

Changes in the water column concentration of the metabolite are determined by adding the amount of metabolite released by cell lysis/excretion and subtracting losses due to microbial degradation. The rate of microbial degradation is given by the equation:

$$\frac{dX}{dt} = -k \cdot T \cdot X \quad (8.8)$$

where k = the degradation rate constant
T = the temperature of the water
X =the concentration of the metabolite

Eagle creek reservoir frequently has a problem with nuisance algae such as *anabaena* growing to populations large enough to cause taste and odour problems for the local water treatment plant. The level of taste and odour compounds in the reservoir was typically the trigger for dosing copper algaecide to kill the nuisance species. For this reason, the production of geosmin was added to the model for *Anabaena* in CAEDYM.

Jones and Korth (1995) reported that in a natural water from the Murrumbidgee River, geosmin had a half life of 2 days due to microbial degradation. To calculate the microbial decay constant (Equation 8.8 was integrated and solved for k at $t_{½}$ (see Equation 8.9 giving a value of $0.01386°C^{-1}day^{-1}$ at 25°C.

$$\begin{aligned} [X]_t &= [X]_0 e^{-ktT} \\ \therefore 0.5 &= e^{-kt_{½}T} \\ \therefore k &= \frac{-\ln(0.5)}{t_{½}T} = \frac{-\ln(0.5)}{2 \cdot 25} \end{aligned} \quad (8.9)$$

The rate of degradation calculated by the CAEDYM model was tested using a simulation where the initial concentration geosmin in the water column was set to 2 ng/L while the initial algae concentration was set to zero i.e. no release of additional geosmin. The result of this simulation was compared with the expected geosmin concentration over time based on Equation 8.9 and the rate constant calculated above. It can be seen that the rate of degradation is similar in both curves and the deviation was attributed to the water column temperature (which was artificially increased to 25°C) fell below 25°C and therefore slowed the rate slightly (Figure 8.16).

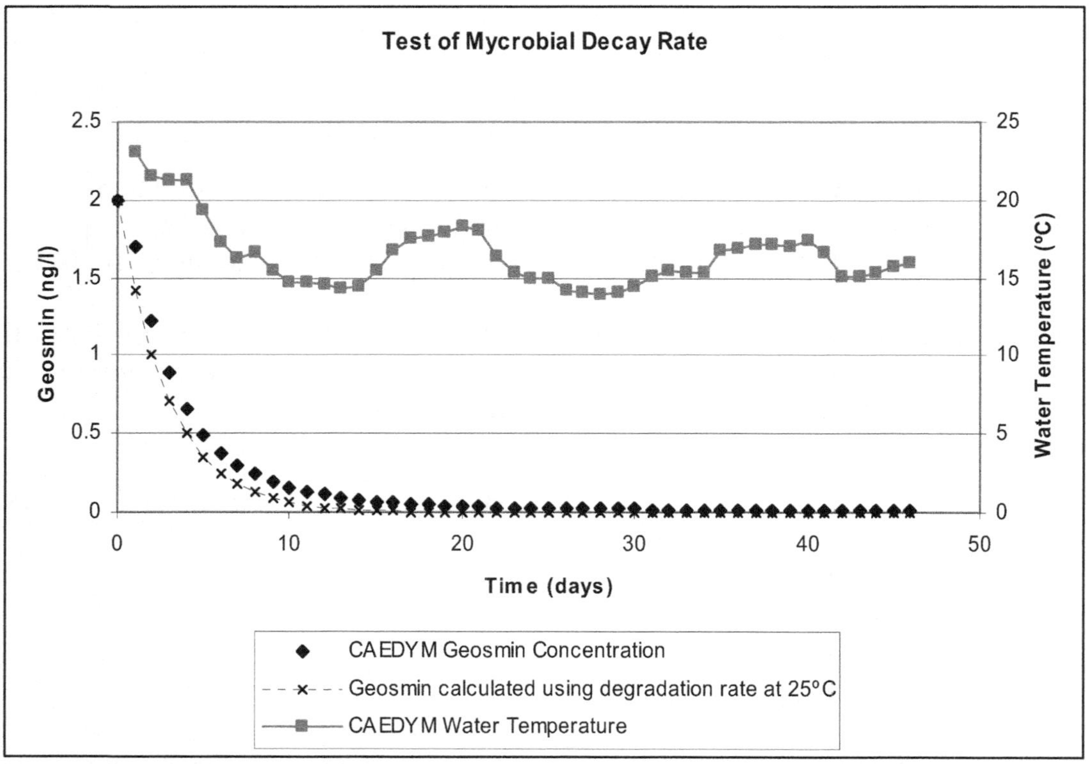

Figure 8.16 Comparison of the average geosmin concentration simulated using CAEDYM and predicted values using the underlying rate equation.

The model (above) that links secondary metabolite production and chlorophyll was proposed after a study of the micocystin content of *M. aeruginosa* cells which showed a strong correlation between the two. A similar trend between geosmin and cellular chlorophyll a concentration in *A. circinalis* was observed by Bowmer et al. (1992) and so it seems likely that the same model could be applied to taste and odour production.

As mentioned above (Equation 8.7) two input values IX_{max} and IX_{min} are required for the model. A value of 20ng/µg Chl*a* was reported by Jones and Korth (1995) for a bloom of *A. circinalis* found in the Murrumbidgee River. In an initial simulation this value (0.02mg/mg Chl*a*) was used for IX_{max} while zero was used for IX_{min}. The simulation using these values gave results two orders of magnitude higher than that observed in the reservoir. There is some conjecture about whether the species of *A. circinalis* found in the Murrumbidgee River was a "super producer" and that the value of this parameter used to describe the *anabaena sp.* found in Eagle

Creek may need to be much smaller. For the scenarios described later, a value of 0.0002mg/mg Chl*a* was used for IX_{max}.

DYREM/CAEDYM models outflows from the reservoir as either withdrawals, where the height of the off-take and the volume removed is given as an input value, or overflows. In each case, the value of all water quality variables such as geosmin concentration are calculated for each outflow and therefore can be used to predict the quality of water entering a treatment plant or flowing down stream from a reservoir. The levels of MIB and geosmin were regularly monitored in the Eagle Creek WTP off-take and so the predicted levels of geosmin calculated for the off-take were compared with the observed values (see Figure 8.17).

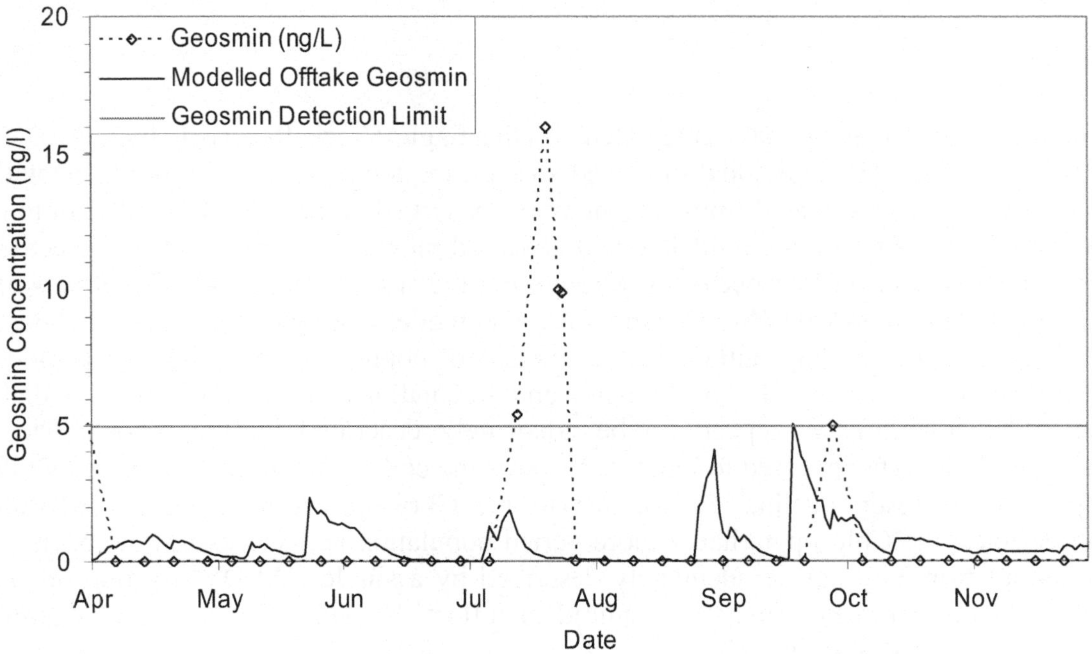

Figure 8.17 Comparison of actual and modelled taste and odour compound concentrations for Eagle Creek.

It should be noted that the secondary metabolite model in CAEDYM only outputs the water column concentration of the metabolite while the analysis for taste and odour includes intra- and extra-cellular material. Although the model actually calculates the internal metabolite concentration there is no facility to output this value in the current version of the model. It is suggested that the model program could be modified to correct this short coming.

Since *Anabaena* is a scum forming species and the water treatment plant has a subsurface off-take it was thought the contribution from cellular material would be minimised in these analyses as opposed to those from the surface of the reservoir and that a useful comparison might be made. One clear discrepancy is apparent for the peak in geosmin that occurred on the 21 July that corresponded with "bloom" of *Anabaena* in the reservoir. The modelled levels of *Anabaena* were not as high as those observed in the reservoir during this event and consequently the predicted geosmin level was less than observed. Other than this bloom, there was only one other sample taken during the simulation period that was above the detection limit for geosmin. During this time period the predicted population of *Anabaena* was closer to what was observed in the reservoir and the simulated geosmin concentration had a peak of similar concentration although it occurred earlier than observed in the reservoir. For the remaining periods where the geosmin concentration was below the detection limit, the simulated concentration was also below the less than the detection limit.

SCENARIO MODELLING

Several scenarios were modelled with CAEDYM to investigate the possible effect of different reservoir management strategies on the algal assemblage and biomass. Firstly, the installation of a bubble plume aerator to provide artificial destratification and also a reduction in the input of total phosphorus from the catchment were modelled.

Unaltered

The simulated algal biomass calculated for the Eagle Creek Reservoir by CAEDYM is shown in Figure 8.18. The individual modelled species are compared (see Figure 8.19) with the corresponding biomass estimated from cells numbers observed in the field. It is difficult to make a direct comparison between the modelled and observed biomass because copper algaecide was dosed into the reservoir on four occasions through out the simulation period. The effects of this were not modelled in CAEDYM. Despite this, the model appears to capture the onset of cyanobaterial growth in July although the modelled population of this initial bloom is significantly less than observed. The biomass and fluctuations in the population of diatoms, represented by *Aulacoseria*, appears to be adequately described by the model. The field population of chloropyta appeared to increase towards the end of the simulation period while the model predicts a steady decline. In contrast to the Chrysophyta which had a persistently dominant species, the Chloropyta had a more varied population and succession of species. This dynamic population may not be adequately described by a single CAEDYM group otherwise, adjustments to the parameters may be required to reflect the general character of Chloropyta found in Eagle Creek Reservoir.

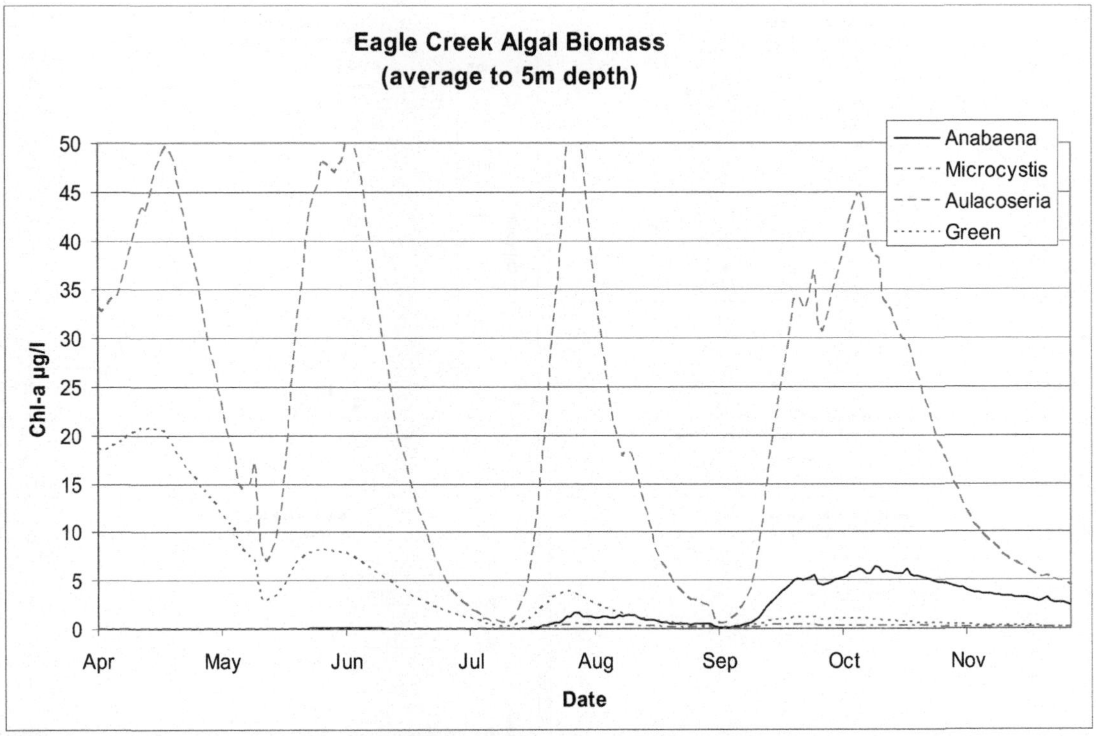

Figure 8.18 Simulated algal biomass for Eagle Creek Reservoir.

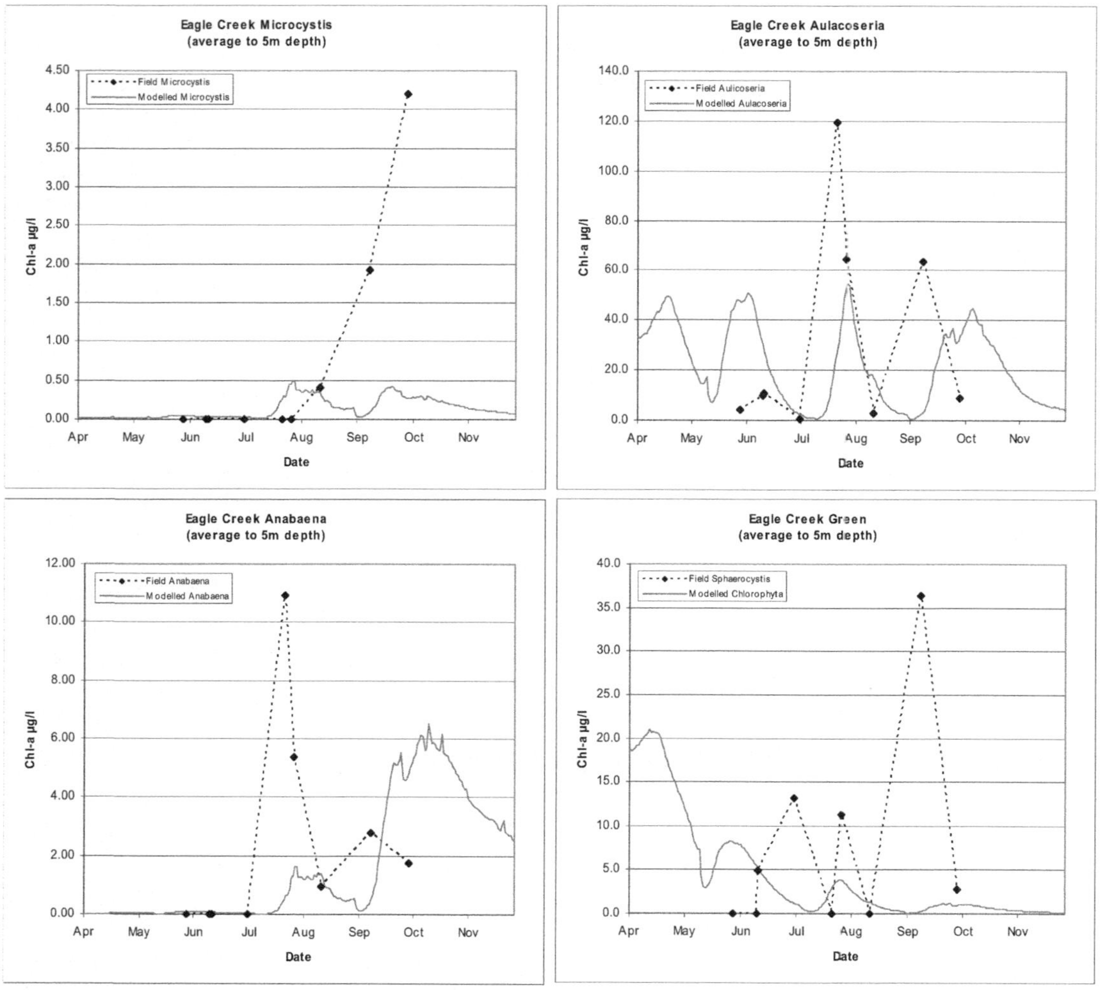

Figure 8.19 Comparison of field and modelled algal biomass.

Aerator

During stratified conditions some cyanobacteria have a competitive advantage over diatoms or green alga by being able to increase their buoyancy and therefore prevent settling out of the euphotic zone of the reservoir. Artificially mixing a reservoir by means of a bubble plume aerator can reduce the incidence of cyanobacterial blooms by increasing the depth of the surface mixed layer below the depth of the euphotic zone. When this occurs all algae have a greater likely hood of experiencing light limiting conditions and therefore the competitive advantage of the cyanobacteria is reduced.

Artificially mixing a reservoir can also reduce the overall productivity of a lake by reducing release of nutrients from the sediment that results from low dissolved oxygen levels in the hypolimnion.

Several scenarios were previously trialled using DYRESM to test different aerator configurations for Eagle Creek Reservoir. The overall conclusion from the modelling results was that an aerator was not likely to increase the depth of the surface mixed layer sufficiently to cause light limitation in cyanobacteria. In addition, since large inflows throughout the summer period add to the pool of available nutrients, any reduction in nutrient release from sediments resulting from mixing the reservoir were not expected to alter the lakes productivity significantly.

To examine this hypothesis, one of the aerator designs trialled previously was modelled using the CAEDYM ecological model linked to DYRESM to simulate the impact of artificial mixing on the algal community.

In contrast to DYRESM, the CAEDYM model has a dynamic model for estimating the penetration of light into the water column and is based on the amount of chlorophyll and suspended solids in the water and also a static term which approximates to the "background" light extinction coefficient for the reservoir (e.g. the amount of colour in the water).

The euphotic depth (Z_{eu}) for simulations with and without the aerator were calculated by determining the depth at which the light level was 1% of the subsurface irradiance. The effect of artificial mixing can be seen by comparing histograms of the euphotic depth for scenarios with and without the aerator (see Figure 8.20). The average Z_{eu} for the simulation with the aerator off was 3.34m and this value did not change significantly with the aerator on (3.38m).

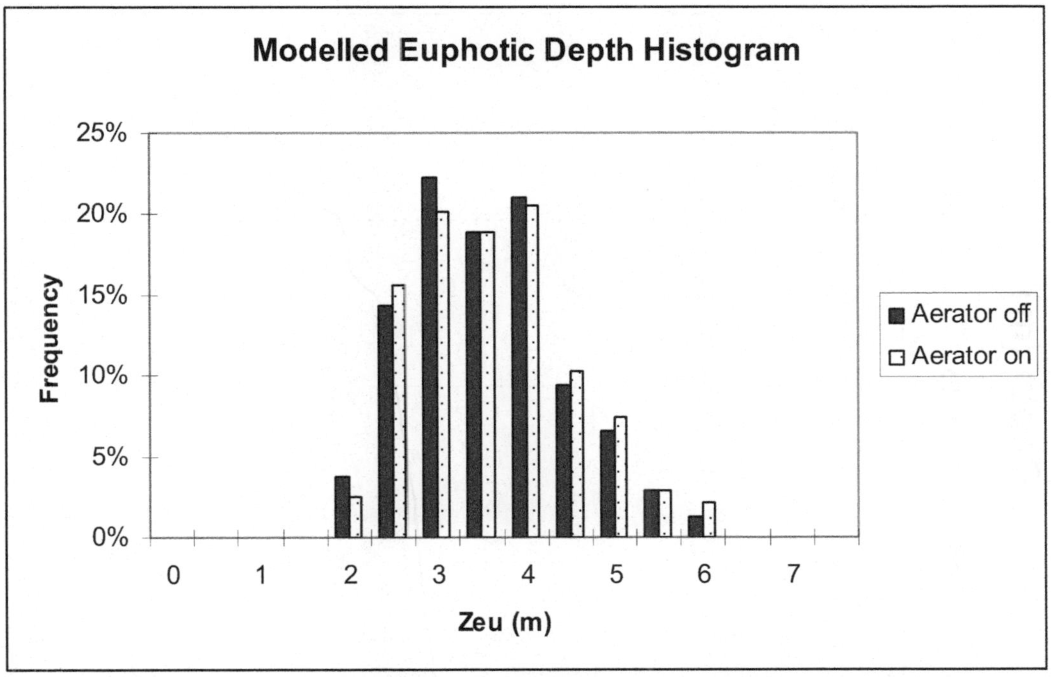

Figure 8.20 Comparison of light penetration depth for scenarios with and without artificial mixing.

For light limitation to occur the depth of the surface mixed layer needs to be greater than the euphotic depth. For the two simulations the surface mixed layer was calculated for each day at midnight and the results compared. The surface mixed layer was defined as the depth at which the temperature changed by more than 0.1°C between values output from the model (~0.11m data interval) which corresponds to ~1°C/m. An example of the model temperature output showing the calculated Z_{mix} & Z_{eu} depths is shown in Figure 8.21.

To test whether artificial mixing would affect the light environment experienced by alga, the ratio Z_{eu}/Z_{mix} was calculated for each day and a histogram of the results was used to compare each set of data (see Figure 8.22). For light limitation the Z_{eu}/Z_{mix} ratio needs to be greater than one and it can be seen that during the summer months there was no real improvement in the potential for light limitation with a mixer operating as the ratio was ≥1 for 80% of the time in both cases. This result suggests that the cyanobacterial population would not be perturbed by artificial mixing.

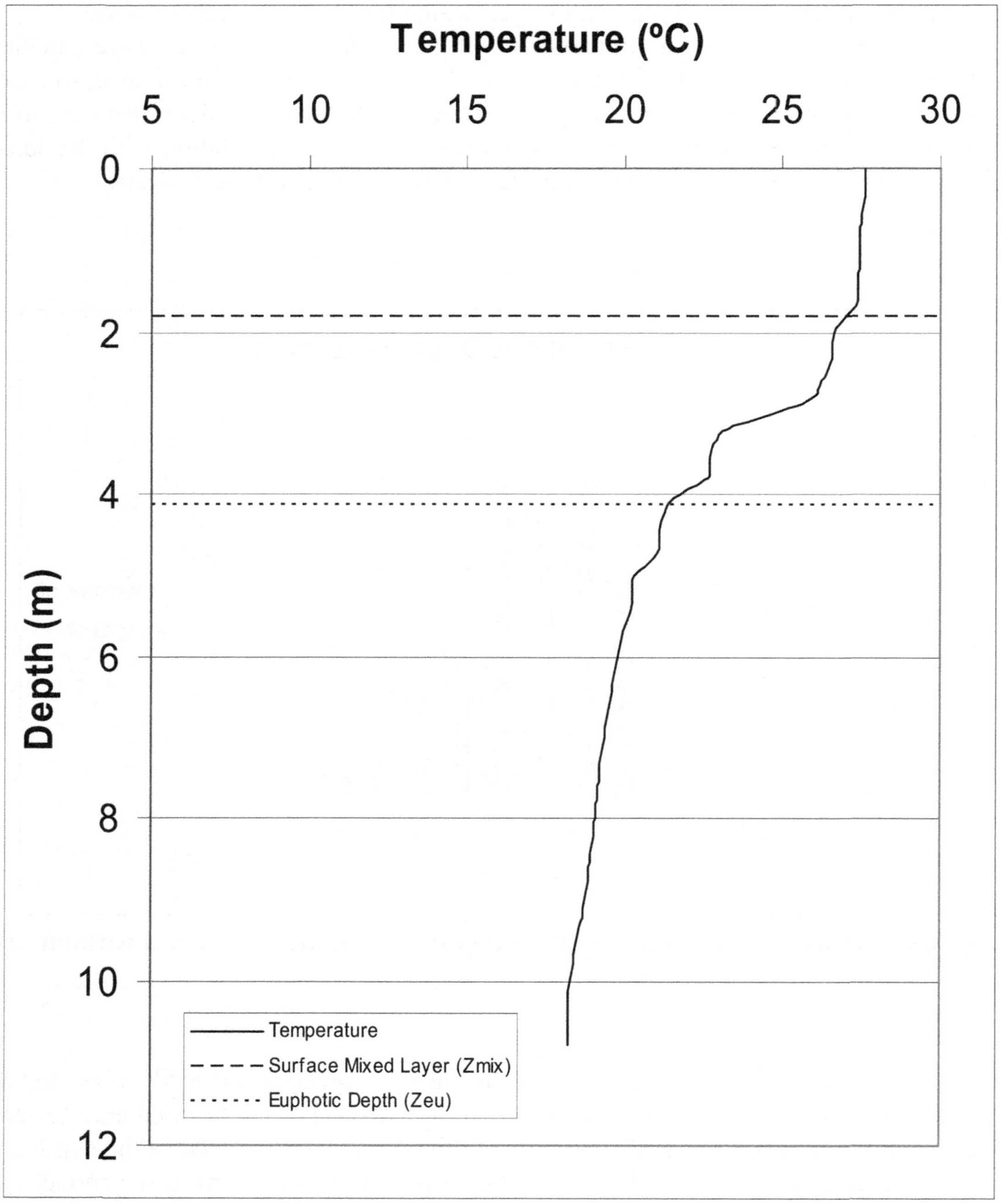

Figure 8.21 Plot of the modelled temperature gradient for Eagle Creek (26/6/03) showing the calculated Z_{mix} & Z_{eu} depths.

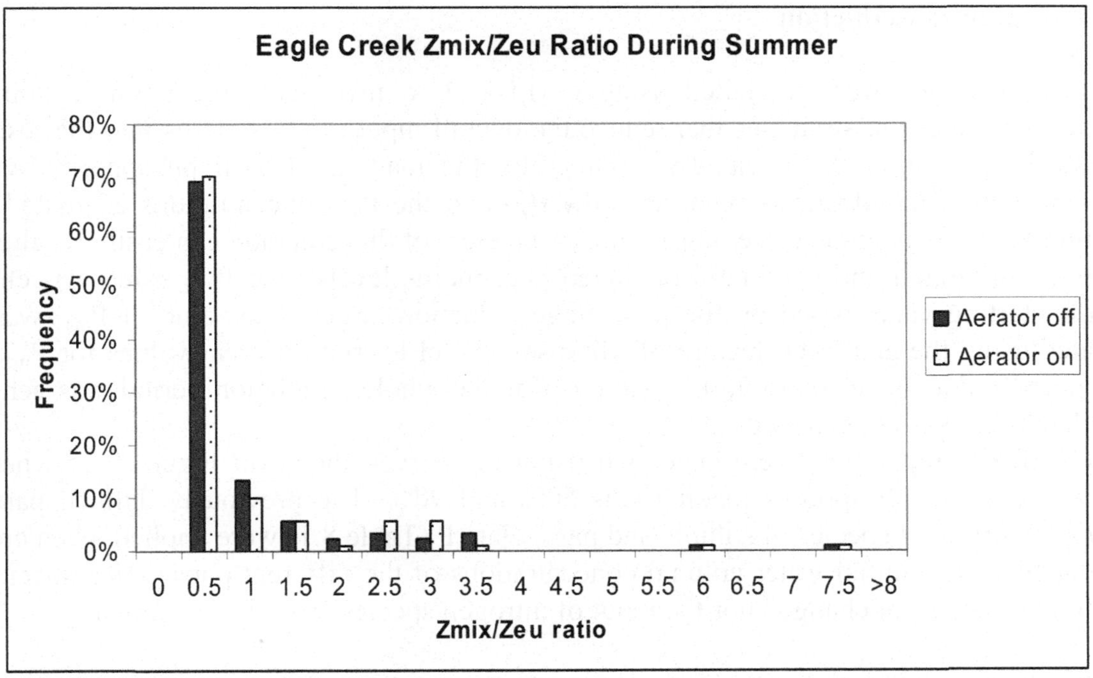

Figure 8.22 Histogram of Z_{eu}/Z_{mix} ratios calculated for model scenarios with and without artificial mixing.

Trends in the biomass estimated by the CAEDYM model are shown in Figure 8.23 for *Anabaena* and *Aulacoseira*. The simulated biomass of the four species of algae showed similar trends in that there was very little difference between simulations with the aerator was on or off except during September. During this period the biomass of all species was less in the simulation with the aerator, however the peak in biomass that occurred after this period was similar to that observed in the scenario with no aerator. This suggests that artificial mixing may have delayed the onset of a bloom by reducing the growth rate of all species.

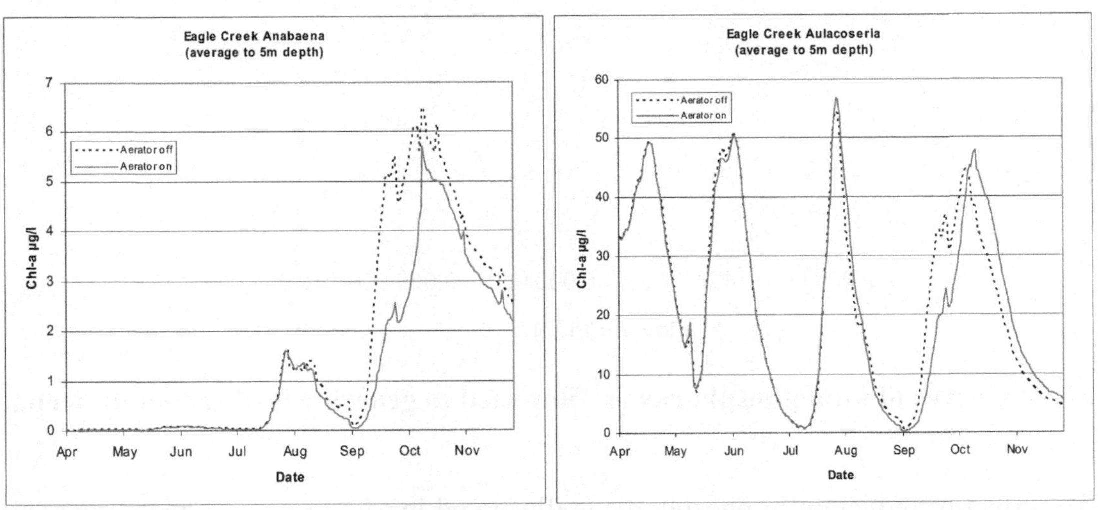

Figure 8.23 Comparison of algal biomass from simulations with and without artificial mixing.

Total Phosphorus Reduction

Two scenarios were modelled using CAEDYM to investigate the possible impact of catchment management strategies that reduce the overall input of phosphorus into the reservoir. A spreadsheet was set up to calculate (integrate) the loading of total phosphorus over the simulation period based on the recorded flow (Q) and the P_{tot} concentrations estimated using (Equation 8.2). As mentioned previously, the parameters of this equation can easily be altered to change the minimum and maximum predicted phosphorus levels. The P_{min} value was changed from the starting value based on the percentage reduction target. The value of P_{max} was then calculated using the goal seek feature of Microsoft Excel to trial different values for P_{max} until the required reduction in phosphorus loading over the whole simulation period was achieved. The value of K_p was not changed.

CAEDYM input files were generated using the curves shown in Figure 8.24 where the target reduction in phosphorus loading was 50% and 90%. The previously defined partitions between the different species of soluble and particulate P (Table 8.1) were applied when creating the input files. The initial water column concentrations of the different phosphorus species and algal biomass were not changed nor the input of nitrogen species from the catchment.

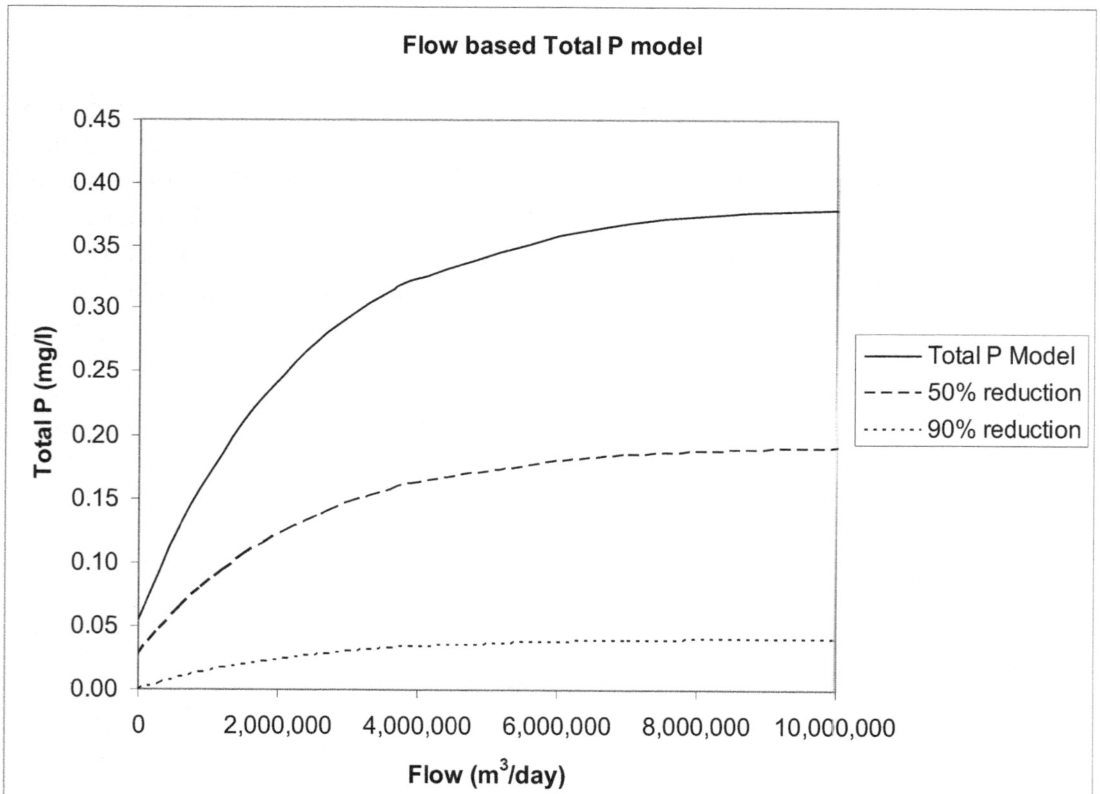

Figure 8.24 Curves of total phosphorus vs. flow used to generate load reduction scenarios.

The effect of reduction in phosphorus is illustrated in Figure 8.25. When compared to the unaltered simulation (Figure 8.18), some trends can be observed in the population of alga. The overall simulated biomass of diatoms and green algae are reduced proportionally with the reduction in nutrients as might be expected. Contrary to this trend, the cyanobacteria are largely unaffected by a 50% reduction in phosphorus while a further 40% reduction causes a decrease in *Anabaena* after September (see Figure 8.25a and 8.25b) and an overall increase in the population of *Microcystis*. This result is probably a reflection of the ability of the different species to compete for limited phosphorus resources. In the model, this ability is represented by the phosphorus half-saturation constant. The lower the value the better the species of algae can compete at low levels of phosphorus. The cyanobacteria in these simulations had the lowest half-saturation constants of the four species represented.

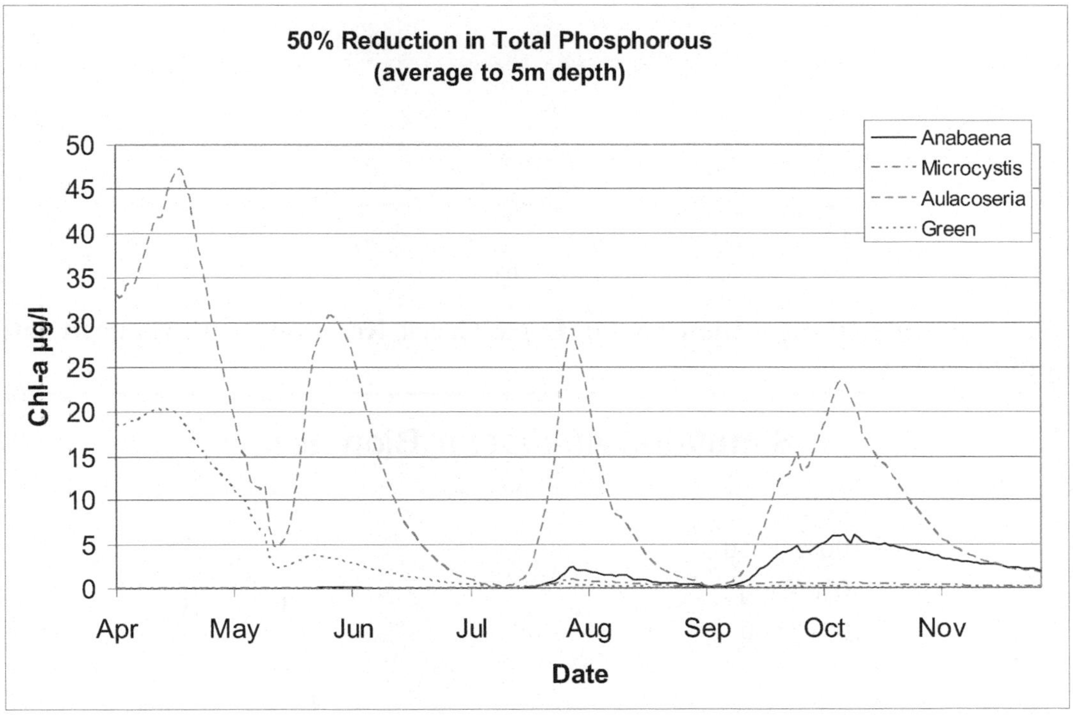

Figure 8.25a Simulated algal biomass for Eagle Creek Reservoir with reduced catchment input of phosphorus.

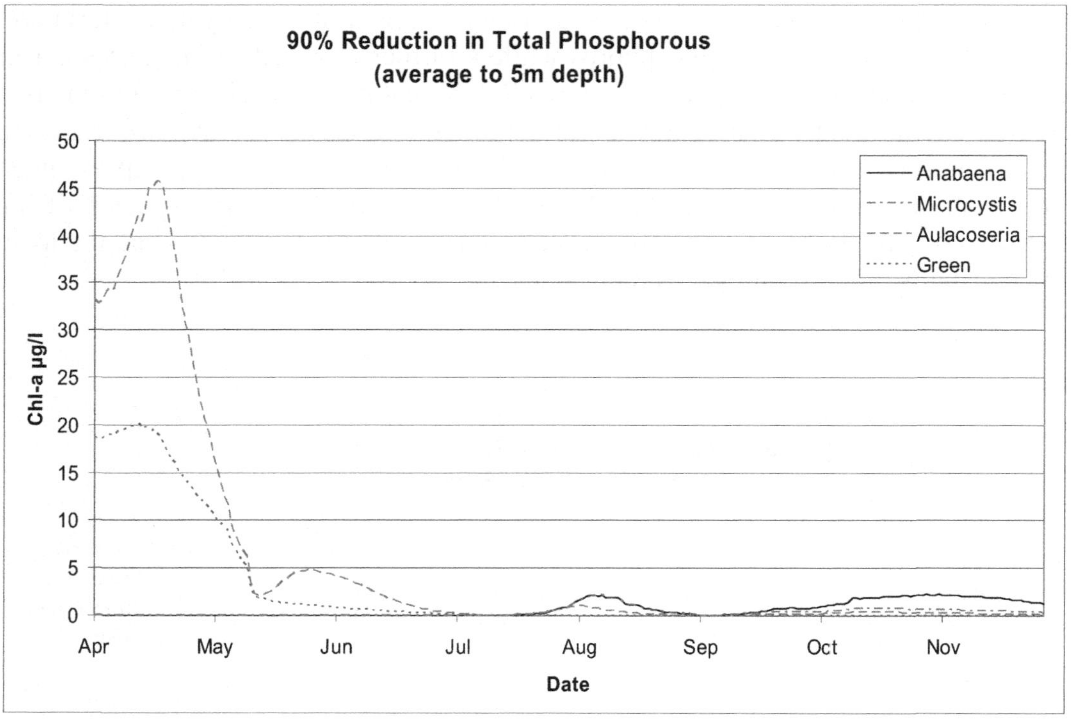

Figure 8.25b Simulated algal biomass for Eagle Creek Reservoir with reduced catchment input of phosphorus.

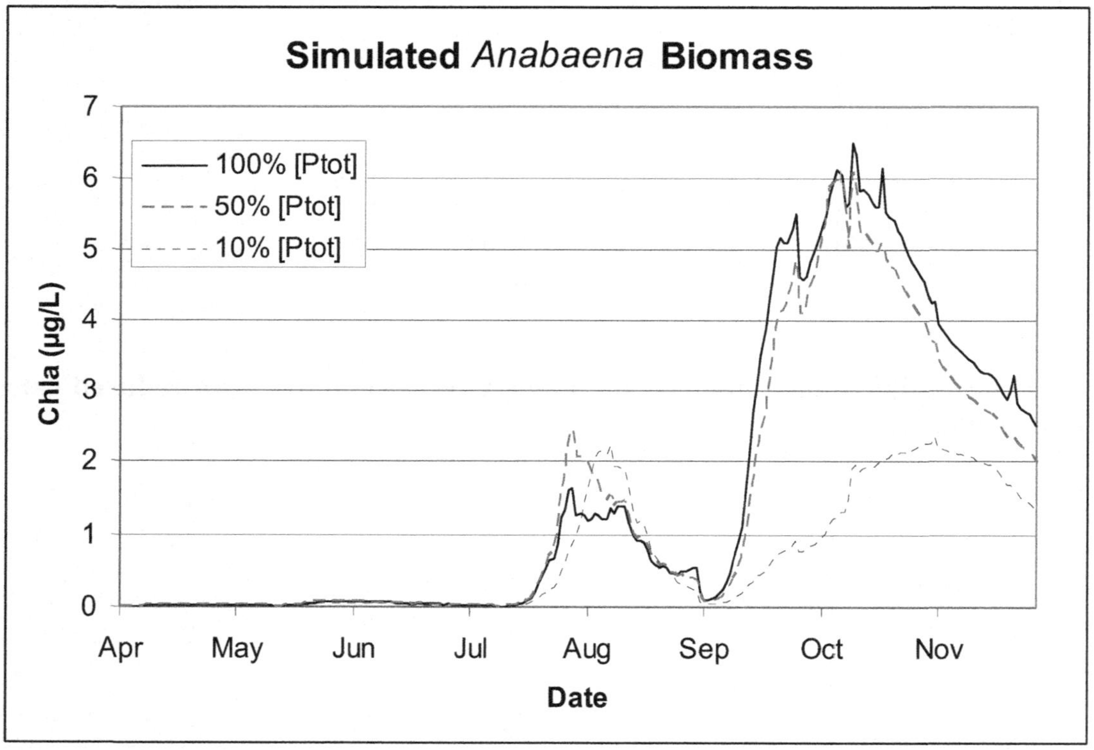

Figure 8.26 Simulated *Anabaena* biomass for Eagle Creek Reservoir with reduced catchment input of phosphorus.

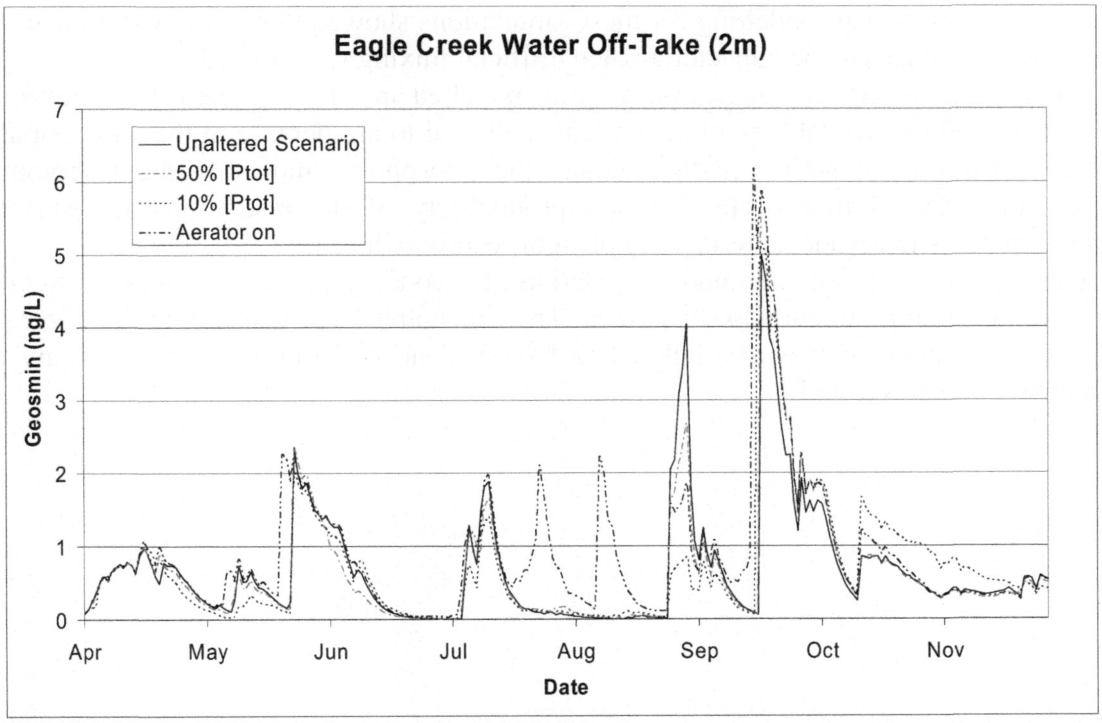

Figure 8.27 Comparison of simulated geosmin levels from reservoir management scenarios.

As might be expected, given the small impact these management strategies had on the population of *Anabaena*, the simulated concentration of geosmin at the plant off-take height was not greatly changed by these measures (Figure 8.27). The scenario which simulated artificial mixing showed the largest differences with the geosmin apparently mixed down to the level of the off-take sooner and more often than the unaltered simulation. Phosphorus limitation appeared to cause a decrease in the levels at the end of August but an increase during October and November.

CONCLUSIONS

The DYRESM reservoir hydrodynamics model of Eagle Creek Reservoir was developed further to allow reservoir management strategies to be simulated using the CAEDYM ecological model. A data set of nutrient input data was created based upon available data and included flow based modelling of the main inflow to the reservoir. A mathematical description of the main groups of algae that populate the reservoir and the other biological processes important to water quality was prepared. The impact of reservoir management using artificial mixing or phosphorus input reduction was simulated in the ecological model and the effect on the algal population examined.

Previous modelling studies have suggested that because of its shallow depth, the Eagle Creek Reservoir is not well suited to artificial mixing by a bubble plume aerator. Large inflows of cold water during periods of strong solar heating reinforces persistent stratification and, due to the low efficiency of aerator mixing, is not easily broken down. The modelling studies presented here support these previous studies and have shown that no appreciable increase in the surface mixed layer or potential for light limitation (Z_{eu}/Z_{mix} ratio) was observed when artificial mixing was simulated. Ecological modelling for these simulations showed that there was no decrease in the population of nuisance cyanobacteria when artificial mixing was applied.

The response of alga to nutrient inflow was modelled and showed that large inflows during summer increased the available pool of nutrients and lead to an increase in the algal population. Catchment management strategies that reduce the phosphorus inputs to the reservoir were simulated and showed that while the total productivity of the reservoir was reduced, the population of nuisance cyanobacteria was not significantly reduced.

A model for the production and degradation of secondary metabolites was configured for the study of geosmin production by *Anabaena*. The concentration of geosmin in the water supply to the local treatment plant was simulated and showed no real improvement with any of the management strategies tested.

CHAPTER 9: EFFECT OF CHLORINE PRETREATMENT ON ALGAL CELL INTEGRITY

INTRODUCTION

There has been a world wide trend towards eutrophication of drinking water supplies (Smith, Tilman and Nekola 1999) and also greater treatment/reuse of waste water. These nutrient rich waters can support large populations of algae and as a result, there has been increased focus on the negative effects that algae can have on water treatment processes. In conventional treatment, algae exert an additional demand on coagulant (Clasen et al. 2000) resulting in higher chemical use and sludge disposal costs. Certain species of algae have been implicated in filter clogging, reducing runtimes and increased back-wash water use (Crittenden 2005). Others can penetrate into the treated water and act as a nutrient source for bacterial regrowth or contribute to disinfectant decay and by-product formation. In addition, several species of algae produce metabolites that cause unpleasant tastes or odours, or are toxic.

A considerable amount of research has been focused on maximising removal of algae in conventional water treatment plants (Mouchet and Bonnélye 1998). In particular, the addition of oxidants such as chlorine, ozone or permanganate to the untreated water prior to coagulation has been shown to increase the removal of algae (Montiel and Welté 1998, Plummer and Edzwald 2002, Chen and Yeh 2005). These 'pre-treatment' chemicals alter the surface charge of algae thus improving their removal during coagulation but can also cause cell lysis (Plummer and Edzwald 2002). Cells lysis is a concern because it could result in the release of unpleasant/toxic metabolites or intracellular material that could promote bacteria growth. The aim of this study was to investigate the effects of chlorine pre-treatment on the cell integrity of *Microcystis aeruginosa* and the release of intracellular microcystin toxin.

Studies have shown that coagulants typically used in water treatment have minimal effect on the cell integrity of cyanobacteria and do not cause release of intracellular material (Velzevoer et al. 1995, Chow et al. 1998, Chow et al. 1999). The whole treatment process may however cause some cells to rupture due to mechanical processes such as pressure gradients and mixing shear etc. A pilot scale study using flocculation filtration to remove a natural population of toxic *Planktothrix rubescens* showed removal efficiencies for intracellular microcystin between 87-99% (Schmidt et al. 2002). Although coagulation chemicals do not cause lysis, cells can start to breakdown in settled flocc and release intracellular material (Drikas et al. 2001).

In contrast chlorine has been shown to cause damage to algal cells. Lam et al. (1995) used transmission electron microscopy to show that chlorine treatment caused damage to individual cells of *M. aeruginosa* and also measured an increase in the amount of extracellular toxin post treatment. Peterson et al. (1995) showed that cells of *Aphanizomenon flos-aquae* released intracellular potassium (indicating cell lysis) and geosmin after exposure to chlorine. Plummer and Edzwald (2002) studied the effect of oxidants on the surface properties of algae and showed that cells of *Scenedesmus* were lysed by chlorine using scanning electron microscopy. Tung et al. (2004) showed that treating cultures of *Oscillatoria tenius* with chlorine caused release of intracellular MIB. In a full scale treatment plant, Jurczak et al. (2005) measured intra- and extra-cellular toxin before and after chlorine-dioxide pre-treatment and showed that the average level of extracellular toxin increased from 16% to 78%. In many cases extracellular toxin was at detectable levels post coagulation while intracellular toxin was not.

While oxidant pre-treatment can cause release of toxin from cells it can also degrade toxin in solution. Many oxidants, such as those used to improve algal coagulation, have been shown to be effective at destroying microcystin (for review see Lawton and Robertson 1999). Chlorine can destroy microcystin and the products have no acute toxicity (Nicholson, Rositano and Burch 1994). However, two factors may limit the effectiveness of toxin destruction by chlorine when applied as a pre-treatment. Firstly, there is likely to be significant chlorine demand in the raw water and therefore chlorine can be consumed by other processes before toxin destruction is complete. Secondly, the rate of degradation is dependant on pH with half lives ranging from ~162 min at pH 9 to ~12.4 min at pH 6 (Acero, Rodrigez and Meriluoto 2005). Since the pH of untreated water can be at the higher end of this range, particularly when the numbers of algae are high (Mukherdee, Pandey and Singh 2002), the time required for complete degradation may be relatevely long.

For this study, the release and decay of microcystin toxin following exposure to chlorine was measured in untreated reservoir water to simulate the conditions of water pre-treatment. Cultures of a toxic *Microcystis aeruginosa* strain were grown in untreated water and then exposed to varying levels of chlorine. Total and extracellular microcystin concentrations were measured following treatment and the integrity of cells was investigated using flow cytometry in conjunction with viability DNA stains.

Flow cytometry measures the optical and florescence properties of individual cells as they pass a laser/detector array as part of a finely "focused" sample stream. It is a useful technique for studying pytoplankton because the optical properties of a large number for cells can be measured very quickly (Collier 2000). Since cells are analysed individually, they can be divided into subpopulations based on differences in their fluorescence and light scattering properties. The use of fluorescent stains allows subpopulations to be characterised and quantified by their physiology or metabolic activity.

EXPERIMENTAL PROCEDURES

Chemicals and Reagents

All chemicals and reagents used were analytical grade or better unless otherwise stated. Solutions were made using Milli Q water (Millipore Pty Ltd, USA).

Toxin spiking experiments were conducted using a purified isolate from a natural bloom of *Microcystis aeruginosa* that occurred in the Torrens Lake in South Australia during the summer of 1998-99. The isolation procedure involved freeze-thawing the bloom material in water and methanol, followed by preparative reverse phase flash chromatography and preparative high performance liquid chromatography (HPLC). The toxin concentrate contained microcystin-LR (m-LR) and microcystin-LA (m-LA). A Microcystin-LR analytical standard was purchased from Sapphire Bioscience Pty Ltd, Australia and was used as a HPLC standard and for extraction efficiency tests.

Water Sources

Water for toxin degradation and release experiments was sourced from the Myponga reservoir South Australia. Samples were collected from the water treatment plant supply prior to any treatment. The water was high in organics and colour (dissolved organic carbon 12 mgL^{-1} as C, true colour 70 Hazen Units) but low in turbidity (3 NTU) and had an alkalinity of 60 mgL^{-1} as $CaCO_3$ and 330 mgL^{-1} total dissolved solids.

Cyanobacterial Culture

The cyanobacteria used in all experiments was a toxic strain of *Microcystis aeruginosa* (strain 309) isolated from a bloom in the Mt.Bold reservoir, South Australia, December 1988 and held as part of the AWQC algal culture collection. The culture was monospecific but not axenic and was grown in ASM-1 medium (Gorham et al. 1964). For our experiments untreated water from Myponga reservoir was fortified with nutrients according to the ASM preparation (except no copper was added), filter sterilised, and then inoculated with an ASM culture in stationary phase (~10^6 cells/mL). The culture was incubated at 25°C under constant light flux.

For counting, algal samples were treated with Lugols iodine, pressurised to 1000 kPa to collapse gas vesicles and then counted at 40× magnification using a 1 mL Sedgwick Rafter chamber.

Chlorination experiments

A chlorine stock solution was prepared by bubbling gaseous chlorine through Milli Q water in a glass flask. The flask was then sealed and stored at 4°C overnight prior to use. Chlorine stock concentrations and free chlorine residuals were determined using the DPD-FAS titration method described in Standard Methods (APHA, AWWA and WEF. 1998). Typical chlorine stock solution concentrations ranged from 3000-5000 mgL^{-1} as free chlorine.

Chlorine decay experiments were conducted in 100 mL Erlenmeyer flasks at room temperature (20±2°C). Chlorine was added from the chlorine stock solution to obtain the desired dose and the flask covered and left for a specified time interval before determining the amount of residual chlorine.

The degradation of microsystin-LR and -LA by chlorine was measured in untreated water following experiments to determine the chlorine demand of the water as described above. A separate volume of water was spiked with cyanobacterial cell extract to give a final concentration of approximately 20 $\mu g\ L^{-1}$ of mLR and mLA and was used for all toxin degradation experiments. Duplicate sub-samples were dosed with chlorine following the same protocol as the decay experiments except that at the end of the prescribed contact time residual chlorine was quenched rather than titrated. Samples were quenched with sodium sulfite and the amount used was calculated using the stoichiometric ratio specified in Standard Methods (APHA, AWWA and WEF 1998) and the results of the decay experiments. Samples were excluded from the protocol based on the persistence of chlorine. The maximum contact time used at each chlorine dose was set at the first point where free chlorine was undetectable in the decay experiment. The quenched sub-samples were then analysed for microcystin.

As an initial investigation into the effect of chlorine on toxic *M.aeruginosa* cells, a batch culture was grown in untreated water and samples removed at intervals throughout the exponential growth phase and treated with a range of chlorine doses (1-20 mgL^{-1}). Cell integrity experiments were conducted in 5 mL cytometer sample tubes with 3 mL of sample. Chlorine was added from a 1:10 diluted stock solution with vigorous stirring (vortex mixer) and was quenched at time intervals with ascorbic acid (200µL, 20gL^{-1}) or sodium sulphite (20µL, 1gL^{-1}). Relative proportions of live and lysed cells were determined at each dose using flow cytometry.

Combined chlorine decay and cell integrity experiments were conducted with 103 mL samples. Samples for cytometric measurement (3 mL) were withdrawn from the bulk sample just prior to the required time interval and injected into a test tube containing quenching agent. The remainder of the sample was used for titrimetric chlorine determination.

Toxin release and degradation experiments with cultured *Microcystis* were carried out in three stages :

1. Small scale cell integrity experiments were performed with a range of chlorine doses 0-20 mgL^{-1}. The results of these experiments were used to set chlorine doses for parts 2 & 3 of the experiment by selecting a dose resulted in ~50% cell lysis and the minimum dose required to lyse 100% of cells.The 30 minute chlorine demand of the culture was measured by dosing 100 mL samples with a range of chlorine doses between 8-20 mgL^{-1}. The residual chlorine of each sample was measured after 30 minutes by FAS titration and subtracted from the initial concentration of chlorine to give the 30 minute chlorine demand. The highest chlorine dose in part 3 of the experiment was set to the 30 min chlorine demand plus 0.5 mgL^{-1}.
2. The three of chlorine doses determined in part 1 of the experiment were added to 100 mL samples of *Microcystis* culture. After a predetermined contact time (0.5, 1, 5, 10, 20 and 30 minutes) the residual free chlorine concentration was analysed using FAS titration.
3. Samples of *Microcystis* culture (1L) were dosed with chlorine as above. After 30 minutes residual free chlorine was quenched with a predetermined amount of Na_2SO_3 solution (calculated from the result of part 2). Samples of the quenched solution (100 mL) were removed, treated with 1g of KI (to reveal combined chlorine) and then titrated with FAS using the same method as for free chlorine. The remaining portion of the treated culture was then treated with a further dose of Na_2SO_3 in order to quench the remaining combined chlorine. Samples were divided in two sub-samples one of which was filtered without delay through GFC to remove cellular material. The remaining sample was subjected to three freeze/thaw cycles before GFC filtration.

ANALYTICAL PROCEDURES

pH

pH was measured using a portable pH meter (pH-340i meter, WTW, Germany) with a combination electrode. The meter was calibrated using a two point calibration at pH 7.00 and pH 4.00.

Trihalomethane Formation Potential

The trihalomethane formation potential analyses were performed by the Australian Water Quality Centre. Buffered water samples (pH 7.4) were heated at 35°C for 4 hours with an excess of chlorine (approximately 20 mgL^{-1} chlorine) and then quenched with ascorbic acid. THM components were determined using a gas chromatograph with a headspace autosampler, coupled with an electron capture detector (similar to standard methods, APHA, AWWA and WEF 1998).

Flow Cytometry

For this study Sytox green stain was used in combination with flow cytometry to measure the impact of chlorination on the cell integrity of *M. aeruginosa*. Sytox green is a nucleic acid probe that diffuses freely into cells with compromised cell membranes but will not cross membranes of living cells. The staining of nucleic acid within compromised cells results in an increase in green fluorescence and thus allows these to be distinguished from uncompromised cells using the green channel of the cytometer. The example shown in Figure 9.1 illustrates this increase in fluorescence with a mixture of live and lysed *M. aeruginosa* cells and shows how the two groups can be resolved. Once the spectral properties of each subpopulation are defined, the relative proportion of each can be determined.

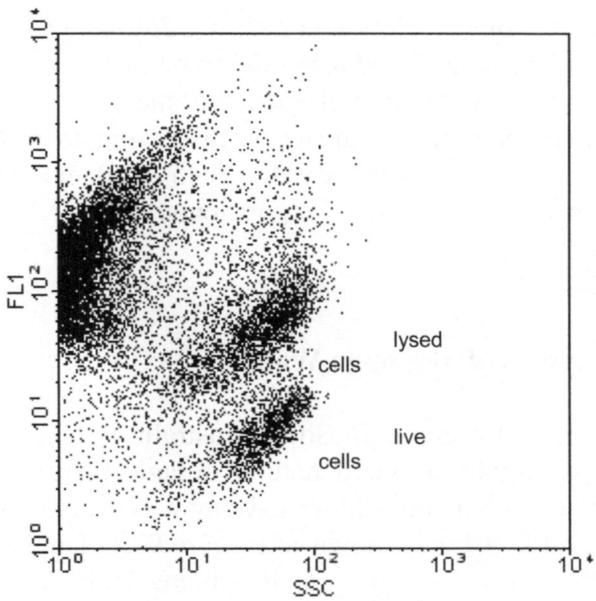

Figure 9.1 Increase in green fluorescence (FL1) observed between Cl_2 treated and untreated microcystis cells stained with Sytox Green. Treated cells were exposed to 15 mgL^{-1} chlorine for 30 min., equal volumes of liveL^{-1}ysed were mixed after Cl_2 quench.

The auto-fluorescence of chlorophyll inside algal cells can also be detected using the red channel of the flow cytometer. This property can be used to distinguish algal cells from other detritus in the sample and can also provide a measure of cell health particularly since chlorine has been shown to decrease autofluorescence in algal cells (Burch, Baker and Beck 1990).

A FACS Calibur Flow Cytometer (Becton Dickinson, San Jose, CA, USA) equipped with an air cooled 15 mW argon laser, emitting at a fixed wavelength of 488 nm was used for all fluorescence measurements. Fluorescent filters and detectors were all standard with green fluorescence collected in channel FL1 (530 nm) and red fluorescence collected in channel FL3. Data were collected and analysed using CellQuest™ software. SYTOX green nucleic acid stain (C_2H_6OS, Molecular probes, S-7020) was used to determine the percentage of viable to nonviable cells in a sample using the method of Regel et al. (2004). A development time of 7 min was used and a final sytox concentration of 0.1 μM. Data collection was stopped when the combined number of events recorded in the viable and nonviable regions reached 5000.

Microcystin Extraction Method

All samples were prepared for solid phase extraction by filtering (0.45 μm) to remove any particulate matter. Each sample was concentrated using a Waters C18 (500 mg, 3cc) solid phase extraction cartridge which had been washed with methanol (10 mL) followed by MQ water (10 mL). Samples were filtered through the cartridge under vacuum at a flow rate of less than 10 mL/min. Cartridges were washed with MQ water (10 mL), 10% methanol (10 mL), 20% methanol (10 mL) and then air dried. The toxin was eluted off with methanol (10 mL) and then evaporated to dryness using a dry heating block (30°C) under a stream of nitrogen.

For each group of extractions a solution containing 250 μL of 5 ppm m-LR standard was prepared and then concentrated and analysed using the same method as with experiment samples. This sample was used to check the extraction efficiency of the C18 cartridges.

Samples were prepared for HPLC analysis by dissolving in methanol (500μL), diluting with water (500 μL) and then filtering (PVDF, 0.45 μm) into HPLC vials. All mLA concentrations were expressed in terms of mLR equivalents.

RESULTS AND DISCUSSION

Chlorine Decay Characteristics of Myponga Water

Water from the Myponga Reservoir in South Australia is characterised as having a high load (12 mgL^{-1} as DOC) of highly coloured natural organic mater (NOM) and as such, was expected to exert a high chlorine demand. Chlorine decay was examined with a range of chlorine doses; 1-20 mgL^{-1} as Cl_2 (Table 9.1 and shown in (Figure 9.2). The untreated water had a very high chlorine demand with doses of 7 mgL^{-1} or less being entirely consumed within the first minute of contact. At higher doses there was a rapid decay of chlorine in the first minute followed by a more gradual decay over the remaining time.

The rapid initial reactivity is attributed to the high concentration of chromophoric compounds present in NOM as these species are very reactive towards free chlorine (Korshin, Li and Benjamin 1997). In similar work by Ho et al. (2006), using treated Myponga water where most of these compounds were removed using enhanced coagulation, the chlorine demand during the first minute of contact was less than 0.75 mgL^{-1}. The addition of algal extract to give an m-LR concentration of 20μgL^{-1} did not significantly change the chlorine decay characteristics of the water (Table 9.1).

Table 9.1
Chlorine decay in untreated Myponga water (pH 7.6-6.8), chlorine doses of 7 mgL^{-1} or less showed no chlorine residual after 1 min.

Contact Time (min)	Free Chlorine Residual (mgL^{-1})						Free Chlorine Residual (mgL^{-1}) spiked with 20μgL^{-1} m-LR		
	8	9	10	12	15	20	12	15	20
1	0.7	1.0	1.6	2.6	5.3	8.4	2.3	5.0	7.9
5	0.0	0.1	0.4	1.3	3.6	7.1	1.5	3.1	6.6
10	0.0	0.0	0.1	1.0	2.8	6.0	1.1	2.6	6.1
20	0.0	0.0	0.0	0.7	2.3	5.9	0.9	2.4	5.7
30	0.0	0.0	0.0	0.5	2.1	5.7	0.5	2.3	5.9
60	0.0	0.0	0.0	0.4	2.0	5.7	0.3	2.3	5.1

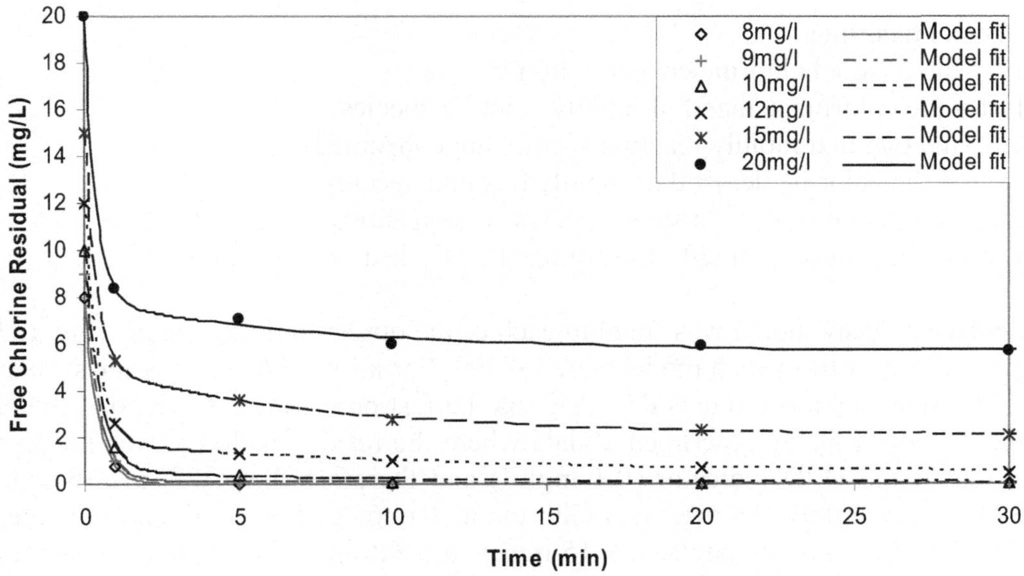

Figure 9.2 Chlorine decay in untreated Myponga water.

Chlorine Decay Rates and Exposure (ct)

The inactivation of pathogens or destruction of contaminants during the chlorination of water is affected by two important parameters, the concentration of chlorine (c) and the contact time (t). In natural and treated waters, the decay of chlorine with time negatively impacts on the efficiency of these processes by reducing both the concentration of chlorine and the time of its effect. In order to account for this phenomenon, a combined term that describes the net chlorine exposure experienced by contaminants in the water is used. This term ct ($mgL^{-1}min$) is derived by integrating the time dependant concentration of chlorine over the time of its action (von Gunten 2003).

Chlorine exposure (ct) values were calculated from chlorine decay data by fitting a model (Equation 9.1)) to the decay curve and then substituting the estimated parameters into the integrated form of the model equation (Equation 9.2). The chlorine decay model assumes that the initial chlorine dose can decay by one of two parallel pathways: fast and slow (Sohn et al. 2004). The fast pathway represents the rapid consumption of chlorine observed within the first minute of contact while the slow represents the slower decay observed following this. These two processes are modelled as a first order decay where the rate of each process is given by a rate constant k and the amount of chlorine removed by Cl_{fast} or Cl_{slow}. An extra term $Cl_{v.slow}$ was added to account for the chlorine that did not decay over the time scale of the experiment.

$$[Cl]_t = Cl_{fast}e^{-k_{fast} \cdot t} + Cl_{slow}e^{-k_{slow} \cdot t} + Cl_{v.slow} \qquad (9.1)$$

$$ct = \int_0^t [Cl]_t \, dt = \frac{Cl_{fast}}{k_{fast}}\left(1 - e^{-k_{fast} \cdot t}\right) + \frac{Cl_{slow}}{k_{slow}}\left(1 - e^{-k_{fslow} \cdot t}\right) + Cl_{v.slow} \cdot t \quad (9.2)$$

where t = the contact time

[Cl]t = the residual chlorine concentration at time t

Clfast = the chlorine demand of rapidly reacting species

kfast = the rate that rapidly reacting species are consumed

Clslow = the chlorine demand of rapidly reacting species

kslow = the rate that slow reacting species are consumed

Clv.slow = the amount of chlorine not reacting by fast or slow routes

The chlorine decay model was fitted to each of the observed decay curves (Figure 9.3), the correlation coefficients from each model were >0.999. For initial chlorine doses less than 8 mgL^{-1} all of the chlorine was modelled as decaying via the fast process (i.e. Cl_{slow}=0, Table 9.2). For many of these, there was no experimental data where the residual chlorine was higher than the detection limit of the DPD indicator (0.1 mgL^{-1}) and therefore there was insufficient data to model the decay accurately. The residual chlorine at 1 minute showed an approximately linear relationship to the initial doses between 8-15 mgL^{-1} and extrapolation of this line suggested that 7 mgL^{-1} would only just be consumed after 1 minute contact time (0.0±0.3 mgL^{-1}, experimental result ≤0.1 mgL^{-1}). The value for this point was artificially set to 0.1 mgL^{-1} (DPD detection limit) and an appropriate decay constant determined, this value was used for all doses less than 7 mgL^{-1}. The inaccuracy in ct values calculated for these doses was thought to have little impact on other modelling work as the values are small compared to those observed for higher doses.

For initial chlorine doses greater than 8 mgL^{-1} the amount of chlorine that decayed by the fast route levelled off at around 10 mgL^{-1} and appeared to represent the waters total fast chlorine demand i.e. the amount of material in the water that will react quickly. Once this material has reacted further addition of chlorine is consumed by the slow decay or persists for the length of the experiment.

Chlorine Degradation of Microcystins in Untreated Water

There is a strong relationship between chlorine exposure and m-LR & m-LA toxin degradation (Figure 9.3). This relationship has been observed previously although typically in the absence of NOM which means the decay of chlorine can effectively be ignored (Zeir, Harrington and Xagoraraki 2003; Acero, Rodrigez and Meriluoto 2005). Under these conditions a pseudo first order rate law (Equation 9.3) adequately describes the degradation of toxin at a given pH.

$$[tox]_{ct} = [tox]_0 . e^{-k_{app} ct} \quad (9.3)$$

where *ct* = the chlorine exposure

$[tox]_{ct}$ = the residual toxin concentration after exposure to chlorine

$[tox]_0$ = the initial toxin concentration

k_{app} = the apparent rate constant

The apparent degradation rate k_{app} is reported to be a combination of three possible reaction pathways, each with a different rate; the relative significance of each pathway is dependant on the pH of the solution (Acero, Rodrigez and Meriluoto 2005). The slowest of these reactions occurs between toxin and hypochlorite (OCl^-) ions and is the predominant reaction at high pH. At lower pH (<7.5) hypochlorous acid (HOCl) is the dominant chlorine species and is more reactive towards microcystins. At very low pH (<5), acid can protonate the toxin molecule and accelerate the reaction with HOCl. Overall the rate of toxin degradation is at a minimum above pH 9 and progressively increases as the pH is lowered by up to three orders of magnitude.

Ho et al. (2006) measured the rate of toxin degradation in treated water with significant chlorine demand and showed that Equation 9.3 appeared to hold using ct values calculated by integrating the observed decay curves (similar to above). The degradation rates observed were different to those measured in pure water and may indicate that solution chemistry, other than pH, may affect the reaction rate.

For the experiment described here, the pH of the untreated water containing toxin was 7.9 and due to the acidity of the chlorine solution was lowered to between 6.8-7.6 depending on the dose of chlorine applied. Under these conditions the dominant chlorine species is HOCl (AWWA 1999) and therefore the degradation rate was expected to be rapid. The results of toxin degradation experiments were $\log_e$ transformed and then linear regression used to determine the apparent rate constant for m-LR and m-LA. The obtained values were 242 $M^{-1}s^{-1}$ for m-LR (R^2 0.93) and 172 $M^{-1}s^{-1}$ for m-LA (R^2 0.98). The model for m-LR is compared with experimental data on a linear scale in Figure 9.3.

A prediction for the m-LR degradation rate in the above experiment can be calculated using the pH relationship developed by Acero, Rodrigez and Meriluoto (2005). For the range of pH conditions used, the rate was predicted to be between 55 and 101 $M^{-1}s^{-1}$ while the actual rate was 2-4 times faster than this. Ho et al. (2006) also measured a faster than expected degradation rate and attributed the difference to the reaction chlorine and NOM producing species that were more reactive towards microcystin than chlorine alone. Since the above reaction also contained NOM, the same explanation could be applied. The rates reported by Ho (2006) are slightly faster ($k_{m\text{-}LR}$: 260, $k_{m\text{-}LA}$: 270 $M^{-1}s^{-1}$) than above although this difference may be explained by the difference in pH between the two experiments.

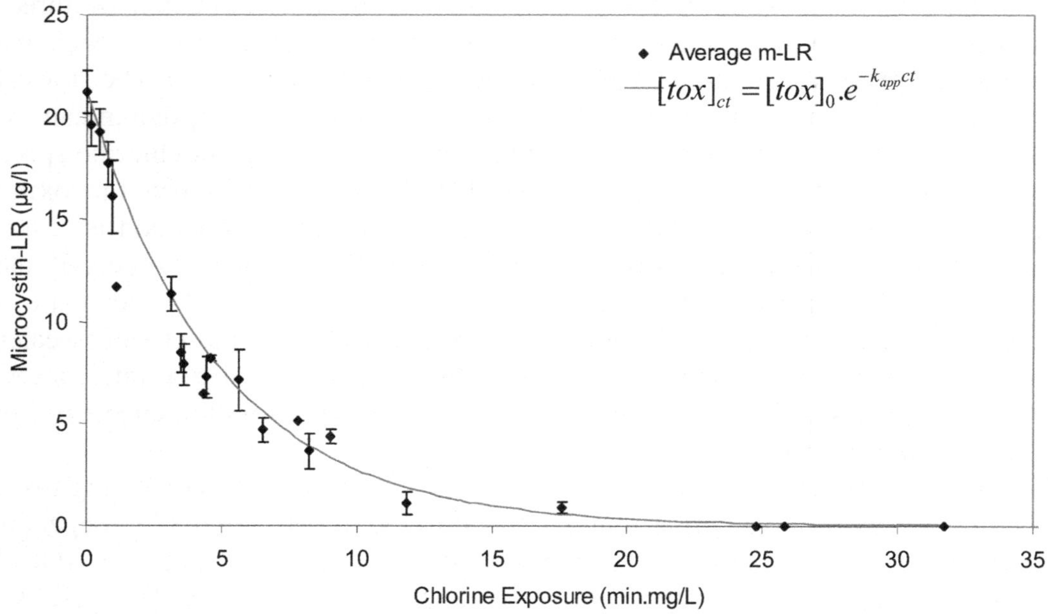

Figure 9.3 Degradation of microcystin-LR in untreated water at pH 7.6-6.8. Error bars show the difference between duplicate sample analyses.

Table 9.2
Chlorination of Microcystin toxins in untreated water (pH 7.6-6.8).

$[Cl]_0$ (mgL^{-1})	Chlorine Decay Constants					Time (min)	ct ($min.mgL^{-1}$)	[m-LR] (μgL^{-1})		[m-LA] (μgL^{-1})	
	k_{slow} min^{-1}	k_{fast} min^{-1}	Cl_{fast} mgL^{-1}	Cl_{slow} mgL^{-1}	$Cl_{v.slow}$ mgL^{-1}						
0							0.0	20.1	22.0	28.2	28.6
								21.7		28.3	
1	-	4.24	1.0	0.0	0.0	1	0.2	20.7	18.6	27.2	25.8
3	-	4.24	3.0	0.0	0.0	1	0.5	20.4	18.1	27.1	24.9
5	-	4.24	5.0	0.0	0.0	1	0.8	18.8	16.7	26.2	23.1
6	-	4.24	6.0	0.0	0.0	1	0.9	14.3	17.9	20.7	23.9
7	-	4.24	7.0	0.0	0.0	1	1.1	-	11.8	-	18.2
8	-	2.29	8.0	0.0	0.0	1	3.1	12.2	10.5	18.2	16.5
8						5	3.5	7.6	9.4	12.4	15.4
9	0.10	2.29	8.9	0.1	0.0	1	3.6	9.0	6.9	15.1	12.7
9						5	4.3	6.5	6.5	12.5	12.4
9						10	4.6	8.1	8.4	13.5	13.6
10	0.10	2.14	9.5	0.5	0.0	1	4.4	8.3	6.3	14.5	11.9
10						5	6.5	5.3	4.1	9.7	9.3
10						10	7.8	5.2	5.2	10.1	10.4
10						20	9.0	4.1	4.7	8.8	10.2
12	0.07	2.35	10.3	1.3	0.4	1	5.6	5.7	8.7	11.5	14.1
12						5	11.8	0.5	1.7	1.9	7.4
12						10	17.6	0.7	1.2	2.6	3.5
12						20	25.8	0.0	0.0	0.0	1.0
12						30	31.7	0.0	0.0	0.0	0.0

(continued)

(continued)

12						60	44.8	0.0	0.0	0.0	0.0
15	0.13	2.72	10.0	3.0	2.0	1	8.3	4.6	2.8	9.7	6.5
15						5	24.8	0.0	0.0	0.9	0.9
15						10	40.6	0.0	0.0	0.0	0.0
15						20	65.5	0.0	0.0	0.0	0.0
15						30	87.0	0.0	0.0	0.0	0.0
15						60	148.4	0.0	0.0	0.0	0.0

Effect of Chlorination on the cell integrity of Microcystis aeruginosa

The response of algal cells to chlorination was evaluated using a flow cytometer to measure changes in cell permeability (sytox green stain) and chlorophyll autofluorescence. Addition of chlorine to the culture gave rise to a second population of algal cells (R2 see Figure 9.4a) with increased green fluorescence and reduced autofluorescence compared to live cells (R1). Chlorine caused cell lysis in this population and the reduction in chlorophyll fluorescence may indicate oxidation of photosynthetic pigments. The population of lysed cells increases with additional chlorine until a point where the R2 population accounted for 100% of the algal cells (Figure 9.4b). The chlorine dose required to achieve 100% lysis of the cells increased with the algal population (see Figure 9.5) and probably reflects the increase in chlorine demand provided by the cells. Additional demand would reduce the persistence of chlorine and therefore lower the chlorine exposure (ct) of cells.

At low cell densities all of the cells were lysed at doses less than 8 mgL^{-1}. Given that similar doses in the untreated water persisted for less than one minute, it would appear that cell lysis occurs rapidly after the addition of oxidant.

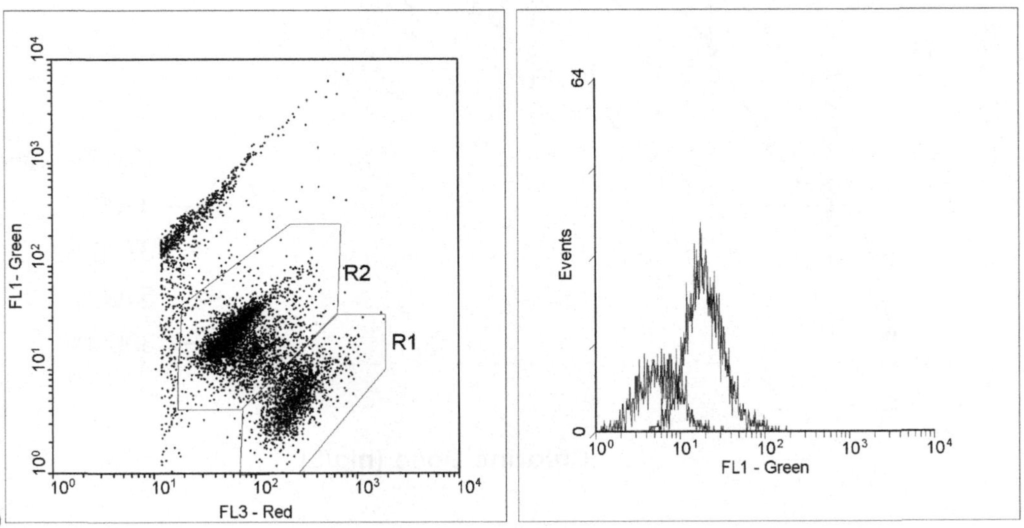

a)

(continued)

(Figure 9.4 continued)

b)

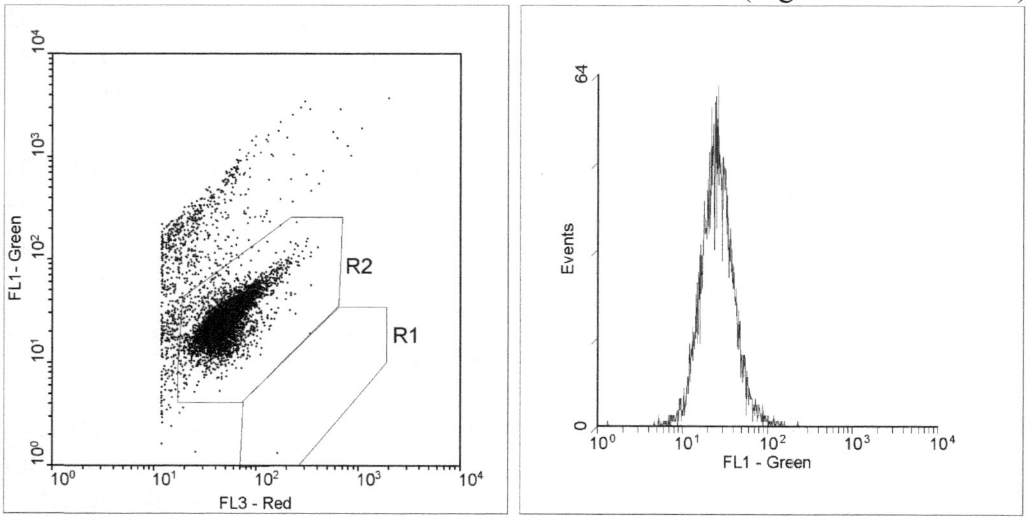

Figure 9.4 Flow cytometry results following chlorination of 37800 cell/mL *M. aeruginosa* with a) 3 mgL^{-1} Cl_2 and b) 6 mgL^{-1} Cl_2; contact time 30 min. R1 & R2 are regions used to define populations as live or lysed (respectively) for histogram plots and % lysis statistics. Axes: FL1-Green corresponds to Sytox green stain, FL3-Red corresponds to chlorophyll autoflourecense.

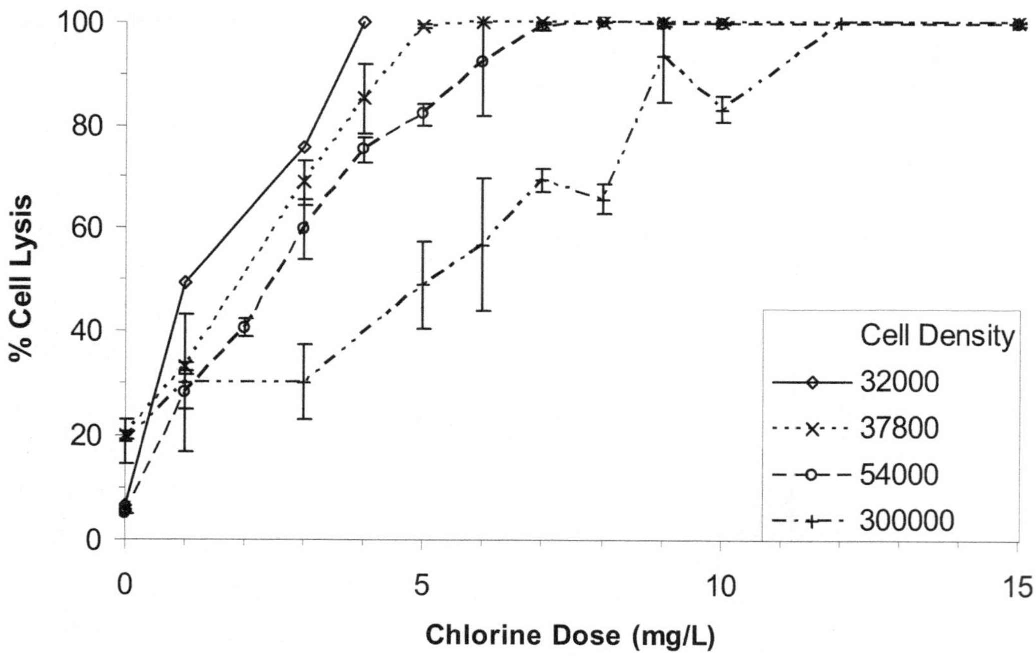

Figure 9.5 Cell lysis statistics for *M. aeruginosa* at different cell densities following treatment with chlorine (30 min contact time).

When the cell population of the culture had reached 300,000 cells/mL combined chlorine decay and cell integrity experiments were conducted over a range of contact times and chlorine doses. Chlorine doses of 12 mgL^{-1} or higher caused lysis in >99% cells within 1 minute of contact (see Table 9.3). Doses less than 7 mgL^{-1} did not persist longer than one minute and resulted in progressively less cell damage as the chlorine concentration was reduced. At dose rates between 8-10 mgL^{-1}, chlorine persisted longer than one minute and a clear difference in the number of lysed cells was evident between samples quenched after 1 minute compared to those quenched after 5 min (Figure 9.6). The fraction of the population lysed after one minute was comparable to lower doses where all of the chlorine had been consumed after this time i.e. samples with similar net exposure. Additional chlorine contact in samples not quenched after one minute resulted in a ~20% increase in cell lysis.

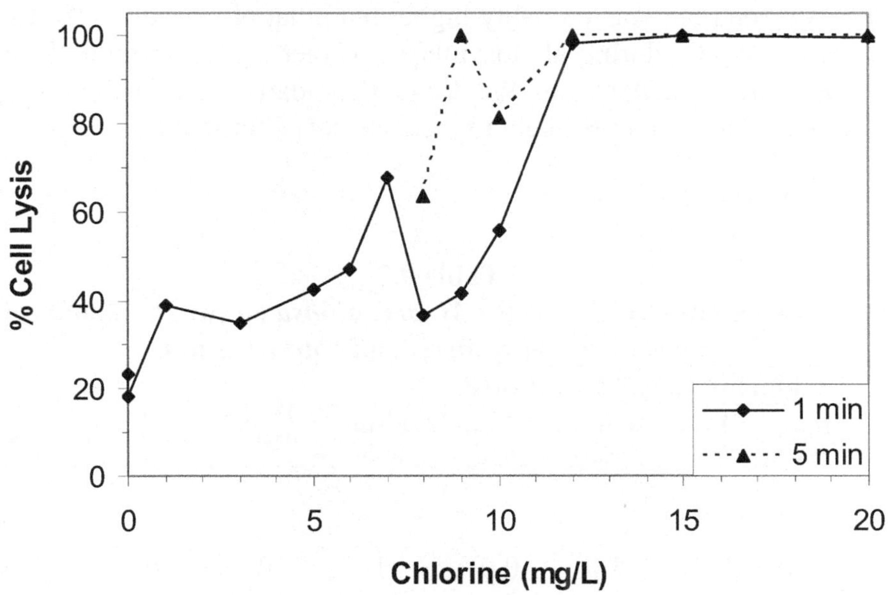

Figure 9.6 Extent of cell lysis for *M. aeruginosa* culture (300,000 cells/mL) for a range of chlorine doses and contact times

Chlorine exposure (ct) values were calculated from chlorine decay data, similar to toxin degradation experiments, by modelling decay curves (Equation 9.1) then using the integrated form of the model Equation 9.2 to calculate the ct value. Parameters for lower doses were not able to be determined accurately using the available data and therefore an estimate based on higher doses was used for these values. However, unlike the toxin experiments where degradation occurred over a range of 0-30 $min.mgL^{-1}$, cell lysis occurs over a smaller range (0-14 $min.mgL^{-1}$) and consequently this estimate (for ct<2 $min.mgL^{-1}$) is a more significant limitation on the accuracy of rate estimates. The results of cell integrity experiments were normalised so that at ct = 0 the proportion of lysed cells was zero, log_e transformed and then linear regression used to estimate the cell lysis rate for chlorine exposure (Equation 9.4).

$$\frac{N_{ct}}{N_0} = e^{-k_{lysis}ct} \quad (9.4)$$

where ct = the chlorine exposure
N_{ct} = the number of uncompromised cells after a given chlorine exposure
N_0 = the number of uncompromised cells at ct=0
k_{lysis} = the rate at which lysis occurs

This is a special form of the Chick-Watson equation which is often used to describe the inactivation of microorganisms in disinfection processes (AWWA 1999). This form was used so that the lysis rate could be directly compared to the toxin degradation rate (dividing k_{lysis} by ln(10) will yield a typical Chick-Watson constant). The relationship between cell lysis and chlorine exposure (Figure 9.7) was not as clear (R^2 0.66) as that observed for toxin degradation although still significant ($p<0.001$). The lysis rate was 670 ± 77 $M^{-1}s^{-1}$, the pH during chlorine exposure was not recorded but was probably higher than that observed for the raw water due to the removal of dissolved CO_2 during photosynthesis. Under these conditions, the lysis rate was approximately three times higher than the toxin degradation rate which suggests that the extracellular concentration of toxin is likely to increase post chlorination.

Table 9.3
Chlorine decay and extent of cell lysis for *M. aeruginosa* culture (300,000 cells/mL) for a range of chlorine doses and contact times.

$[Cl]_0$ (mgL^{-1})	Chlorine Decay Constants				Time (min)	$[Cl]_t$ (mgL^{-1})	ct ($min.mgL^{-1}$)	% Lysed Cells
	k_{slow} min^{-1}	k_{fast} min^{-1}	Cl_{fast} mgL^{-1}	Cl_{slow} mgL^{-1}				
0								23.4
0								18.0
1	-	0.08	1.0	0.0	1	0	0.2	39.2
3	-	0.08	3.0	0.0	1	0	0.7	35.1
5	-	0.08	5.0	0.0	1	0	1.2	42.8
6	-	0.08	6.0	0.0	1	0	1.4	47.5
7	-	0.10	7.0	0.0	1	0.28	2.1	67.6
8	-	0.10	8.0	0.0	1	0.65	2.9	36.8
8					5	0	3.2	63.5
9	-	0.10	9.0	0.0	1	1.21	3.9	41.9
9					5	0	4.5	99.9
10	2.10	0.10	9.8	0.2	1	1.36	4.3	55.7
10					5	0.15	5.4	81.4
12	2.23	0.08	11.2	0.8	1	1.93	5.2	98.1
12					5	0.53	8.3	100.0
12					10	0.37	10.5	100.0
12					20	0.13	13.0	99.8
12					30	0.11	14.1	99.9
15	2.26	0.05	12.8	2.2	1	3.47	7.2	100.0

(continued)

(continued)

15					5	1.69	15.6	100.0
15					10	1.55	23.5	100.0
15					20	0.71	34.4	100.0
15					30	0.58	41.2	100.0
20	2.54	0.03	14.2	5.8	1	6.75	10.9	99.7
20					5	4.75	32.6	100.0
20					10	4.65	55.9	100.0
20					20	3.41	93.5	100.0
20					30	2.18	121.7	100.0

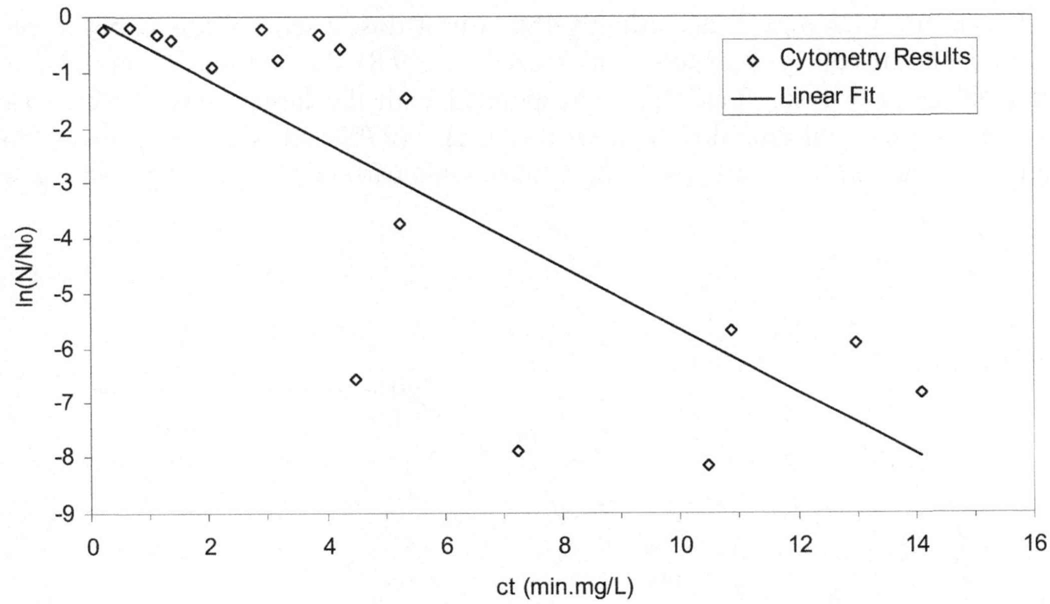

Figure 9.7 Extent of cell lysis for *M. aeruginosa* culture (300,000 cells/mL) relative to chlorine exposure.

Quenching agents

The effect of the chlorine quenching agents, ascorbic acid and sodium sulphite, on the cell integrity of *M.aeruginosa* was examined using flow cytometry to ensure that this had no effect that would confound the effect of chlorine alone. Samples of cyanobacteria mixed with an aliquot of quenching agent showed no increase in cell lysis over control samples. Similarly, cyanobacteria added to a mixture of the quenching agent and a typical dose of chlorine showed no increase in cell lysis indicating that the products of the quenching reaction do not cause cell lysis.

Toxin Release and Degradation

The total and extracellular levels of toxin in several *M. aeruginosa* cultures were measured after exposure to three different chlorine doses to assess the release of toxin from cells compared to the overall degradation of toxin by chlorine. The maximum chlorine dose was chosen to provide a residual free chlorine concentration of 0.5 mgL^{-1} after 30 minutes of contact. This level was suggested by Nicholson, Rositano and Burch (1994) to provide sufficient chlorine exposure to completely degrade soluble toxin provided the pH was below 8. Treatment of cyanobacterial cells under these conditions would allow this recommendation to be tested in relation to intracellular toxin which would be afforded some protection from chlorine exposure by the cell wall. The remaining two doses were set based on the extent of cell lysis measured at different doses using flow cytometry. The lowest dose was chosen to coincide with lysis of approximately 50% of algal cells while the intermediate dose was based on >99% cell lysis.

Experiment 1

A culture containing 300,000 cells/mL was used for this experiment and had a 30 minute chlorine demand of 11.5 mgL^{-1}; accordingly the largest dose used for this experiment was 12 mgL^{-1}. The results of cell lysis experiments (see Figure 9.8) showed that 12 mgL^{-1} Cl_2 caused 99.9% of algal cells to lyse, since this corresponded with the largest dose based on chlorine decay, the intermediate chlorine dose was set to 9 mgL^{-1} (87% cell lysis). The lowest dose used was 5 mgL^{-1} and caused 55% cell lysis. The total organic carbon content of the culture was 16.9 mgL^{-1}.

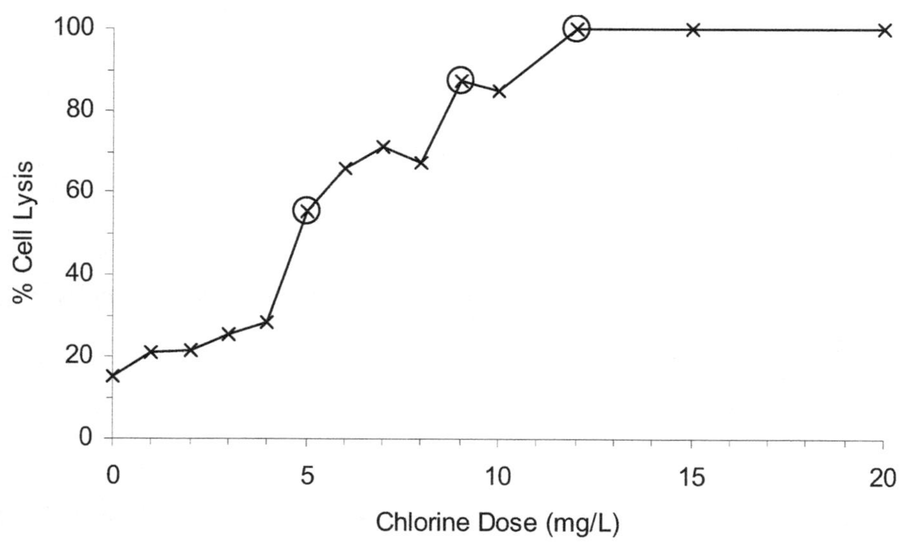

Figure 9.8 Extent of cell lysis for *M. aeruginosa* culture (300,000 cells/mL) as a function of chlorine dose. Contact time = 30 min. Circled results indicate doses used for toxin release and degradation experiment.

The amount of extracellular toxin remaining after 30 minutes contact with chlorine is displayed in Figure 9.9 together with the total amount of toxin. The algal culture before exposure to chlorine contained 6.8 μgL^{-1} m-LA toxin and only a very small proportion (6%) was found outside of the algal cells. The lowest chlorine dose persisted for between 30-60 sec. (Table 9.4) and resulted in a 335% increase in the amount of extracellular toxin while only 9% of the total toxin was degraded. The intermediate dose persisted between 5-10 min. and degraded 58% of the total toxin; there was a 40% net increase in the amount of extracellular toxin although it is clear that degradation of extracellular toxin had occurred. The total amount of toxin was reduced by 93% using the highest dose and was at a similar concentration to that found outside the cells indicating that most of the toxin had been released.

After the 30 min contact time, quenched samples were analysed for combined chlorine, the measured amounts were 3.1, 2.7 & 1.5 mgL^{-1} as Cl_2 for doses 5, 9 & 12 mgL^{-1}, respectively. The combined nitrogen and chlorine species (such as chloramine) are not reactive towards microcystins (Nicholson, Rositano Burch 1994) however the organic amines associated with cyanobacteria could contribute a significant free chlorine demand.

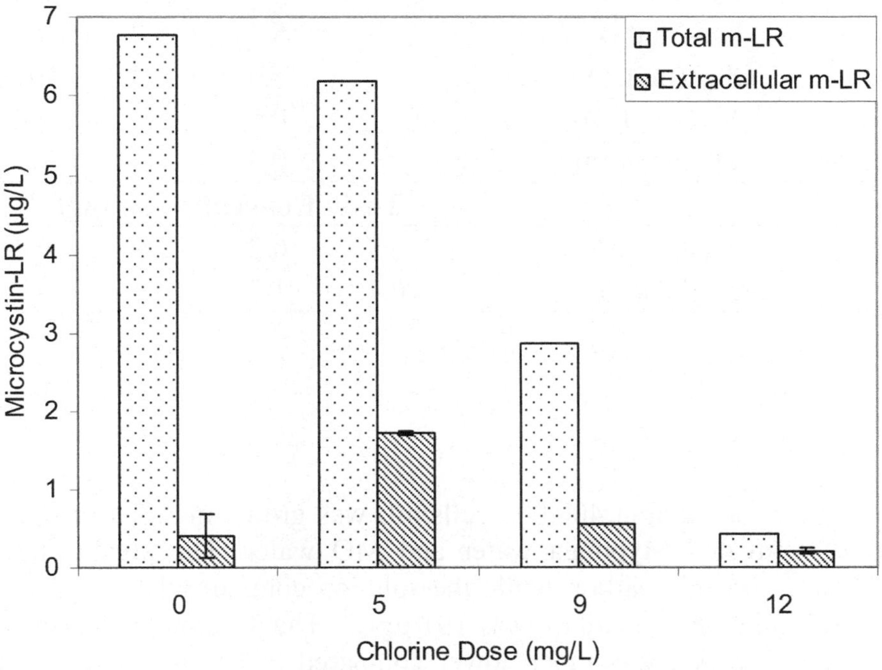

Figure 9.9 Microcystin concentration of *M. aeruginosa* culture (300,000 cells/mL) and media (extracellular) after treatment with chlorine (30 min contact time). Results are from single samples for total m-LR, extracellular samples were analysed in duplicate.

Table 9.4
Chlorine decay and toxin analysis results for *M. aeruginosa* culture (300,000 cells/mL).

Time (min)	**Residual Chlorine (mgL^{-1})**			
0	0	5.0	9.0	12.0
0.5		0.3	2.2	-
1		0.0	1.6	3.1
5		0.0	0.5	1.5
10		0.0	0.0	1.0
20		0.0	0.0	0.7
30		0.0	0.0	0.5
	***ct* Model Parameters**			
k_{slow} (min^{-1})		-	0.31	0.046
k_{fast} (min^{-1})		5.6	6.0	2.0
Cl_{fast} (mgL^{-1})		5.0	6.8	10.2
Cl_{slow} (mgL^{-1})		0.0	2.2	1.8
Calculated 30 min *ct* ($mgL^{-1}min$)	0	0.9	8.1	34.2
	Toxin Concentration (μgL^{-1})			
Total m-LR	6.8	6.2	2.9	0.4
Extracellular m-LR	0.4	1.7	0.6	0.2

Experiment 2

A second culture containing 400,000 cells/mL was grown in ASM media prepared using equal amounts of untreated Myponga water and MQ water. The algal cells in the culture contributed 2.8 mgL^{-1} organic carbon while the solution contained 10.1 mgL^{-1} DOC; the total THM formation potential for the culture was 191 μgL^{-1}. The 30 minute chlorine demand of this culture was 3.7 mgL^{-1} which was much lower compared to previous experiments due to the reduced amount of NOM present. As a consequence of this, complete cell lysis occurred at much lower chlorine doses since less was required for a persistent residual. The largest dose used for this experiment was 4 mgL^{-1} and resulted in 99.9% cell lysis (see Figure 9.10). An intermediate dose of 3 mgL^{-1} Cl_2 was used and corresponded to 93% cell lysis, while the lowest dose was set to 1.5 mgL^{-1} and caused approximately 45% cell lysis. The pH of this culture was 10 and accordingly the rate of toxin degradation by chlorine was expected to be at a minimum.

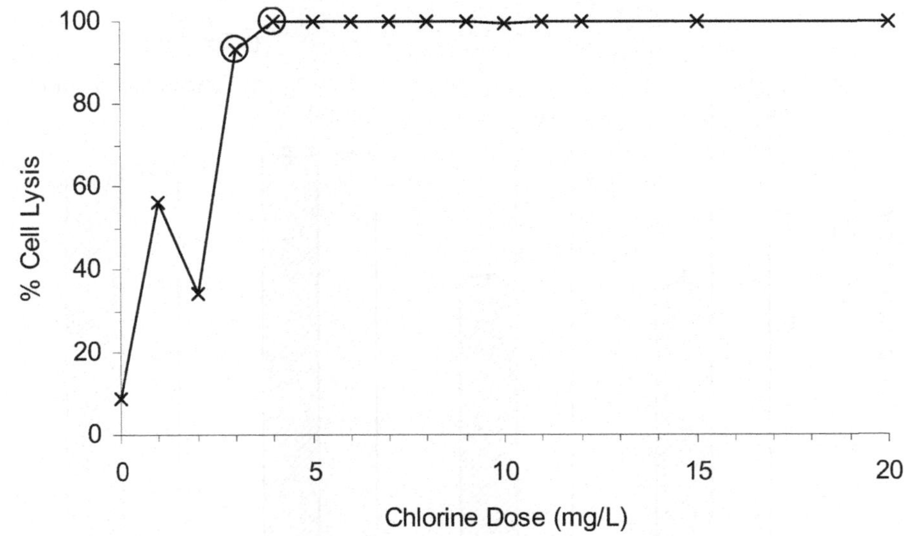

Figure 9.10 Extent of cell lysis for *M. aeruginosa* culture (400,000 cells/mL) as a function of chlorine dose. Contact time = 30 min. Circled results indicate doses used for toxin release and degradation experiment.

The amount of extracellular toxin remaining after 30 minutes contact with chlorine is displayed in Figure 9.11 together with the total amount of toxin. The algal culture before exposure to chlorine contained 34.7 μgL^{-1} m-LA toxin and in contrast to the previous experiment a large proportion (63%) was found outside of the algal cells. The lowest chlorine dose persisted for between 30-60 seconds. (Table 9.5) and resulted in no net increase in the amount of extracellular toxin however, the amount of intracellular toxin decreased by 1.6 μgL^{-1} (13%). This may indicate that toxin release was balanced by degradation since there also a 1.6 μgL^{-1} drop in the total amount of toxin recovered. The intermediate dose persisted between 20-30 minutes and caused all of the intracellular toxin to be released; a further 8% of total toxin was degraded. The highest dose resulted in only 14% degradation of total toxin and also caused complete release of intracellular toxin.

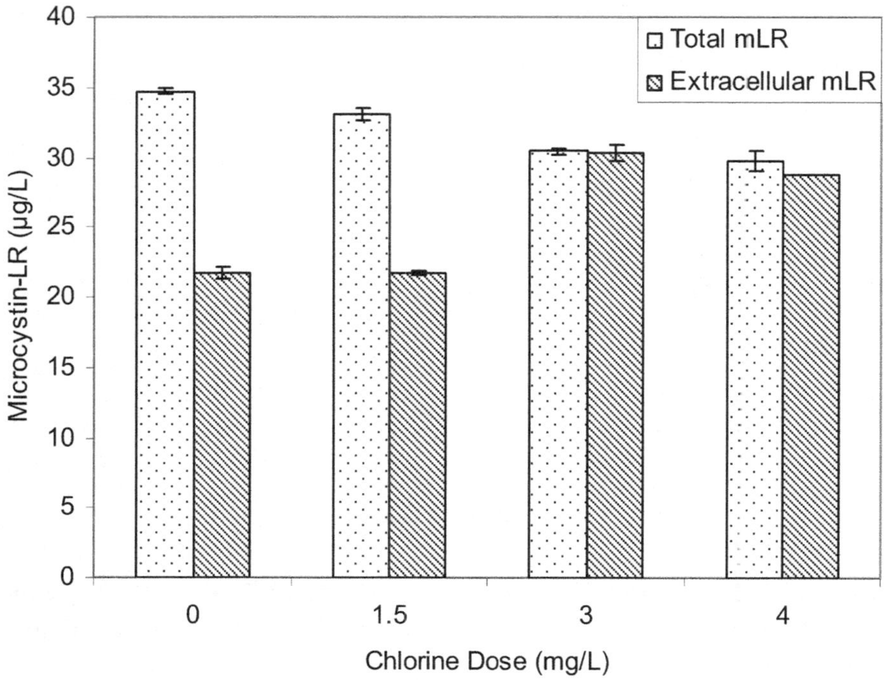

Figure 9.11 Toxin recovered from cultured *M. aeruginosa* (400,000 cells/mL) after treatment with chlorine (30 min contact time). Results are from single samples analysed in triplicate.

Table 9.5
Chlorine decay and toxin analysis results for *M. aeruginosa* culture (400,000 cells/mL).

Time (min)	Residual Chlorine (mgL^{-1})			
0	0.0	1.5	3.0	4.0
0.5		0.1	0.6	1.5
1		0.0	0.6	1.5
5		0.0	0.5	1.2
10		0.0	0.4	0.9
20		0.0	0.2	0.5
30		0.0	0.0	0.3
	***ct* Model Parameters**			
k_{slow} (min^{-1})		-	0.066	0.057
k_{fast} (min^{-1})		5.6	11.6	11.6
Cl_{fast} (mgL^{-1})		1.5	2.4	2.5
Cl_{slow} (mgL^{-1})		0.0	0.6	1.5
Calculated 30 min *ct* ($mgL^{-1}min$)	0.0	0.3	8.6	22.2
	Toxin Concentration (μgL^{-1})			
Total m-LR	34.7	33.1	30.4	29.8
Extracellular m-LR	21.7	21.7	30.3	28.7

Experiment 3

A batch culture containing 1,050,000 cells/mL was grown in untreated Myponga water. The algal cells in the culture contributed 2.8 mgL^{-1} organic carbon, 4.38 mgL^{-1} organic nitrogen and 0.058 mgL^{-1} ammonia. The bulk solution contained 27.8 mgL^{-1} DOC, 1.97 mgL^{-1} organic nitrogen and 0.011 mgL^{-1} ammonia. The culture had a 30 minute chlorine demand of 10.1 mgL^{-1} and a total THM formation potential of 594 μgL^{-1}. When filtered to remove algal cells, the culture medium consumed 3.7 mgL^{-1} less chlorine after 30 min (at the same dose rate) and the THMFP was reduced by 179 μgL^{-1}. The pH of the culture was 8.8 and was adjusted using hydrochloric acid (15 mL, 0.1 M) to obtain a final pH of 7.9 to match the conditions used when measuring the toxin degradation rate above.

The cell lysis experiments using chlorine gave surprising results with 100% cell lysis occurring at a dose of 5 mgL^{-1} (Figure 9.12). These results are comparable to the previous experiment despite 3 fold higher chlorine demand. This result may be partly explained by the observation that, in this culture, the chlorine persisted for longer during the fast decay processes. However in terms of chlorine exposure, 100% lysis occurred at 7.1 $min.mgL^{-1}$ which was low compared to the previous experiments. Lowering the pH of the culture may also have affected the cell lysis rate as it would lead to an increase in the proportion of chlorine present as hypochlorous acid (HOCl), which is a stronger oxidant.

The largest dose used for this experiment was 10.3 mgL^{-1} which was more than double the amount required to lyse all of the cells and represented the recommended dosing regime for complete toxin destruction. As discussed above, 100% lysis occurred at 5 mgL^{-1} Cl_2; this dose was used as the intermediate dose while the lowest dose used was 2 mgL^{-1} and resulted in 56% cell lysis.

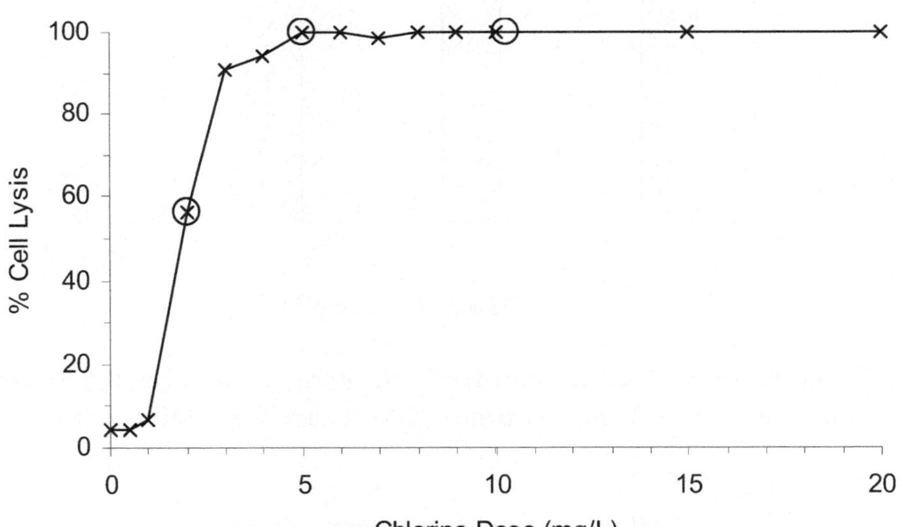

Figure 9.12 Extent of cell lysis for *M. aeruginosa* culture (1,050,000 cells/mL) as a function of chlorine dose. Contact time = 30 min. Circled results indicate doses used for toxin release and degradation experiment.

The amount of extracellular toxin remaining after 30 minutes contact with chlorine is displayed in Figure 9.13 together with the total amount of toxin. The algal culture before exposure to chlorine contained 36.1 μgL^{-1} m-LA toxin and a large proportion (61%) was found outside of the algal cells. The lowest chlorine dose persisted for between 1-5 min. (Table 9.6) and resulted in a 47% increase in the amount of extracellular toxin. This corresponds to release of 84% of the intracellular toxin. Under these conditions the total toxin level was reduced by 4%. The large difference between the amount of toxin released and the amount degraded, demonstrates that release occurs much faster than degradation. The intermediate dose persisted between 5-10 min. and resulted in no significant change in the amount extracellular toxin compared to untreated cells. The toxin release was offset by degradation with the total level of toxin being reduced by 30%. The highest dose resulted in 89% degradation of total toxin and caused complete release of intracellular toxin.

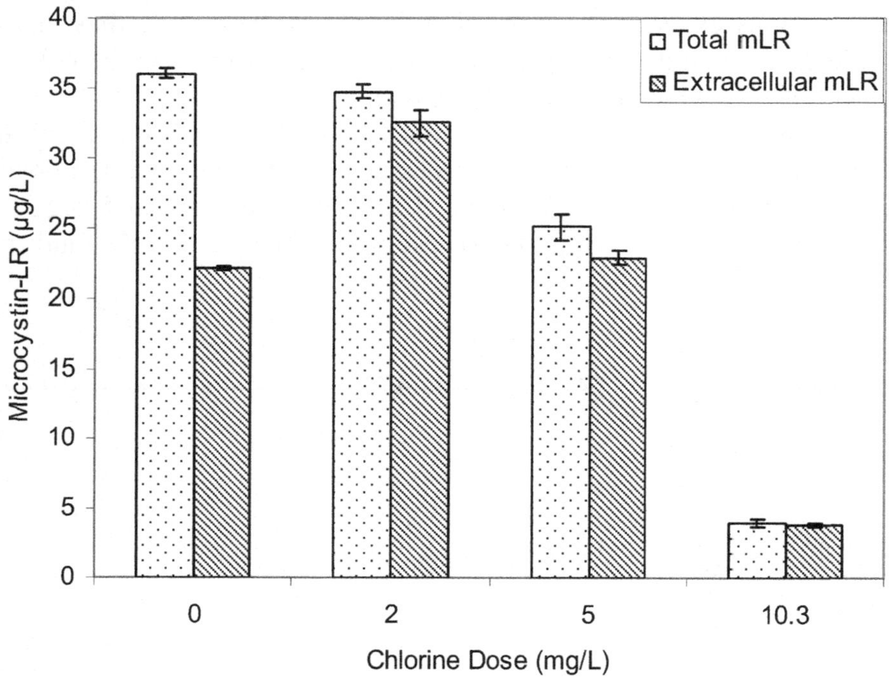

Figure 9.13 Toxin recovered from cultured *M. aeruginosa* (1,050,000 cells/mL) after treatment with chlorine (30 min contact time). Results are from triplicate samples.

Comparing the results of all three toxin experiments (Figure 9.14), toxin release appears to correlate with cell lysis. Between 85 and 100% of intracellular toxin was released when a chlorine dose that caused 100% cell lysis was applied. Complete cell lysis occurred at ct values between 7 and 29 $min.mgL^{-1}$. Compared with the range of chlorine exposure values needed to inactivate other water contaminants, *M. aeruginosa* lies midway between viruses (3-6 $min.mgL^{-1}$) and Giadia cysts (2-log inactivation, 69 $min.mgL^{-1}$) (AWWA 1999). Accordingly, *Microcystis* cells present in water would be lysed under disinfection practices aimed at reducing the risk from all pathogens and would result in the release of toxin. However, chlorine pre-treatment is usually targeted at achieving treatment outcomes other than water disinfection and so the applied doses may result in much less exposure. This may limit the release of toxin although at doses where chlorine persisted for longer than 30 seconds significant cell lysis occurred.

Table 9.6
Chlorine decay and toxin analysis results for *M. aeruginosa* culture (1,050,000 cells/mL).

Time (min)	Residual Chlorine (mgL^{-1})			
0	0.0	2.0	5.0	10.3
0.5		0.3	2.4	7.0
1		0.2	1.8	5.5
5		0.0	0.3	2.3
10		0.0	0.0	1.7
20		0.0	0.0	0.7
30		0.0	0.0	0.2
	***ct* Model Parameters**			
k_{slow} (min^{-1})		-	0.10	0.084
k_{fast} (min^{-1})		3.5	1.4	1.2
Cl_{fast} (mgL^{-1})		2.0	4.6	6.7
Cl_{slow} (mgL^{-1})		0.0	0.4	3.6
Calculated 30 min *ct* ($mgL^{-1}min$)	0.0	0.6	7.1	45.3
	Toxin Concentration (μgL^{-1})			
Total m-LR	36.1	34.8	25.1	4.0
Extracellular m-LR	22.2	32.5	22.9	3.9

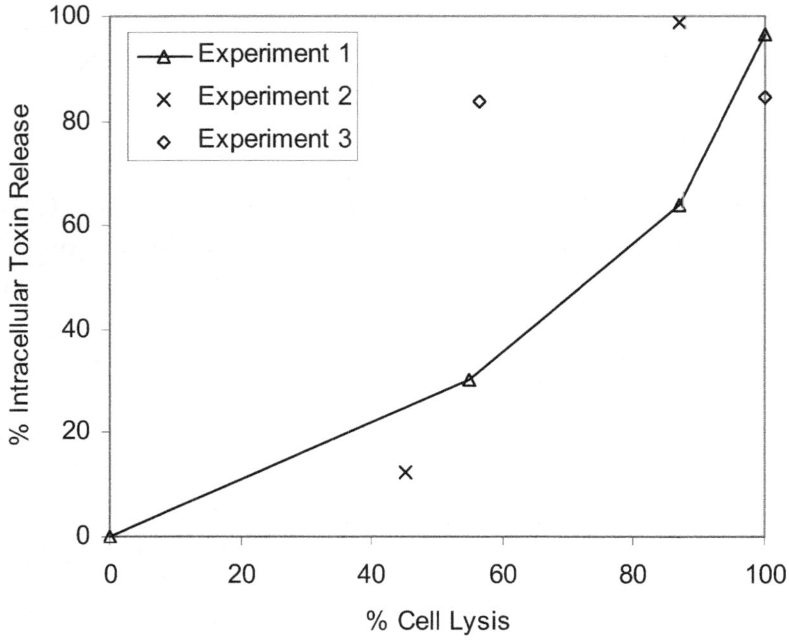

Figure 9.14 Comparison of intracellular toxin release relative to cell lysis in three cultures of *M. aeruginosa*. Experiment 1 - Myponga water, 5-12 mgL^{-1} Cl_2, 3×10^5 cells/mL; Experiment 2 - Myponga/MQ water, 1.5-4 mgL^{-1} Cl_2, 4×10^5 cells/mL; Experiment 3 - Myponga water, 2-10.3 mgL^{-1} Cl_2, 10^6 cells/mL.

There was a net increase in the amount of extracellular toxin following chlorine treatment in all three toxin release and degradation experiments. This suggests that intracellular toxin is released from damaged cells at a rate faster than it is degraded. In support of this, the cell lysis rate measured using flow cytometry was three times faster than the fastest observed degradation rate. In a system of consecutive reactions such as this, the rate of slowest reaction can often be used to describe the rate of the whole process. The degradation of total toxin in these three experiments was well described by first order decay relative to chlorine exposure (Table 9.7 & Figure 9.15) and therefore this approximation appears to be valid.

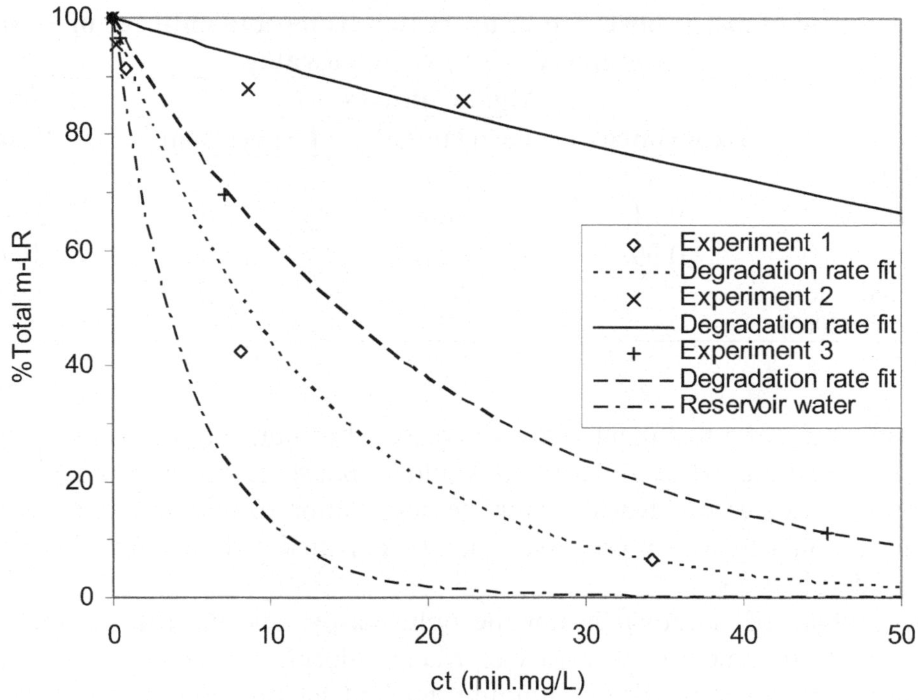

Figure 9.15 Comparison of microcystin-LR degradation relative to chlorine exposure in untreated water and three cultures of *M. aeruginosa*.

The rate of total toxin degradation was different in each experiment where it was measured. Experiment 2 had the slowest degradation rate; this was because, at pH 10, all of the chlorine was present as the less reactive hypochlorite species. The fastest degradation rate occurred in untreated water that was free from cyanobacterial cells. The observed rate was much faster than expected based on published rates measured in pure water (Acero, Rodrigez and Meriluoto 2005). Additional solution chemistry that occurred in the presence of natural organic matter was thought to have accelerated the degradation reaction in this example. Water from the same source was used to grow the algal culture used in experiment 3 and the pH was adjusted to be the same as the cell free water. Despite similar conditions, the rate of degradation was much slower than the experiment without cells and was closer to the value measured in pure water. It is possible that the process that accelerated degradation in the cell free water was diminished by competition with other molecules similar to microcystin (such as peptides) from within lysed algal cells. Experiment 1 also used cyanobacteria grown in reservoir water but had a much lower cell concentration than experiment 3. Interestingly, the rate was in-between the values recorded for cell-free and high cell number experiments and could be the result of less competition between cell derived material and toxin. While this is one possible explanation, the pH of experiment 1 was not recorded and so rate changes due to this factor cannot be ruled out.

Table 9.7
Degradation rate of m-LR observed in untreated water and cultures of *M. aeruginosa* compared with predicted values

	Algal Cultures			Reservoir
	Experiment 1	**Experiment 2**	**Experiment 3**	**Water**
Initial pH	n.r.	10	9.7	9.7
k_{app} ($min.mgL^{-1}$)	96.1	9.6	57.6	241.8
r^2	0.99	0.58	1.00	0.93
Predicted k_{app} Acero et al. (2005)	-	7.1	38.0	38.0

While chlorine pre-treatment may increase the number of cells removed during coagulation, any resulting extracellular toxin would be poorly removed by coagulation/filtration. If insufficient pre-oxidant is added for complete degradation of released toxin, it would need to be removed using other treatment processes such as activated carbon, ozonation or by additional chlorine before distribution.

Enhanced algal cell removal is not the only reason chlorine pre-treatment is used. For situations where pre-treatment is necessary or where chlorine disinfection is the only treatment, toxin degradation appears to be the rate limiting step and can effectively be used to describe total toxin destruction. In the presence of NOM and algal cells, the observed degradation rates were at least as fast as those predicted using the pH relationship described by Acero, Rodrigez and Meriluoto (2005) and were often faster. The recommendation of Nicholson, Rositano and Burch (1994) that complete toxin destruction can be achieved by maintaining a chlorine residual >0.5 mgL-1 for 30 min at pH less than 8 was also tested. Under these conditions 90-100% of total toxin was removed.

It should be noted that the particular strain of *M. aeruginosa* used in this study was non-colonial. Many strains of microcystis do form colonies and may behave differently when exposed to chlorine. Cells at the centre of a colony may be protected from chlorine exposure by outer layers of cells or mucilage. This would reduce the rate of toxin release following pre-treatment, but likewise slow the rate of total toxin degradation by chlorine disinfection.

The use of chlorine as a pre-treatment step to conventional water treatment should be avoided when a bloom of toxic Microcystis affects the source water. This is because chlorine causes intact cells to lyse and release intracellular toxin into solution where it is poorly removed by coagulation. Soluble toxin can be destroyed by chlorine, although in uncoagulated water the pH and chlorine demand may reduce the effectiveness of this reaction. Alternatively, coagulation without pre-treatment is effective at removing toxin inside intact cells and is likely to reduce the total amount of toxin post filtration compared to the previous scenario. In both scenarios chlorine disinfection post filtration could be used to degrade any remaining toxin however, there would be increased reliance on this process for the pre-treatment scenario.

For other scenarios where chlorine pre-treatment may be necessary, such as a bloom of filter clogging algae or soluble manganese in the source water, the risk of microcystin release would need to be assessed in terms of the amount toxin present and the dose of chlorine required to degrade the toxin to acceptable levels. For relatively low concentrations of *Microcystis*, released toxin may be effectively degraded by chlorine pre-treatment. Under these situations particular attention would need to be applied to the treatment conditions in order to maximise toxin degradation. This would include ensuring the pH was less than 8 and that the toxin receives sufficient chlorine exposure throughout the treatment process. A conservative estimate for the exposure nessesary for toxin destruction can be calculated using the pH relationship described by Acero et al. (2005).

CHAPTER 10: FACTSHEETS

FACT SHEET OBJECTIVE

The Framework for Management of Drinking Water Quality contained in Chapter 2 of the Australian Drinking Water Guidelines (ADWG), (http://www7.health.gov.au/nhmrc/publications/synopses/eh19syn.htm) outlines the methodology for providing safe drinking water by managing the complete catchment to tap water supply system. This document is achieving global recognition as the best way to manage our drinking waters as we move into the 21st Century and is being incorporated into National and State Health Guidelines.

It is important to understand the level of risk that the different cyanobacteria and toxins pose to drinking water. This allows managers of catchments and urban water utilities to focus their efforts on policies, works and operational practices to not only lower risks to public health but also improve the environmental health of these waters.

These fact sheets present the findings of a major research program carried out by the Australian Cooperative Research Centre (CRC) for Water Quality and Treatment into areas such as understanding cyanobacterial growth, detection methods for cyanobacterial toxins and water treatment options for cyanobacterial cells and toxins.

FACT SHEET CONTENTS

These fact sheets are derived from the following CRC for Water Quality and Treatment research projects

- 1.0.0.2.5.1 Destratification for Control of Cyanobacteria in Reservoirs
- 2.0.2.2.1.4 Reservoir Management Strategies for the Control and Degradation of Algal Toxins
- 1.0.0.2.6.1 ARMCANZ National Algal manager
- 1.0.0.3.2.6 Optimisation of Adsorption Processes – Stage II
- 2.0.2.4.0.5 Biological Filtration Processes for the Removal of Algal Metabolites
- 2.0.2.4.1.3 Management Strategies for Toxic Blue-green Algae: A Guide for Water Utilities

The fact sheets are comprised of four sections: Research Findings, Implementation, More Information and Contact Details.

Summary of key points

Cyanobacteria present a particular challenge to water utilities because of both aesthetic issues associated with taste and odour compounds and human health concerns surrounding cyanobacterial toxins. The research undertaken by the CRC for Water Quality and Treatment has examined cyanobacterial control strategies from catchment to tap. The research has revealed that with a multi-barrier approach to the cyanobacteria hazards in the catchment, reservoir and treatment plant can reduce the risks associated with cyanobacteria.

An important outcome from the research has been the development of practical tools for identifying cyanobacterial risks, operation response guides for reservoir managers and appropriate treatment technologies to control tastes, odours and toxins. These tools are transferable to any catchment, reservoir or treatment plant. The fact sheets provide the initial information on cyanobacteria and a list of further information where a detailed understanding can be gained.

Factsheet topics

- FS 1 The Ecology of Cyanobacteria
- FS 2 The Cyanobacterial Toxins
- FS 3 Sampling Waterbodies for the Detection of Cyanobacteria
- FS 4 Microscopic Identification of Potentially Toxic Cyanobacteria
- FS 5 Molecular Identification of Cyanobacteria
- FS 6 Treatment of Cyanobacterial Toxins
- FS7 Taste and Odour Removal
- FS 8 Biodegradation of Microcystin Toxins
- FS 9 Biodegradation of Cylindrospermopsin
- FS 10 Artificial Destratification for Control of Cyanobacteria
- FS 11 Algicides as a Management Tool to Control Cyanobacteria
- FS 12 The Alert Levels Framework for Cyanobacteria in Drinking Water
- FS 13 Modelling Tools for Predicting Cyanobacterial Growth
- FS 14 Nutrient Control

In Australia, drinking water quality management is undertaken in the context of the Framework for Management of Drinking Water Quality contained in the Australian Drinking Water Guidelines. In Table 10.1, the salient research findings are presented within the Framework to aid in their implementation by the Australian water industry.

Table 10.1
Summary of factsheet findings and relationship with ADWG Framework elements.

ADWG Framework Elements		Key research findings and reference to Factsheet No.
Assessment of the drinking water supply system	Water Supply System Analysis	All factsheets provide information necessary for control and management of cyanobacteria
	Review of Water Quality Data	FS 6 Data sets used to determine what toxins are likely to occur and the appropriate treatment technology to apply FS 9 Cell count data provides the context for cyanobacteria risk assessment FS 12 Historical data provides a validation for modelling studies
	Hazard Identification and Risk Assessment	FS 1 Sampling for cyanobacteria FS 2 Cyanobacteria are identified and counted by microscopy FS 3 Molecular techniques are becoming available for rapid detection of cyanobacteria FS 9 Protocol for operational response to cyanobacterial blooms

(continued)

Table 10.1 (Continued)

ADWG Framework Elements		Key research findings and reference to Factsheet No.
Planning-preventative Strategies for Drinking Water Quality Management	Multiple Barriers	FS 13 Nutrients exported from catchments can be reduced with soil amending chemicals FS 10 Destratification can control nutrient release from sediment and can promote mixing to light limit cyanobacteria FS 6 Coagulation and the removal of intact cell is the first treatment barrier FS 6 Residual toxin can be degraded with the oxidants chlorine or ozone FS 7 Bio-filters can be very effective for microcystin removal
	Critical Control Points	FS 6 Management and maintenance of treatment technologies is a critical control prior to releasing water for consumption FS 11 Algicides can be applied in response to cyanobacterial blooms
Verification of Drinking Water Quality	Drinking Water Quality Monitoring	FS 1 Appropriate sampling is critical to obtain a representative overview of water quality FS 2 Accurate identification of problem algae is achievable with microscopy FS 9 Monitoring data can be evaluated in the context of the alert levels framework to determine the appropriate response FS 10 Hydrodynamic and ecological models are useful for prediction of water quality
Research and Development	Investigative Studies and Research Monitoring	FS 3 Rapid genetic tests are being developed to improve identification of cyanobacteria and reduce operational response time. FS 7 Rapid degradation of microcystin in biofilters shows promise as a low cost alternative for treatment of toxins FS 12 Reservoir and cyanobacterial growth models transfer knowledge from science to operations and allows the outcome of management options to be predicted

FACTSHEET # 1: THE ECOLOGY OF CYANOBACTERIA

- Cyanobacteria have existed on earth for 3 billion years, however there is a general opinion that 'cyanobacterial blooms' are increasing in frequency due to anthropogenic eutrophication.
- The building of dams and river regulation has also created more habitats suitable for cyanobacteria.
- There are three general constraints on cyanobacterial growth; light, nutrients and temperature.

Research Findings

Light

- The amount of light available to a colony of cyanobacteria is determined by the latitude, the time of year and the degree of mixing relative to the depth of light penetration.
- Species such as *Microcystis aeruginosa* and *Anabaena circinalis* have maximal growth rate when cells are mixed to the depth of the euphotic zone (1% of surface irradiance). Deeper mixing causes light limitation to growth.

Nutrients

- There is a direct correlation between the amount of phosphorus in a lake and the quantity of phytoplankton the lake can support. Therefore, any intervention that reduces the load of nutrients entering a lake will eventually reduce the magnitude of the algal bloom.
- Concentrations of filterable reactive phosphorus less than 0.01 mgL^{-1} are considered to be limiting for growth.
- A concentration of 0.1 mgL^{-1} of soluble inorganic nitrogen is considered the minimum concentration to maintain growth during the growing season

Buoyancy regulation

- The success of some cyanobacteria is, in part, attributable to the presence of gas vesicles that provide buoyancy.
- During stratified conditions the ability of some cyanobacteria to float provides the opportunity to be in the illuminated surface layers and access the light required for productivity, nitrogen fixation and growth.
- The classic hypothesis is that cyanobacteria migrate to access the vertically separated resources, light and nutrients (Figure 10.1).

Implementation

Limiting light and nutrients can create conditions that limit cyanobacteria. This can include destratification (see FS 10) and catchment management (see FS14).

Figure 10.1 A day in the life of *Anabaena* is a cartoonist's impression of the buoyancy regulation mechanism used by cyanobacteria to migrate vertically and overcome the vertical separation of light (near the surface) and nutrients near the sediment. Cells near the surface photosynthesize and accumulate carbohydrate (CHO) which makes them heavy and they sink. Away from light the CHO is respired and buoyancy is restored.

More Information

Oliver, R.L. 1994. Floating and sinking in gas-vacuolate cyanobacteria. *Journal of Phycology*, 30:161-173.

Chorus, I., and J. Bartram. 1999. Toxic Cyanobacteria in Water. World Health Organization. London: E&FN Spon.

Reynolds, C.S. 1997. Vegetation processes in the pelagic: a model for ecosystem Theory. Germany: Ecology Institute, Oldendorf, Luhe.

Contact details

Justin Brookes: justin.brookes@adelaide.edu.au
Mike Burch: mike.burch@sawater.com.au

FACTSHEET # 2: THE CYANOBACTERIAL TOXINS

Research Findings

The main cyanobacterial toxins of concern in Australia are cylindrospermopsins, microcystins, and saxitoxins (paralytic shellfish toxins). In Australia cylindrospermopsins are produced by *Cylindrospermopsis raciborskii* and *Aphanizomenon ovalisporum*, microcystins by various *Microcystis* species but predominantly by *Microcystis aeruginosa*, and the saxitoxins are produced by *Anabaena circinalis*.

Cylindrospermopsins (3 types currently known) are alkaloid toxins that inhibit protein synthesis and can disrupt the structure of DNA. The first of these actions can cause death in exposed animals but the second also suggests the possibility of cancer initiation.

Microcystins (>80 types currently known) are cyclic peptides that inhibit enzymes called protein phosphatases, which are involved in the regulation of many important cellular processes. This can lead to rapid death if the dose is high enough. At lower does, the effects on cell regulation may allow cancers to escape normal controls, increasing their growth rate.

Saxitoxins (~30 types currently known) are also alkaloids. These toxins block nerve transmission and so cause death by inhibiting the muscles required for respiration. There are no known long-term effects of a non-lethal dose.

Implementation

The CRC has concentrated its toxicological research efforts on:

- Explaining the mechanisms of action of the toxins.
- Conducting animal studies of toxic effects.
- Developing and assessing toxicity based assays for detection of cyanotoxins in source and drinking waters.

The outcomes from the research have been implemented by:

- Providing the data to WHO along with recommendations for guideline safety values for microcystin and cylindrospermopsin (both at 1 μg/L). A guideline value for microcystin has been published by WHO and the relevant committee on chemical contaminants in water is currently considering our data as the basis for a guideline value for cylindrospermopsin.
- Development of an assay for cylindrospermopsin based upon its mechanism of toxic action, that is, protein synthesis inhibition. This provides a sensitive assay that detects all three known cylindrospermopsin analogues, plus any others that are yet to be discovered, providing a measure of total cylindrospermopsin-like toxicity in a water sample.
- Evaluation of cell and antibody based assays for detection of saxitoxins. The neuroblastoma cell-based assay was found to be specific for these toxins and was able to detect toxin analogues that are yet to be chemically characterised. Specific recommendations are contained in the report for AWAARF Project 2789.

More information

Falconer, I.R., and A.R. Humpage. 2001. Preliminary evidence for in vivo tumour initiation by oral administration of extracts of the blue-green alga *Cylindrospermopsis raciborskii* containing the toxin cylindrospermopsin. *Environmental Toxicology*, 16:192-195.

Falconer, I.R. 2005. Cyanobacterial Toxins of Drinking Water Supplies: Cylindrospermopsins and Microcystins. Boca Raton, Florida: CRC Press.

Falconer, I.R., and A.R. Humpage. 2005. Health risk assessment of cyanobacterial (blue-green algal) toxins in drinking water. *International Journal of Environmental Research and Public Health*, 2:43-50.

Froscio, S.M., A.R. Humpage, P.C. Burcham, and I.R. Falconer. 2001. Cell-free protein synthesis inhibition assay for the cyanobacterial toxin cylindrospermopsin. *Environmental Toxicology*, 16:408-412.

Froscio, S.M., A.R. Humpage, P.C. Burcham, and I.R. Falconer. 2003. Cylindrospermopsin-induced protein synthesis inhibition and its dissociation from acute toxicity in mouse hepatocytes. *Environmental Toxicology*, 18:243-251.

Humpage, A.R., J. Rositano, A.H. Bretag, R. Brown, P.D. Baker, B.C. Nicholson, and D.A. Steffensen. 1994. Paralytic shellfish poisons from Australian cyanobacterial blooms. *Australian Journal of Marine and Freshwater Research*, 45:761-771.

Humpage, A.R., and I.R. Falconer. 1999. Microcystin-LR and liver tumour promotion: Effects on cytokinesis, ploidy and apoptosis in cultured hepatocytes. *Environmental Toxicology*, 14:61-75.

Humpage, A.R., S.J. Hardy, E.J. Moore, S.M. Froscio, and I.R. Falconer. 2000. Microcystins (cyanobacterial toxins) in drinking water enhance the growth of aberrant crypt foci in the mouse colon. *Journal of Toxicology & Environmental Health*, 61:155-165.

Humpage, A.R., M. Fenech, P. Thomas, and I.R. Falconer. 2000. Micronucleus induction and chromosome loss in transformed human white cells indicate clastogenic and aneugenic action of the cyanobacterial toxin, cylindrospermopsin. *Mutation Research*, 472:155-161.

Humpage, A.R., and I.R. Falconer, 2003. Oral toxicity of the cyanobacterial toxin cylindrospermopsin in male Swiss albino mice: Determination of no observed adverse effect level for deriving a drinking water guideline value. *Environmental Toxicology*, 18:94-103.

Humpage, A.R., F. Fontaine, S. Froscio, P. Burcham, and I.R. Falconer. 2005. Cylindrospermopsin genotoxicity and cytotoxicity: Role of cytochrome P-450 and oxidative stress. *Journal of Toxicology and Environmental Health*, 68:739-753.

Norris, R.L., A.A. Seawright, G.R. Shaw, M.J. Smith, R.K. Chiswell, and M.R. Moore. 2001. Distribution of 14C cylindrospermopsin in vivo in the mouse. *Environmental Toxicology*, 16:498-505.

Norris, R.L., A.A. Seawright, G.R. Shaw, P. Senogles, G.K. Eaglesham, M.J. Smith, et al. 2002. Hepatic xenobiotic metabolism of cylindrospermopsin in vivo in the mouse. *Toxicon*, 40:471-476.

Contact Details

Andrew Humpage: andrew.humpage@sawater.com.au
Suzanne Froscio: suzanne.froscio@sawater.com.au

FACT SHEET # 3: SAMPLING WATERBODIES FOR THE DETECTION OF CYANOBACTERIA

Recommended procedures based upon best practice

- The type of sampling adopted depends upon the aim(s) of the monitoring program, the water body type and on the current health alert status of the waterbody.
- Collecting samples to determine the 'true' cyanobacterial population in a waterbody to detect population changes is difficult. Samples need to be representative of the whole waterbody remembering that cyanobacteria can have patchy distribution.
- The recommended sample method to detect population size is the depth-integrated sample. This integrates vertical variation and is regarded as generally providing good representation of the cyanobacterial population.
- To assess the potential cyanobacterial contamination of the drinking water system, the samples should be taken adjacent to the water offtake tower and at the same depth as the offtake.
- Where toxin monitoring is required, it is recommended that toxin analysis be performed on the same sample used for cyanobacterial identification and enumeration
- Different sampling techniques are required to assess for benthic cyanobacteria.
- For toxicity assessment it may be necessary to collect a highly concentrated sample of cells or scum rather than a water sample
- To prevent any changes to a sample from when it was taken to when it is analysed, transport in the dark (eg in an esky with a lid - cooler) and on ice, unless other transport requirements are indicated.

Implementation

Table 10.2
Scale of sampling effort and procedures for monitoring cyanobacteria (for operators)

Water Body Type	Priority	Sampling Access	Sample Type (method)*	Number of Samples†	Frequency of Sampling‡
Reservoirs & lakes	High	Open water by boat	Integrated depth	Multiple sites	Weekly or bi-weekly
	High	Supply offtake	Offtake depth or integrated depth	Single site	Weekly or bi-weekly
	Moderate-Low	Shoreline	Surface	Single or multiple sites	Weekly or fortnightly
Rivers and weir pools	High	Midstream by boat or bridge or weir	Integrated depth	Multiple sites	Weekly or bi-weekly
	High	Supply offtake	Offtake depth or integrated depth	Single site	Weekly or bi-weekly
	Moderate-Low	River bank	Surface	Single site or multiple sites	Weekly or fortnightly

(continued)

Table 10.2 (Continued)

Water Body Type	Priority	Sampling Access	Sample Type (method)*	Number of Samples†	Frequency of Sampling‡
All water bodies	Low	Near to offtakes, bank or shorelines	Visual inspection for water discolouration or surface scums	Single or multiple sites	Fortnightly to daily depending on season and frequency of use

*Integrated depth samples are collected with a flexible or rigid hosepipe, depth (2-5m) depending on mixing depth; surface or depth samples collected with a van Dorn or Niskin sampler; shoreline or bank samples collected with a 2m sampling rod and bottle.

†Multiple sites should be suitably spaced (eg. a minimum of 100m apart, except in smaller water bodies such as farm dams), and should include one near the offtake. Multiple samples can also be pooled and one composite sample obtained. River monitoring should include upstream sites for early warning. Samples from recreational waters should be collected within the water contact area.

‡Sampling should be programmed at the same time of day for each location. Visual inspection for surface scums should be conducted in calm conditions, early in the morning.

More information

Burch, M.D., F.L. Harvey, P.D. Baker, and G. Jones, (In review) National Protocol for the Monitoring of Cyanobacteria and their Toxins in Surface Fresh Waters. ARMCANZ.

Contact Details

Mike Burch: mike.burch@sawater.com.au

FACTSHEET # 4: MICROSCOPIC IDENTIFICATION OF POTENTIALLY TOXIC CYANOBACTERIA IN AUSTRALIAN FRESHWATERS

Background

Cyanobacterial (blue-green algae) blooms are a public health concern due to their ability to produce potent toxins. These toxins have been implicated in episodes of human illness in Australia and deaths overseas. Monitoring of cyanobacterial taxa, cell numbers (or equivalent cell biovolume) and toxin concentrations provides an excellent basis for assessing health risks associated with toxic blooms.

Implementation

Phase contrast light microscopy, at magnifications greater than 100 times, is typically used for the identification and enumeration of cyanobacteria and provides the first indication of potentially toxic species in source waters. Toxicity is species specific and considerable taxonomic expertise is required to differentiate the potentially toxic species.

The following describes the morphological characteristics of the potentially toxic cyanobacterial species recognised from Australia's freshwater habitats.

Anabaena circinalis (order Nostocales)

A. circinalis is a planktonic cyanobacterium morphologically presented as open spirally coiled trichomes, greater than 50 μm in diameter. The vegetative cells are spherical or compressed at the poles, breadth 7-8.5 μm. Vegetative cells contain aerotopes (gas vesicles). Heterocytes (nitrogen fixing cells) are spherical, breadth 7.5-9 μm. Mature akinetes (resting cells or spores) are cylindrical, slightly curved and remote from the heterocytes, length 27 μm, breadth 14 μm.

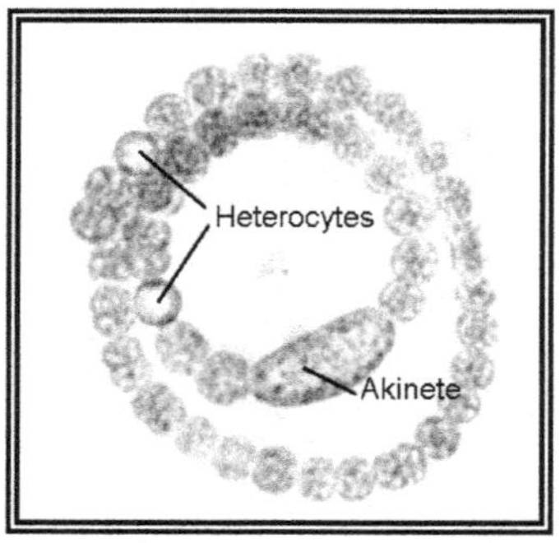

Figure 10.2 ***Anabaena circinalis.***

Aphanizomenon ovalisporum (order Nostocales)

A. ovalisporum is a planktonic cyanobacterium morphologically presented as straight or slightly curved trichomes. Trichomes taper towards the ends. The vegetative cells are clearly constricted at the cross walls, cylindrical, length 3.1-9.8 µm, breadth 2.3-5.2 µm. Vegetative cells contain aerotopes. Apical cells (end cells) are distinctly narrowed and elongated, largely hyaline (clear), 6.7-17.3µm long, 1.2-4.3 µm broad. Heterocytes are solitary, spherical or ellipsoid, length 3.4 - 8.2 µm, breadth 2.2 - 6.5µm. Mature akinetes are solitary or commonly 2 - 3 in series, oval, usually removed from the heterocytes by one or more cells, length 6.2-16.8 µm, breadth 5.3 -11.8 µm.

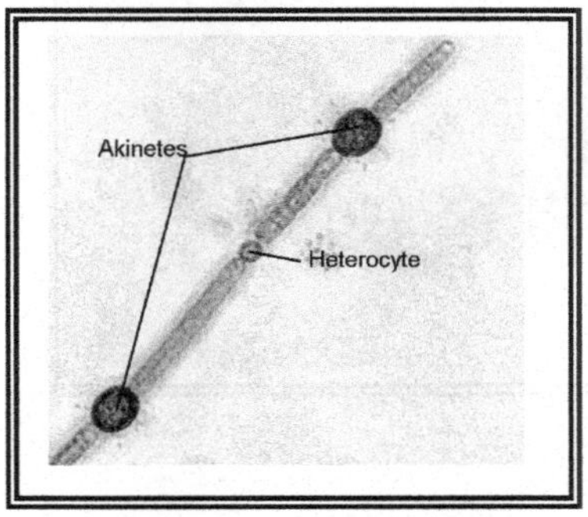

Figure 10.3 ***Aphanizomenon ovalisporum.***

Cylindrospermopsis raciborskii (order Nostocales)

C. raciborskii is a planktonic cyanobacterium morphologically presented as either straight, slightly curved or spirally coiled trichomes. Trichomes are cylindrical or slightly narrowed or tapered towards the ends. The vegetative cells are cylindrical or slightly barrel-shaped when constricted, length 2.0-8.5 µm, breadth 2.5-4.0 µm. Vegetative cells contain aerotopes. Heterocytes, which develop **only** from terminal narrowed cells, are short to conical and pointed to rounded at the ends, length 3.5-10.5 µm, breadth 2.5-4.0 µm. Akinetes, which develop beside or slightly distant from the heterocytes, can be present solitary, in pairs or in a short series. Mature akinetes are oval, length, 7.5-16.0 µm, breadth 3.5-4.5 µm.

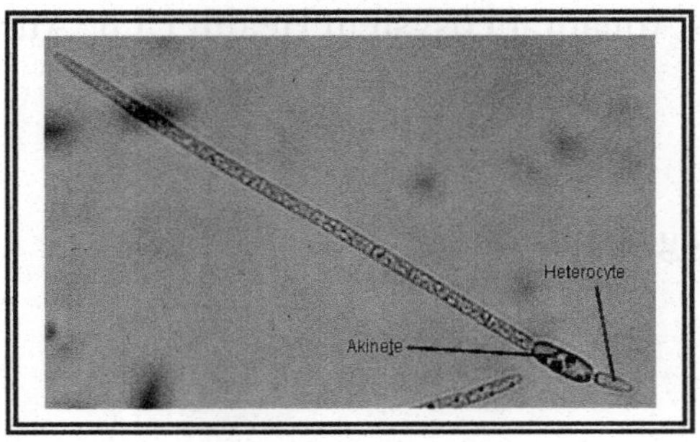

Figure 10.4 ***Cylindrospermopsis raciborskii.***

Microcystis aeruginosa (order Chroococcales)

M. aeruginosa colonies are variable in morphology, commonly more or less spherical or elongate, later irregularly lobate, elongated or clathrate (net-like) or composed of sub-colonies. Mucilage broad forming a margin around colony 10-50μm wide. Cells spherical, 5-7μm in diameter, densely aggregated in the common mucilage. Cells contain numerous aerotopes. No specialised cells present.

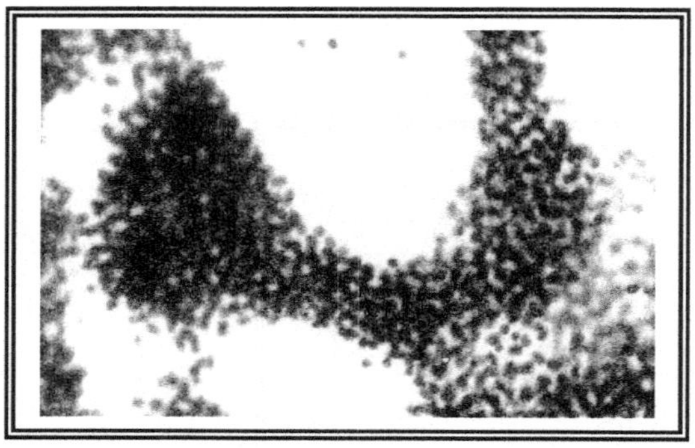

Figure 10.5 ***Microcystis aeruginosa.***

More information

Komárek, J., and A. Konstantinos. 1999. *Cyanoprokaryota: the Chroococcales*, Germany.

Baker, P., and L. Fabbro. 1999. *A guide to the Identification of Common Blue-Green Algae (Cyanoprokaryotes) in Australian Freshwaters*, CRCFE, Australia

Baker, P. 1991. *Identification of Common Noxious Cyanobacteria. Part I – Nostocales*, Research Report No.29, Urban Water Research Assoc. of Aust, Adelaide.

Baker, P. 1992. *Identification of Common Noxious Cyanobacteria. Part II – Chroococcales, Oscillatoriales*, Research Report No.46, Urban Water Research Assoc. of Aust, Adelaide.

McGregor, G., and L. Fabbro. 2001. *A Guide to the Identification of Australian Freshwater Planktonic Chroococcales (Cyanoprokaryota/Cyanobacteria),* CRCFE, Australia.

McGregor, G., and J. Lobegeiger. 2004. *Harmful freshwater micro algae information Sheet No. HAB01, Aphanizomenon ovalisporum*, Department of Natural Resources, Mines & Energy, Natural Resource Sciences - Aquatic Ecosystem Health Unit, Australia.

Contact Details

Maree Smith: maree.smith@griffith.edu.au

FACT SHEET # 5: DETECTION OF TOXIGENIC CYANOBACTERIA USING GENETIC METHODS

Research Findings

Toxigenic cyanobacteria that produce cylindrospermopsin, microcystin, and nodularin can be detected by using the polymerase chain reaction (PCR) which identifies the genes responsible for toxin production. Tests for saxitoxin producing cyanobacteria are in progress.

Using real-time PCR the amplification of the toxin genes can be followed as the reaction proceeds, providing rapid assay turnaround (eg 1-2 hr).

Real-time PCR can be used for screening water samples to detect the presence of toxigenic cyanobacteria. However, where standards are used, analysis is semi-quantitative and will indicate both the presence of the toxin gene and the approximate number of copies present in the starting sample (Table 10.3).

Table 10.3
An example of real-time PCR results (toxin gene copies/ml) for the detection of the gene associated with cylindrospermopsin production by *Cylindrospermopsis raciborskii.* Note the uneven distribution across the reservoir.

Sample points at one reservoir	Toxin gene (copies.mL^{-1})	Cell count (cells.mL^{-1})	Toxin (µg.L^{-1}) detected by LCMS
Site 1	50028	27607	0.2
Site 2	43149	27950	0.8
Site 3	52793	57855	0.4
Site 4	58004	73670	0.9
Site 5	92866	77290	0.8
Site 6	54598	73625	0.9
Site 7	90401	66550	1

Table 10.3 shows that many sites around a reservoir can be sampled and rapidly analysed giving an overall assessment of the distribution of potentially toxic algae for the whole reservoir. This information can then be used for further management and water treatment decisions.

Environmental samples that may contain toxigenic cyanobacteria can be prepared for PCR analysis using simple and rapid methods that include the possibility of performing the test on the site where the sample was collected giving rapid results.

Due to natural variation several toxin genes may need to be targeted to reduce a low level of false positives, this can be achieved by multiplex PCR.

Implementation

- Conventional or real-time PCR assays should be used to supplement existing monitoring strategies that include microscopic counts and toxin analysis by chemical methods (eg HPLC etc)
- PCR results should be used for the rapid detection of negative samples to save further labour and analysis costs
- Real-time PCR should be made available in emergency, rapid-response or time critical situations to rapidly assess potentially toxic cynaobacterial blooms

More information

Rinta-Kanto, J.M., A.J.A. Ouellette, G.L. Boyer, et al. 2005. Quantification of toxic *Microcystis* spp. during the 2003 and 2004 blooms in western Lake Erie using quantitative real-time PCR. *Environmental Science & Technology*, 39(11):4198-4205.

Dittmann, E., and T. Borner. 2005. Genetic contributions to the risk assessment of microcystin in the environment. *Toxicology and Applied Pharmacology*, 203(3):192-200.

Kurmayer, R., and T. Kutzenberger. 2003. Application of real-time PCR for quantification of microcystin genotypes in a population of the toxic cyanobacterium *Microcystis* sp. *Applied and Environmental Microbiology*, 69: 6723-6730.

Becker, S., M. Fahrbach, P. Boger, et al. 2002. Quantitative tracing, by Taq nuclease assays, of a *Synechococcus* ecotype in a highly diversified natural population. *Applied and Environmental Microbiology*, 68(9):4486-4494.

Foulds, I.V., A. Granacki, C. Xiao, et al. 2002. Quantification of microcystin-producing cyanobacteria and *E-coli* in water by 5 '-nuclease PCR. *Journal of Applied Microbiology*, 93(5):825-834.

Fergusson, K.M., and C.P. Saint. 2003. Multiplex PCR assay for *Cylindrspermopsis raciborskii* and cylindrospermopsin-producing cyanobacteria. *Environmental Toxicology*, 18: 120-125.

Schembri, M.A., B.A., Neilan, and C.P. Saint. 2001. Indentification of genes implicated in toxin production in the cyanobacterium *Cylindrospermopsis raciborskii*. *Environmental Toxicology*, 16:413-421.

Contact details

Paul Rasmussen: paul.rasmussen@sawater.com.au
Chris Saint: chris.saint@sawater.com.au

FACT SHEET #6: TREATMENT OF CYANOBACTERIAL TOXINS

Research Findings

- The toxins of most concern to water suppliers in Australia are saxitoxins, microcystins and cylindrospermopsin.
- Conventional treatment processes such as coagulation, flocculation, sedimentation and filtration will remove up to 90% of the total toxin present if it is contained within healthy cyanobacterial cells (less for cylindrospermopsin).
- Dissolved toxin (ie toxin that has been released from the cells) must be removed using additional treatment such as powdered activated carbon (PAC), granular activated carbon (GAC), ozone or chlorine.
- The doses of oxidant and PAC, and the type of activated carbon required for treatment is dependent on the type of toxin and the water quality.

Implementation

Table 10.4 outlines the appropriate treatment techniques for treatment of cyanobacterial toxins.

More information

Chow, C.W.K., M. Drikas, J. House, M.D. Burch, and R.M.A. Velzeboer. 1999. The impact of conventional water treatment processes on cells of the cyanobacterium *Microcystis aeruginosa*. *Water Research*, 33(15):3253-3262.

Chow, C.W.K., J. House, R.M.A. Velzeboer, M. Drikas, M.D. Burch, and D.A. Steffensen. 1998. The effect of ferric chloride flocculation on cyanobacterial cells. *Water Research*, 32(3):808-814.

Cook, D., and G. Newcombe. 2002. Removal of microcystin variants with powdered activated carbon. *Water Science & Technology: Water Supply*, 2(5/6):201-207.

Cook, D., G. Newcombe, and J. Morrison. 2000. Tastes, odours and algal toxins, which PAC is best? *Water- Journal of the Australian Water Association*, 27(2):28-31.

Donati, C., M. Drikas, R. Hayes, and G. Newcombe. 1993. Adsorption of microcystin-LR by powdered activated carbon. *Water*, 20(3):25-28.

Drikas, M., C.W.K. Chow, J. House, and M.D. Burch. 2001. Using coagulation, flocculation and settling to remove toxic cyanobacteria. *Journal of the American Water Works Association*, 93(2):100-111.

Grützmacher, G., G. Böttcher, I. Chorus, and H. Bartel. 2002. Removal of microcystins by slow sand filtration. *Environmental Toxicology*, 17(4):386-394.

Lam, A.K-Y., E.E. Prepas, D. Spink, and S.E. Hrudey. 1995. Chemical control of hepatotoxic phytoplankton blooms: Implications for human health. *Water Research*, 29(8):1845-1854.

Newcombe, G., and M. Burch. 2004. Toxic blue-green algae, coming soon to a neighbourhood near you? *Water*, November, 57-59.

Newcombe, G., and B. Nicholson. 2004. Water treatment options for dissolved cyanotoxins. *Aqua*, 53(4):227-239.

Contact details

Gayle Newcombe: gayle.newcombe@sawater.com.au

Table 10.4

Techniques for treatment of cyanobacterial toxins.

Treatment process	Cyanobacteria/toxin	Treatment efficiency
		Intact cells
Coagulation sedimentation	cyanobacterial cells	Very effective for the removal of intracellular toxins provided cells accumulated in sludge are isolated from the plant
Rapid filtration	cyanobacterial cells	Very effective for the removal of intracellular toxins provided cells are not allowed to accumulate on filter for prolonged periods
Slow sand filtration	cyanobacterial cells	As for rapid sand filtration, with additional possibility of biological degradation of dissolved toxins
Combined coagulation/ sedimentation/ filtration	cyanobacterial cells	Extremely effective for the removal of intracellular toxins provided cells accumulated in sludge are isolated from the plant cells and any free cells are not allowed to accumulate on filter for prolonged periods
Membrane processes	cyanobacterial cells	Very effective for the removal of intracellular toxins provided cells are not allowed to accumulate on membrane for prolonged periods
Dissolved Air Flotation (DAF)	cyanobacterial cells	As for coagulation/sedimentation
Oxidation processes	cyanobacterial cells	Not recommended as a treatment for cyanobacterial cells as this process can lead to cell damage and lysis and consequent increase in dissolved toxin levels
Dissolved Toxins		
ADSORPTION		
Adsorption - powdered activated carbon (PAC) (doses required vary with water quality)	Microcystins (except m-LA)	Wood-based, chemically activated carbon is the most effective, or similar, 60 minutes contact time recommended
	Microcystin LA	High doses required
	Cylindrospermopsin	Wood-based, chemically activated carbon is the most effective, or similar, 60 minutes contact time recommended
	Saxitoxins	A microporous carbon (coconut or coal based) 60 minutes contact time recommended effective for the most toxic of the variants

(continued)

Table 10.4 (continued)

Treatment process	Cyanobacteria/toxin	Treatment efficiency
Adsorption -granular activated carbon (GAC)	All dissolved toxins	GAC adsorption displays a limited lifetime for all toxins. This can vary between 2 months to more than one year depending on the type of toxin and the water quality
Biological filtration	All dissolved toxins	When functioning at the optimum this process can be very effective for the removal of most toxins. However, factors affecting the removal such as biofilm mass and composition, acclimation periods, temperature and water quality cannot be easily controlled
OXIDATION		
Ozonation	All dissolved toxins	Ozonation is effective for all dissolved toxins except the saxitoxins. A residual of at least 0.3 mg L^{-1} for 5 minutes will be sufficient. Doses will depend on water quality
Chlorination	All dissolved toxins	If a residual of 0.5 mg L^{-1} is maintained for at least 30 minutes, and a dose of at least 3 mg L^{-1} is applied, most microcystins and cylindrospemopsin should be destroyed. Microcystin LA may require a higher residual. Limited data suggest chlorination is only effective at elevated pH for saxitoxins
Chloramination	All dissolved toxins	Ineffective.
Chlorine dioxide	All dissolved toxins	Not effective with doses used in drinking water treatment
Potassium permanganate	All dissolved toxins	Effective for microcystin, limited or no data for other toxins
Hydrogen peroxide	All dissolved toxins	Not effective on its own
UV Radiation	All dissolved toxins	Capable of degrading microcystin-LR and anatoxin-a and cylidrospermopsin, but only at impractically high doses or in the presence of a catalyst
EXCLUSION		
Membrane Processes	All dissolved toxins	Depends on membrane pore size distribution

FACT SHEET # 7: TASTE AND ODOUR REMOVAL

Research Findings

- Cyanobacteria can produce the taste and odour compounds geosmin and 2 methyl isoborneol (MIB) and these are a significant water quality issue (not a health issue).
- Conventional water treatment (coagulation/sedimentation/filtration) is an effective process for the removal of these compounds provided they are contained within the algal cells, however, in dissolved form, alternate treatment options are required.
- Ozone and activated carbon, in powdered and granular forms, are treatment options which are able to remove extracellular MIB and geosmin. The extent of removal is dependent upon the water quality, in particular, the natural organic matter (NOM) concentration and character.
- Procedures have been established to determine the best activated carbon for the removal of MIB and geosmin under treatment plant conditions, and to predict the PAC doses required to reach a satisfactory concentration of these compounds in finished water
- Recently, biological treatment of geosmin has been explored. A number of projects within the CRC for Water Quality and Treatment are currently focussed on identifying the bacteria capable of degrading MIB and geosmin and applying this in both source waters and biologically active filters.

Implementation

Powdered activated carbon (PAC) is the major treatment option employed in the Australian water industry for the removal of geosmin and MIB. Figure 10.6 shows the results of laboratory tests revealing the extent of removal of MIB by four activated carbons as a function of time. This information can be used to determine the most cost effective carbon to purchase. The table gives a practical example of the doses required of each carbon and the costs associated with dosing PAC for 10 days. In this case the most expensive carbon produces the best results in terms of water quality, at only slightly higher cost overall than the cheapest carbon, and with the great advantage of lower amounts of PAC (8 tonnes compared with 15.5).

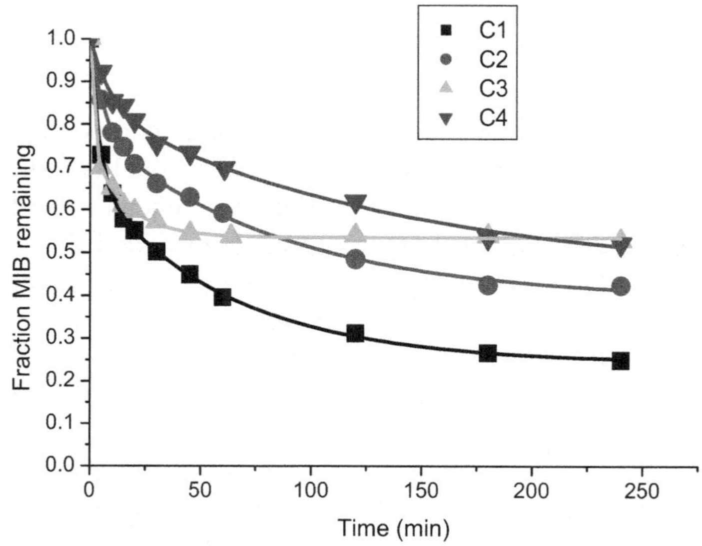

Figure 10.6 Removal of MIB by four activated carbons as a function of time

Table 10.5
Doses required for the removal of MIB from 20 to 10 ngL^{-1} for four activated carbons, and costs associated with dosing at a 50 ML per day flow for 10 days in Myponga Reservoir water.

	C1	C2	C3	C4
PAC dose (mgL^{-1})	16	31	26	38
PAC required for 10 days dosing (tonne)	8.0	15.5	13.0	19.0
Cost for 10 days dosing (AUS$)	28 000	24 800	41 600	28 500

More information

Cook, D., G. Newcombe, and P. Sztajnbok. 2001. The application of powdered activated carbon for MIB and geosmin removal: Predicting PAC doses in four raw waters. *Water Research*, 35(5):1325-1333.

Ho, L., J-P. Croué, and G. Newcombe. 2004. The effect of water quality and NOM character on the ozonation of, MIB and geosmin. *Water Science & Technology*, 49(9):246-255.

Newcombe, G., and D.Cook. 2002. Influences on the removal of tastes and odours by PAC. *Journal of Water Supply: Research and Technology – Aqua*, 51(8):463-474.

Newcombe, G., and D. Cook. 2002. Removing tastes and odours: Is your cheap PAC costing you too much? *Water*, 29(1):24-26.

Contact details

Gayle Newcombe: gayle.newcombe@sawater.com.au
Lionel Ho: lionel.ho@sawater.com.au

FACT SHEET # 8: BIODEGRADATION OF MICROCYSTIN TOXINS

Research Findings

- The microcystin toxins are the most commonly reported of algal toxins world wide.
- Studies have shown that the microcystins biodegradable by micro-organisms in both source waters (Figure 10.7) and through biologically active filters (biofilters).
- To date only a few bacterial species of the genus *Sphingomonas* have shown the ability to effectively degrade microcystin. While these organisms are widespread, they are not present in all waters.
- Recently, genetic methods have identified the genes responsible for the degradation of microcystin.
- Degradation by bacteria is not known to produce any harmful by-products.

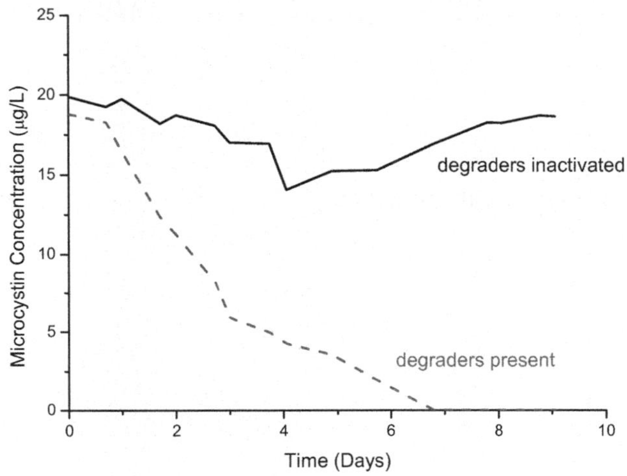

Figure 10.7 Degradation of microcystin toxin by bacteria

Implementation

- Kinetic parameters extracted from data like that shown in Figure 10.7 can be used to estimate degradation of microcystins in reservoirs and lakes. Degradation to below detection could usually be expected in 2-3 weeks in the presence of sufficient numbers of degrading bacteria
- An effective biofilter, with the appropriate biomass and microbial community, can reduce high levels of microcystin to below detection
- By using genetic methods, such as the polymerase chain reaction, it will be possible to screen water sources and biofilters to determine whether they contain the bacteria which are able to degrade microcystin.
- In situations where the bacteria are not present it may be possible to "artificially seed" the degrading bacteria into the system to remove the toxin.

- Researchers within the CRC for Water Quality and Treatment are working to determine the feasibility of practical application of these techniques.

More information

Jones, G.J., and P.T. Orr. 1994. Release and degradation of microcystin following algicide treatment of a *Microcystis aeruginosa* bloom in a recreational lake, as determined by HPLC and protein phosphatase inhibition assay. *Water Research*, 28(4):871-876.

Bourne, D.G., P. Riddles, G.J. Jones, W. Smith, and R.L. Blakely. 2001. Characterisation of a gene cluster involved in bacterial degradation of the cyanobacterial toxin microcystin LR. *Environmental Toxicology*, 16(6):523-534.

Ho, L., T. Meyn, A. Keegan, J. Brookes, C. Saint, and G. Newcombe. 2005. Biological treatment of microcystin toxins. In *Proceedings of the AWA Ozwater Convention and Exhibition,* May 8-12, 2005, Brisbane, Australia.

Saito, T., K. Okana, H-D. Park, T. Itayama, Y. Inamori, B.A. Neilan, B.P. Burns, and N. Sugiura. 2003. Detection and sequencing of the microcystin LR-degrading gene, *mlrA*, from new bacteria isolated from Japanese lakes. *FEMS Microbiology Letters*, 229(2):271-276.

Contact details

Lionel Ho: lionel.ho@sawater.com.au

FACTSHEET # 9: BIODEGRADATION OF CYLINDROSPERMOPSIN TOXINS

Research Findings

- Microbial degradation of soluble cylindrospermopsin (CYN) was confirmed in natural waters with a previous history of toxic *Cylindrospermopsis raciborskii* blooms.
- Degradation of CYN typically began 3-11 days (lag phase) after addition to surface water samples. Once degradation had begun, CYN was completely removed within 10-33 days.
- The lag phase and hence the time taken for complete removal of CYN was reduced upon re-addition of the toxin to induced surface water samples.
- Degradation of CYN in a range of waterbodies with no recorded history of toxic *C. raciborskii* blooms begun approximately 64-125 days after the addition of CYN. Once degradation was detected, CYN was only 16-59% removed within 212 days.

Implementation

- The lag phase and the time taken for total degradation of CYN to occur varied between studies and water bodies.
- The length of time required for complete removal of CYN may vary depending on a number of factors such as the local bacteria present, the presence of essential growth factors for degrading organisms and perhaps the presence of more readily metabolisable substrates in the waterbody.
- Thus, it is not yet possible to accurately predict a safe withholding period for a particular waterbody and each toxic bloom incidence must be treated separately.
- CYN levels must be monitored (refer to fact sheet on sampling) to ensure that degradation has occurred.
- The rate of CYN degradation is generally proportional to the concentration of CYN present, therefore it could be possible to estimate the time taken for complete CYN removal in some waterbodies.
- The reduction of the lag phase upon re-addition of CYN to water samples suggests that biodegradation may commence more rapidly in water bodies that have frequent exposure to CYN producing cyanobacterial blooms.
- The induction of biodegradation in waterbodies with no previous recorded history of toxic *C. raciborskii* blooms is probable, however, the lag period and time taken for complete removal of toxin is likely to be lengthy.
- This study has shown that biological activity has a role in the reduction of soluble CYN concentrations in water bodies containing active CYN degrading organisms. Therefore, if water samples for CYN analysis are not correctly preserved and stored, there may be a reduction in the concentration of CYN by the time the sample is analysed.
- The most common method used for sample preservation to reduce microbial activity is to lower the temperature of the sample to approximately 4ºC as soon as practical after collection. It is recommended that the sample remain chilled until analysis can begin. If analysis is delayed it is recommended that the sample is frozen.

More information

Smith, M. 2005. Biodegradation of the cyanotoxin cylindrospermopsin. PhD Thesis, School of Medicine, Brisbane, Queensland University.
Chiswell, C. 1999. Investigations into the toxicological properties and environmental fate of the cyanobacterial toxin cylindrospermopsin. School of Public Health, Brisbane, Griffith University.

Contact details

Maree Smith: maree.smith@pmhc.nsw.gov.au
Glen Shaw: g.shaw@griffith.edu.au

FACT SHEET #10: ARTIFICIAL DESTRATIFICATION FOR CONTROL OF CYANOBACTERIA

Research Findings

- Thermal stratification is the layering of water of different temperatures in lakes and reservoirs. Thermal stratification promotes stable conditions suitable for cyanobacterial growth.
- During stratification the hypolimnion (bottom waters) are effectively separated from the atmosphere and become depleted in oxygen. Under these conditions contaminants such as ammonia, phosphorus, iron and manganese are released from sediment.
- An effective bubble plume aerator for destratification requires a suitable number of bubble plumes to impart sufficient energy to destratify the water column. The mechanical efficiency of a bubble plume is determined by the depth of the water column, the degree of stratification and the air flow rate. The number of plumes, plume interaction and the feasible length of aerator hose must also be considered in aerator design.
- As a general rule, bubble plumes are more efficient in deeper water columns. In shallow water columns (<5.0m depth) the individual air flow rates of the plumes must be very small to maintain efficiency.

Implementation

- Reservoirs can be artificially destratified to disrupt thermal stratification, control cyanobacteria and reduce sediment release of contaminants.
- The most common and effective artificial destratification devices are bubble plume aerators, consisting of a submerged perforated pipe through which air is pumped from a land-based compressor (Figure 10.8).
- The rising air bubbles expand and entrain water from throughout the water column into the bubble plume.
- When the plume reaches the surface, the bubbles are released to the atmosphere and the entrained water plunges to the depth of equivalent density (temperature) and moves through the reservoir.
- This flow displaces water and creates a return flow, generating circulation and weakening the prevailing stratification. This enables deeper mixing generated by wind and these deeper mixing events can induce light limitation in cyanobacteria (see FS 1).

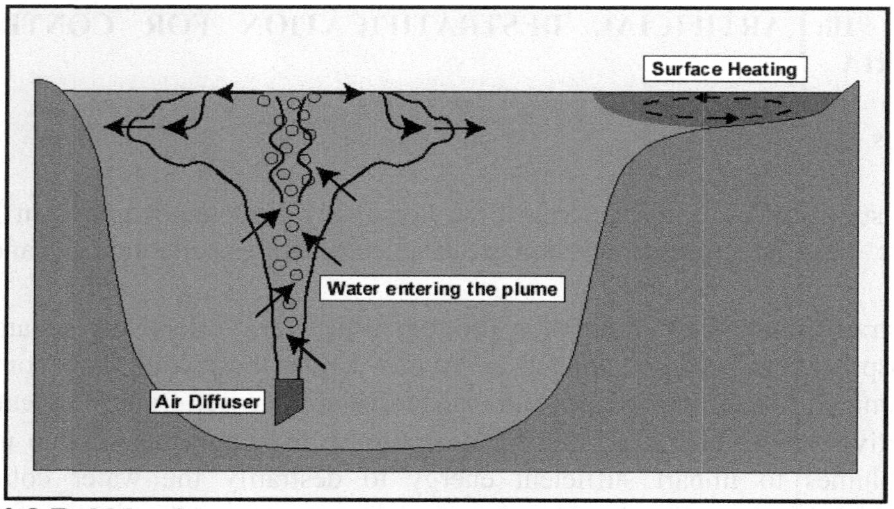

Figure 10.8 Bubble plume aerators consist of a submerged perforated pipe through which air is pumped. Mixing is generated and stratification is weakened. Surface heating and stratification can still occur away from the immediate impact of the bubble plume.

More information

Brookes, J.D., M.D. Burch, and P. Tarrant. 2000. Artificial destratification: Evidence for improved water quality. *Water,* 27 (3):18-22.

McAuliffe, T.F., and R.F. Rosich. 1990. The triumphs and tribulations of artificial mixing in Australian waterbodies. *Water*, Aug p 22-23.

Schadlow, S.G. 1992. Bubble plume dynamics in a stratified medium and the implications for water-quality amelioration in lakes. *Water Resources Research*, 28:313-321.

Visser, P.M., B. Ibelings, B. van der Veer, J. Koedods, and L.R. Mur. 1996. Artificial mixing prevents nuisance blooms of the cyanobacterium Microcystis in Lake Nieuwe Meer, The Netherlands. *Freshwater Biology*, 36:435-450.

Sherman, B.S., J. Whittington, and R.L. Oliver. 2000. The impact of destratification on water quality in Chaffey dam. *Archiv für Hydrobiology. Spec Issues Advance Limnol.*, 55:15-29.

Contact details

Justin Brookes: justin.brookes@adelaide.edu.au
Jason Antenucci: antenucc@cwr.uwa.edu.au

FACT SHEET # 11: *ALGICIDES AS A MANAGEMENT TOOL TO CONTROL CYANOBACTERIA*

Research Findings

- Algicides have been widely used in many countries used to control cyanobacterial blooms. They can provide immediate and cost effective control of algae and cyanobacteria. Use is usually confined to small to medium service reservoirs.
- The most common algicides used in Australia are copper compounds, mainly copper sulphate ($CuSO_4.5H_2O$).
- The chemical character of the receiving water particularly pH, alkalinity and dissolved organic carbon (DOC) control copper speciation and complexation, and this greatly affects copper toxicity.
- When treating cyanobacteria with algicides, it is important that application occurs in the early stages of bloom development when cell numbers are low. This reduces the potential release of high concentrations of toxins or odour compounds into the reservoir water. It may be necessary to withhold the water to allow toxins and odours to degrade if appropriate water treatment is not available (see FS 8, 9).
- There may be local regulations in place to control the use of algicides, due to their potential adverse environmental impacts.

Implementation

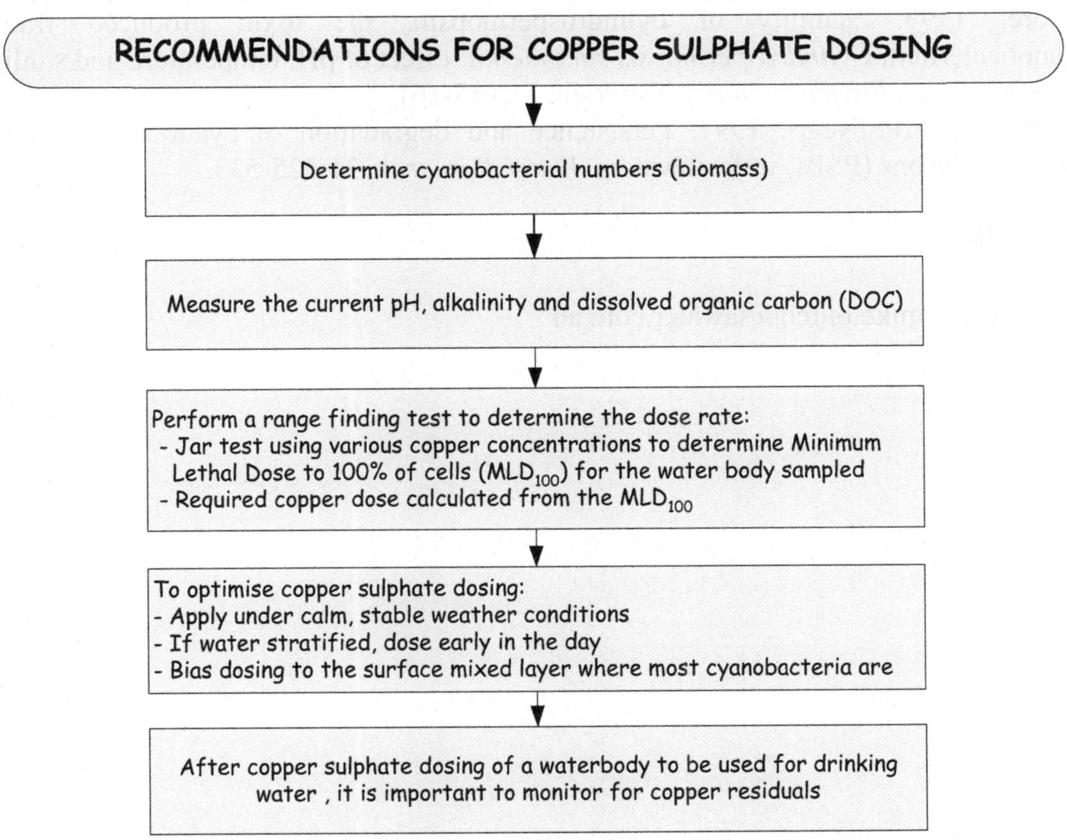

Figure 10.9 Flow diagram for copper sulphate dosing recommendations

More information

Burch, M.D., R.M.A. Velzeboer, C.W.K. Chow, H.C. Stevens, C.M. Bee, J. House. 1998. Evaluation of copper algicides for the control of algae and cyanobacteria. *Urban Water Research Association of Australia, Research report, No. 130*, April, 1998. UWRAA, Melbourne.

Burch, M., C.W.K. Chow, and P. Hobson. 2001. Algicides for control of toxic cyanobacteria. In: *Proceedings of the American Water Works Association Water Quality Technology Conference*, November 12-14, 2001, Nashville, Tennessee. CD-ROM.

Hobson, P., C.W.K. Chow, and M.D. Burch. 2002. Evaluation of the effectiveness of copper algicides against *Anabaena circinalis* from Happy Valley reservoir. Report of investigations into copper sulphate treatment in Happy Valley Reservoir prepared for Water Operations, SA Water. Australian Water Quality Centre.

House, J., and M.D. Burch. 2002. Using algicides for the control of algae in Australia. Registered products for use against algae and cyanobacteria in dams, potable water and irrigation water supply systems, in Australia. CRC for Water Quality and Treatment. Technical memorandum.

Brookes, J., R. Regel, G. Shaw, M. Burford, M. Burch, L. Linden, T. Meyer, K. McNeale, G. Newcombe, S. Rinck-Pfieffer, M. Smith, and R. Hall. 2004. Reservoir management strategies for control and degradation of algal toxins. Awwa Reasearch Foundation Project 2976 First Periodic Report.

Jones, G. 1990. Biodegradation and removal of cyanobacterial toxins in natural waters. Water Board Blue-Green Algal Seminars, November 21st and 22nd: 33.

Chiswell, R.K., G.R. Shaw, G. Eaglesham, M. Smith, R.L. Norris, A.A. Seawright, and M.M. Moore. 1999. Stability of cylindrospermopsin, the toxin produced from the cyanobacterium, *Cylindrospermopsis raciborskii*: effect of pH, temperature and sunlight on decomposition. *Environmental toxicology*, 14:155-161.

Jones, G.J., and A.P. Negri. 1997. Persistence and degradation of cyanobacterial paralytic shellfish poisons (PSPs) in freshwaters. *Water Research*, 31:525-533.

Contact details

Mike Burch: mike.burch@sawater.com.au

FACT SHEET # 12: THE ALERT LEVELS FRAMEWORK FOR DRINKING WATER

Research Findings

- The Alert Levels Framework (ALF) is a monitoring and action sequence framework for a graduated management response to the development of a potentially toxic cyanobacterial bloom in source (raw) waters used for drinking water supply.
- The ALF uses cyanobacterial cell counts and equivalent cell biovolumes from monitoring programs in conjunction with the Australian Drinking Water Guidelines as a situation assessment tool.
- The cell counts/biovolumes are conservative triggers in the management plan and are supplements and surrogates for toxin measurements which may or may not be required.
- Alert levels frameworks also exist for recreational waters and livestock drinking waters.
- The ALF can be adapted to develop a specific local protocol if desired.

Implementation

See Figure 10.10

1. The cell numbers that define the Alert Levels are from samples that are taken from the location adjacent to, or as near as possible to, the water supply offtake (ie. intake point). It must be noted that if this location is at depth, there is potential for higher cell numbers at the surface at this or other sites in the waterbody.
2. The actual numbers for a cell count estimate of 2,000 cells/mL are likely to be in the range 1,000 - 3,000 cells/mL. This is based upon a likely minimum precision of +/-50% for counting colonial cyanobacteria such as *Microcystis aeruginosa* at such low cell densities. For counting filamentous cyanobacteria such as *Anabaena circinalis* the precision is likely to be much better at these cell densities (~+/-20%), giving an actual cell density in the range of 1,600-2,400 this count.
3. This biovolume (>0.4 mm^3/L) is approximately equivalent to $\geq$ 5,000 cells/mL of *M. aeruginosa* for Level 2.
4. This biovolume (> 4 mm^3/L) is approximately equivalent to $\geq$ 50,000 cells/mL of *M. aeruginosa* for Level 3.

More information

Burch, M.D., F.L. Harvey, P.D. Baker, and G. Jones. In review. National Protocol for the Monitoring of Cyanobacteria and their Toxins in Surface Fresh Waters. ARMCANZ.

Newcombe, G., M. Burch, J. House, and L. Ho. In review. Strategies and Practices for Management of Toxic Blue-Green Algae: A Guide. CRC for Water Quality and Treatment.

House, J., Ho, L., Newcombe, G., and Burch, M. 2004. Management strategies for toxic blue-green algae: Literature review. CRC for Water Quality and Treatment.

Contact details

Mike Burch mike.burch@sawater.com.au

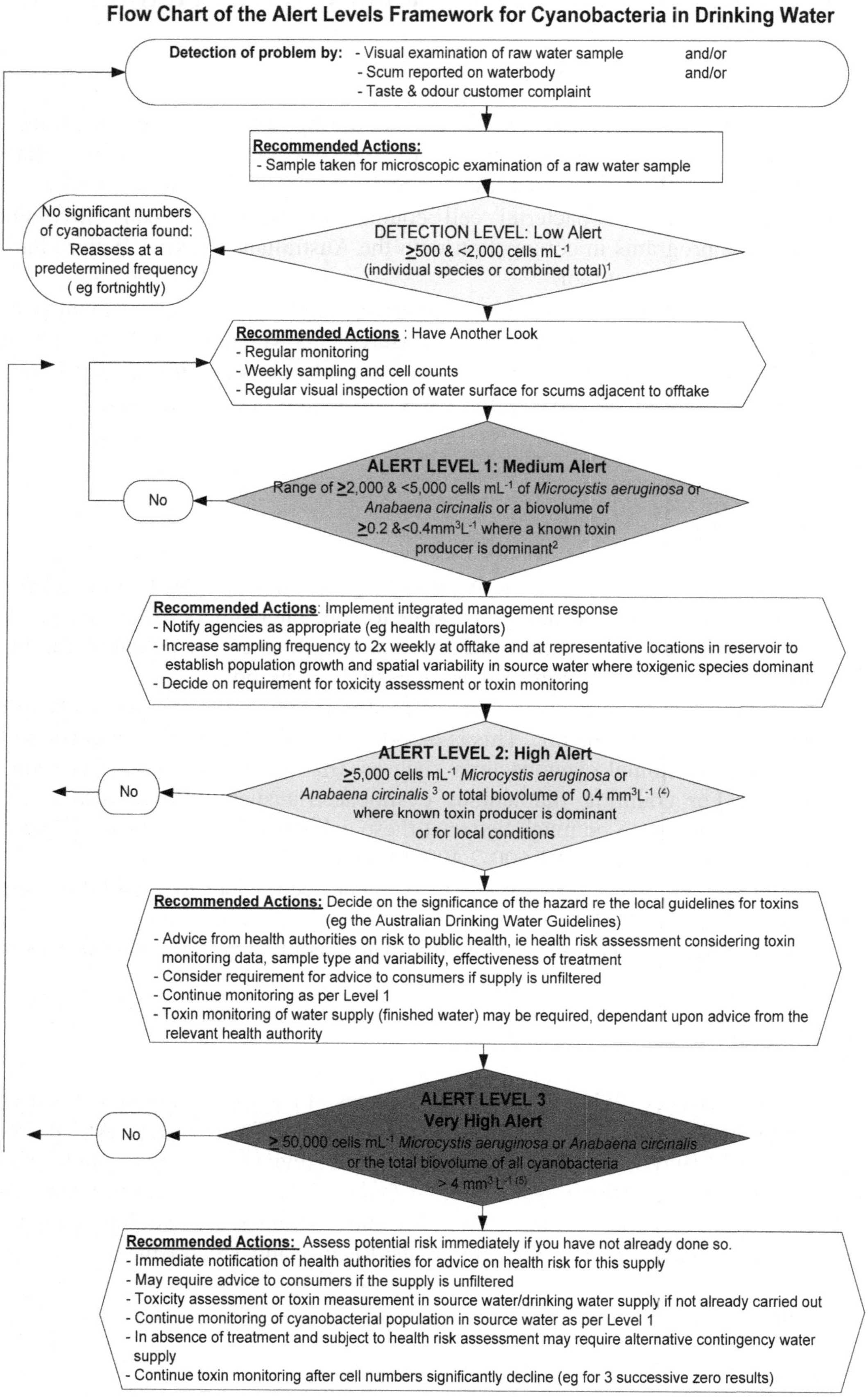

Figure 10.10 Flow chart on the alert levels framework for cyanobacteria in drinking water.

FACT SHEET # 13: *MODELLING TOOLS FOR PREDICTING CYANOBACTERIAL GROWTH*

Background

- Computer simulation of reservoir hydrodynamics, biogeochemistry and ecology enables prediction of reservoir behaviour in response to environmental factors and management intervention.
- DYRESM (DYnamic REservoir Simulation Model) is a one dimensional hydrodynamic model for predicting the vertical distribution of temperature, salinity and density in lakes and reservoirs. The model was developed at the Centre for Water Research at the University of Western Australia and is available from their website www.cwr.uwa.edu.au.
- CAEDYM (Computational Aquatic Ecosystem Dynamics Model) is an aquatic ecological model that is used for investigations involving biological and chemical processes.
- CAEDYM consists of a series of mathematical equations describing the major biogeochemical processes influencing water quality.
- The model can be run in isolation or coupled to DYRESM for studies of the seasonal, annual or decadal variation in water quality.

Implementation

- The growth and vertical distribution of cyanobacteria can be modelled along with other groups of algae including diatoms, dinoflagellates and green algae.
- Implicit in the model are parameters describing phytoplankton response to light and nutrients, and algorithms describing their vertical migration or sinking velocity.
- Inputs to the hydrodynamic model include short wave radiation (Wm^{-2}), incident long wave radiation (Wm^{-2}), air temperature (°C), vapour pressure (hPa), wind (ms^{-1}) and rainfall. The volume and temperature of inflowing water is also required along with the volume of water drawn from the reservoir.
- Management scenarios such as changes in nutrient loading or destratification can be simulated with the models to enable prediction of the most appropriate management strategy and risk reduction that can be achieved.
- As an example, modelling of cyanobacterial growth in Myponga Reservoir South Australia suggests that the cyanobacterial population would be reduced by 75% with a bubble plume aerator (See FS 10).
- Modelling of *Anabaena* growth at Myponga Reservoir with no artificial mixing showed that the population could reach a maximum concentration of 7 μgL^{-1}. However, if artificial destratification using a bubble plume aerator was undertaken the model predicts that the maximum abundance of the *Anabaena* population would be reduced to less than 2 μgL^{-1}, a reduction of 75%.

More information

Antenucci, J.P., R. Alexander, J.R. Romero, and J. Imberger. 2003. Management strategies for a eutrophic water supply reservoir – San Roque Reservoir (Argentina). *Water Science and Technology*, 47(7-8):149-155.

Lewis, D.M., J.D. Brookes, and M.F. Lambert. 2004. Numerical models for management of *Anabaena circinalis*. *Journal of Applied Phycology*, 16(6):457-468.

Romero, J.R., J.P. Antenucci, and J. Imberger. 2004. One- and three-dimensional biogeochemical simulations of two differing reservoirs. *Ecological Modelling*, 174:143-160.

Contact details

Justin Brookes: justin.brookes@adelaide.edu.au
Jason Antenucci: antenucc@cwr.uwa.edu.au

FACT SHEET # 14: *NUTRIENT CONTROL*

Background

- Nutrient levels, particularly nitrogen and phosphorus, have increased within water impacted by human development, leading to 'artificial' or 'cultural' eutrophication.
- Increased nutrient levels are a result of increased inputs to the surrounding catchment. In many cases, fertilisers that are used to increase soil fertility are the major additional nutrient input.
- Most additional nutrients are supplied to water bodies from surface run-off, particularly during high rainfall events.
- When hydrological conditions are favourable, increased nutrient levels result in increased phytoplankton biomass (reflected by chlorophyll and cell concentrations). Eventually, cell numbers may reach 'bloom' levels, which can be detrimental for ecosystems and water treatment processes.

Research Findings

- A reduction in nutrient levels can reduce phytoplankton biomass
- Management practices include:
  - A reduction in nutrient (fertiliser) application to reduce nutrient inputs. This may achieved by most closely matching application with crop nutritional requirements.
  - Alteration to farming practices to reduce surface run-off, soil erosion, and dissolved nutrient concentrations.
  - The addition of nutrient-sorbing substances to farm-land to retain nutrients.
  - Rehabilitation of riparian vegetation, known as buffer zones, to increase nutrient utilisation before nutrients enter water bodies.
  - The creation of artificial wetlands to remove dissolved and particulate nutrients from water.
  - Rehabilitation of aquatic vegetation to compete with phytoplankton for nutrients.

Implementation

- The most efficient management strategies at reduced nutrient levels are not limited to one of these management practices, but will use a combination of all management practices.
- In addition, alterations to hydrology (see Factsheet #10) can be used in combination with these practices to reduce phytoplankton biomass within water bodies.

More Information

Callahan, M.P., Kleinman, P.J.A., Sharpley, A.N. and Stout, W.L. 2002. Assessing the efficacy of alternative phosphorus sorbing soil amendments. *Soil Science* 167:539-547.

Dougherty, W.J., Fleming, N.K., Cox, J.W. and Chittleborough, D.J. (2004) Phosphorus Transfer in Surface Runoff from Intensive Pasture Systems at Various Scales: A Review. *J. Environ. Qual.*, 33:1973–88

Nash, D. and Halliwell, D.J. 1999. Fertilisers and phosphorus loss from productive grazing systems. *Australian Journal of Soil Research*, 37:403-429.
Carpenter, S.R, Christensen, D.L, Cole, J.J., Cottingham, K.L., He, X., Hodgson, J.R, Kitchell, J.F, Knight, S.E.; Pace, M. L., Post, D.M, Schindler, D.E., Voichick, N. 1995. Biological control of eutrophication in lakes. *Environmental Science and Technology,* 29(3):784-786.

Contact details

Justin Brookes: justin.brookes@adelaide.edu.au

CHAPTER 11: RECOMMENDATIONS

This study has led to the development of the following recommendations:

MANAGING TOXIN PRODUCTION IN RESERVOIRS

The best way to control cyanotoxins is to control the growth of cyanobacteria. This can be achieved by reducing nutrients and light. Light limitation can be induced by mixing with artificial destratifiers. Nutrient reduction is more difficult to achieve but must be part of a long term strategy for water quality improvements.

ALGICIDE TREATMENT AND TOXIN RELEASE

If algicides are to be used it is recommended that they be used to arrest growth when cell numbers are low (<5,000 cells/mL) to minimize the concentrations of toxins and geosmin released into the water. The use of algicides when cell numbers are high, and the subsequent cell lysis, can lead to high concentrations of toxins and taste and odour compounds which are difficult to remove. Algicides can also destroy bacteria responsible for the degradation of toxins and so released toxins will persist for longer in water following algicide treatment than they would naturally.

CHLORINE PRE-TREATMENT AND TOXIN RELEASE AND DECAY

Pre-treatment with oxidants such as chlorine, ozone or potassium permanganate can improve the removal of contaminants but can also lead to cell lysis. If cyanobacteria are in high numbers it is recommended that pre-treatment not be used so that the majority of toxins, taste and odour compounds can be removed during coagulation/flocculation. If pre-treatment must be used, a chlorine residual of 0.5 mg/L must be maintained for 30 minutes at pH <8 to ensure toxin degradation. This will vary depending upon the dissolved organic carbon concentration and should not be used as a rule of thumb without testing on specific water.

TOXIN DEGRADATION IN WATER TREATMENT PLANTS

Conventional treatment processes such as coagulation, flocculation, sedimentation and filtration will remove up to 90% of the total toxin present if it is contained within healthy cyanobacterial cells (less for cylindrospermopsin) and should be used in supplies that have cyanobacterial issues. Dissolved toxin (ie toxin that has been released from the cells) must be removed using additional treatment such as powdered activated carbon (PAC), granular activated carbon (GAC), ozone or chlorine. The doses of oxidant and PAC, and the type of activated carbon required for treatment is dependent on the type of toxin and the water quality. More information is provided in factsheet #6.

BIODEGRADATION AND TOXINS IN RESERVOIRS

We cannot rely on biodegradation to control the cyanobacterial toxins, microcystin or cylindrospermopsin in drinking water reservoirs. Toxins can be present in the water column without degrading bacteria being present. Extracellular cylindrospermopsin can persist for months suggesting that biodegradation does not always occur. Additionally, screening with PCR for microcystin degrading organisms revealed that these organisms were not always present in toxic blooms of *Microcystis aeruginosa*.

When a toxic *C. raciborskii* bloom occurs in a water body that has had previous exposure to toxic blooms of CYN producing cyanobacteria it is highly probable that biodegradation of the soluble fraction of toxin will be biodegraded. If no previous history of toxic CYN producing organisms has been recorded it is unlikely, in the first instance, that biodegradation will occur.

REFERENCES

Acero, J. L., E. Rodrigez, and J. Meriluoto. 2005. Kinetics of Reactions Between Chlorine and the Cyanobacterial Toxins Microcystins. *Water Research,* 39:1628-1638.

Alexander, M. 1973. Biotechnology report: Nonbiodegradable and Other Recalcitrant Molecules. *Biotechnology and Bioengineering*, 15:611 -647.

Alexander, M. 1981. Why Microbial Predators and Parasites do not Eliminate Their Prey and Hosts. *Annual Review Microbiology*, 35:113-133.

Alexander, M. 1994. *Biodegradation and Bioremediation*. San Diego: Academic Press.

Antenucci, J., and A. Imerito. 2000. The CWR Dynamic Reservoir Simulation Model DYRESM: Science Manual. Perth: Centre for Water Research, University of Western Australia.

APHA, AWWA, and WEF (American Public Health Association, American Water Works Association, and Water Environment Federation). 1995. *Standard Methods for the Examination of Water and Wastewate*r. 20th ed. Washington, D.C.:APHA.

APHA, AWWA, and WEF (American Public Health Association, American Water Works Association and Water Environment Federation). 1998. *Standard Methods for the Examination of Water and Wastewater*. Washington, D.C.: APHA.

Antenucci, J., A. Ghanouani, M. Burford, and J. Romero. 2005. The Impact of Artificial Destratification on Phytoplankton Species Composition in a Sub-Tropical Reservoir. *Freshwater Biology,* 50:1081-1093.

Atlas, R., and R. Bartha. 1972. Biodegradation of Petroleum in Seawater at Low Temperatures. *Canadian Journal of Microbiology*, 18(12):1851-1855.

AWWA (American Water Works Association). 1999. *Water Quality and Treatment: A Handbook for Community Water Supplies*. New York: McGaw-Hill.

Baldwin, D.S. (1996). The Phosphorus Composition of a Diverse Series of Australian Sediments. *Marine and Freshwater Research*, 35:63-73.

Baldwin, D.S., and A.M. Mitchell. 2000. The Effects of Drying and Re-flooding on the Sediment and Soil Nutrient Dynamics of Lowland River-Floodplain Systems: A Synthesis. *Regulated Rivers: Research and Management*, 16:457-467.

Banker, R., S. Carmeli, O. Hados, B. Teltsch, R. Porat, and A. Sukenik. 1997. Identification of Cylindrospermopsin in *Aphanizomenon ovalisporum* (Cyanophyceae) Isolated from Lake Kinneret, Israel. *Journal Phycology*, 33:613-616.

Benoufella, F., A. Laplanche, V. Boisdon, and M.M. Bourbigot. 1994. Elimination of Microcystis Cyanobacteria (Blue-green Algae) by an Ozoflotation Process: A Pilot Plant Study. *Water Science and Technology*, 30(8):245-257.

Bishop, C.T., E.F.L.J. Anet, and P.R. Gorham. 1959. Isolation and Identification of the Fast-death Factor in *Microcystis aeruginosa* NRC-1. *Canadian Journal of Biochemistry and Physiology*, 37:453-471.

Boethling, R., and M. Alexander. 1979. Effect of Concentration of Organic Chemicals on Their Biodegradation by Natural Microbial Communities. *Applied and Environmental Microbiology*, 37(6):1211-1216.

Boon, P.I., S.E. Bunn, J. D. Green, and R.J. Shiel. 1994. Consumption of Cyanobacteria by Freshwater Zooplankton: Implications for the Success of "Top-down" Control of Cyanobacterial Blooms in Australia. *Australian Journal of Marine and Freshwater Research*, 45:875-887.

Bormans, M., P.W. Ford, L. Fabbro, and G. Hancock. 2004. Onset and Persistence of Cyanobacterial Blooms in a Large Impounded Tropical River. Australia. *Marine and Freshwater Research*, 55:1-15.

Bourke, A., R. Hawes, A. Neilson, and N. Stallman. 1983. An Outbreak of Hepato-Enteritis (the Palm Island Mystery Disease) Possibly Caused by Algal Intoxication. *Toxicon,* 3:45-48.

Bourne, D.G., G.J. Jones, R.L. Blakeley, A. Jones, A.P. Negri, and P. Riddles. 1996. Enzymatic Pathway for the Bacterial Degradation of the Cyanobacterial Cyclic Peptide Toxin Microcystin LR. *Applied & Environmental Microbiology*, 62(11):4086-4094.

Bourne, D.G., P. Riddles, G.J. Jones, W.Smith, and R.L. Blakely. 2001. Characterisation of a Gene Cluster Involved in Bacterial Degradation of the Cyanobacterial Toxin Microcystin LR. *Environmental Toxicology*, 16(6):523-534.

Bouvy, M., R. Molica, S. de Oliveria, M. Marinho, and B. Beker. 1999. Dynamics of a Toxic Cyanobacterial Bloom (*Cylindrospermopsis raciborskii*) in a Shallow Reservoir in the Semi-arid Region of Northeast Brazil. *Aquatic Microbial Ecology*, 20:285-297.

Bouvy, M., D. Dalcão, M. Marinho, M. Pagano, and A. Moura. 2000. Occurrence of *Cylindrospermopsis raciborskii* (Cyanobacteria) in 39 tropical reservoirs during the 1998 drought. *Aquatic Microbial Ecology*, 23:13-27.

Bowmer, K.H.; A. Padovan; R.L. Oliver, W. Korth, and G.G Ganf. 1992. Physiology of Geosmin Production by *Anabaena Circinalis* Isolated from the Murumbidgee River, Australia. *Water Science and Technology*, 25(2):259-267.

Branco, C.W.C., and P.A.C. Senna. 1994. Factors Influencing the Development of *Cylindrospermopsis raciborskii* and *Microcystis aeruginosa* in Paranoa Reservoir, Brazil. *Algological Studies*, 75:85-96.

Briand, J.F., C. Robillot, C. Quiblier-Llobyeras, J.F. Humbert, A. Coutye, and C. Bernard. 2002. Environmental Context of *Cylindrospermopsis raciborskii* (cyanobacteria) Blooms in a Shallow Pond in France. *Water Research*, 36:3183-3192.

Briand, J.F., C. Leboulanger, and J.F. Humbert. 2004. *Cylindrospermopsis raciborskii* (Cyanobacteria) Invasion at Mid-latitudes: Selection, Wide Physiological Tolerance, or Global Warming. *Journal of Phycology*, 40:231-238.

Brock, T., and M. Madigan. 1991. *Biology of Microorganisms.* Englewood Cliffs, New Jersey: Prentice-Hall International Inc..

Brookes, J.D., M.D. Burch, and P. Tarrant. 2000. Artificial Destratification: Evidence for Improved Water Quality. *Water: Official Journal of the Australian Water and Wastewater Association*, 27 (3):18-22.

Brookes, J.D., G.G. Ganf, D. Green, and J. Whittington. 1999. The Influence of Light and Nutrients on Buoyancy, Filament Aggregation and Flotation of *Anabaena circinalis. Journal of Plankton Research,* 21(2):327-341.

Brookes, J.D., R.H. Regel, and G.G. Ganf. 2003. Changes in the Photo-chemistry of *Microcystis aeruginosa* in Response to Light and Mixing. *New Phytologist*, 158:151-164.

Burch, M., P. Baker, and K. Beck. 1990. The Effect of Chlorine and Chloramine on Algae. In *Chloramination of Water Supplies*. Edited by P.M. Thomas. UWRAA, Melbourne: Melbourne and Metropolitan Board of Works.

Burch, M., C. Chow, and P. Hobson. 2001. Blue-Green Algae: Their Significance and Management, Algicides for Control of Toxic Cyanobacteria. *In Proceedings of AWWA Water Quality Conference*, Nashville: AWWA.

Burch, M., P. Hobson, and C. Chow. 2002. Copper Sulphate – Does It Really Work? *In Proceedings of AWWA Water Quality Technology Conference*. AWWA.

Button, K., and B. Robertson 1985. Effect of Toluene Exposure Time and Concentration on Induction of High Affinity Values for Toluene Oxidation by Bacteria of Estuarine Seawater Samples. *Marine Ecology Progress Series*, 26:187-193.

Byth, S. 1980. Palm Island Mystery Disease. *Medical Journal Australia*, 2:40-41.

Campbell, D., V. Hurry, A.K. Clarke, P. Gustafsson, and G. Oquist. 1998. Chlorophyll Fluorescence Analysis of Cyanobacterial Photosynthesis and Acclimation. *Microbiology and Molecular Biology Reviews*, 62:667-683.

Carmichael, W. 1992. Cyanobacteria Secondary Metabolites - The Cyanotoxins. *Journal of Applied Bacteriology*, 72:445-459.

Carmichael, W. 1994. The Toxins of Cyanobacteria. *Scientific Americans*, 170(1):78-86.

Carmichael, W. 2001. Health Effects of Toxin-Producing Cyanobacteria: The CyanoHABs. *Human and Ecological Risk Assessment*, 7:1393-1407.

Carmichael, W., S. Azevedo, J.S. An, R. Molica, E. Jochimsen, S. Lau, K. Rinehart, G. Shaw, and G. Eaglesham. 2001. Human Fatalities from Cyanobacteria: Chemical and Biological Evidence for Cyanotoxins. *Environmental Health Perspectives*, 109:663-668.

Carmichael, W., D. Briggs, and P. Gorham. 1975. Toxicology and Pharmacological Action of *Anabaena flos-aquae* Toxin. *Science*, 187:542-544.

Chapman, A.D., and C.L. Schelske. 1997. Recent Appearance of *Cylindrospermopsis* (Cyanobacteria) in Five Hypertrophic Florida lakes. *Journal of Phycology*, 33:191-195.

Chen, J.-J., and H.-H. Yeh. 2005. The Mechanisms of Potassium Permanganate on Algae Removal. *Water Research,* 39:4420-4428.

Chiswell, R. 1999. Investigations into the Toxicological Properties and Environmental Fate of the Cyanobacterial Toxin Cylindrospermopsin. Ph.D thesis. Griffith University, Brisbane.

Chiswell, R., G. Shaw, G. Eaglesham, M. Smith, R. Norris, A. Seawright, and M. Moore. 1999. Stability of Cylindrospermopsin, The Toxin from the Cyanobacterium, *Cylindrospermopsis raciborskii:* Effect of pH, Temperature and Sunlight on Decomposition. *Environmental Toxicology*, 14:155-161.

Chong, M., B. Wong, P. Lam, G. Shaw, and A. Seawright. 2002. Toxicity and Uptake Mechanism of Cylindrospermopsin and Lophyrotomin in Primary Rat Hepatocytes. *Toxicon*, 40:205-211.

Chorus, I., and J. Bartram. 1999. *Toxic Cyanobacteria in Water. A Guide to Their Public Health Consequences, Monitoring and Management*. London: E and FN Spon.

Chow, C.W.K., M. Drikas, J. House, M.D. Burch, and R.M.A. Velzeboer. 1999. The Impact of Conventional Water Treatment Processes on Cells of the Cyanobacterium *Microcystis aeruginosa*. *Water Research*, 33(15):3253-3262.

Chow, C., J. House, R. Velzeboer, M. Drikas, M. Burch, and D. Steffensen. 1998. The Effect of Ferric Chloride Flocculation on Cyanobacterial Cells. *Water Research,* 32(3):808-814.

Christoffersen, K., S. Lyck and A. Winding. 2002. Microbial Activity and Bacterial Community Structure During Degradation of Microcystins. *Aquatic Microbial Ecology*, 27(2):125-136.

Clasen, J., U. Mischke, M. Drikas, and C. Chow. 2000. An Improved Method for Detecting the Electrophoretic Mobility of Algae During the Destabilisation Process of Flocculation: Flocculant Demand of Different Species and the Impact of DOC. *Journal of Water SRT–Aqua*, 49:89-101.

Codd, G.A. 1995. Cyanobacterial Yoxins: Occurrence, Properties and Biological Significance. *Water Science & Technology*, 32(4):149-156.

Coles, J. F., and C. J. Jones. 2000. Effect of Temperature on Photosynthesis-light Response and Growth of Four Phytoplankton Species Isolated from a Tidal Freshwater River. *Journal of Phycology*, 36:7-16.

Collier, J. L. 2000. Flow Cytometry and the Single Cell in Pycology. *J of Phycology*, 36:628-644.

Cook, D., G. Newcombe, and P. Sztajnbok. 1998. Optimising PAC Dosing to Remove MIB and Geosmin in Four Adelaide Metropolitan Water Treatment Plants. Presented at 61st Annual Water Engineers and Operators' Conference. Shepparton, September 2-3, 1998.

Cousins, I.T., D.J. Bealing, H.A. James, and A. Sutton. 1996. Biodegradation of Microcystin-LR by Indigenous Mixed Bacterial Populations. *Water Research*, 30(2):481-485.

Cox, L., M.C. Hermosin, R. Celis, and J. Cornejo. 1997. Sorption of Two Polar Herbicides in Soils and Soil Clays Suspensions. *Water Research*, 33:1309-1316.

Crittenden, J. 2005. Water Treatment Principles and Design. New York: John Wiley and Sons.

Daubaras, D., and A. Chakrabarty. 1992. The Environment, Microbes and Bioremediation: Microbial Activities Modulated by the Environment. *Biodegradation*, 3:125-135.

Dawson, R. M. 1998. The Toxicology of Microcystins. *Toxicon*, 36:953-962.

Denman, K. L., and A.E. Gargett. 1983. Time and Space Scales of Vertical Mixing and Advection of Phytoplankton in the Upper Ocean. *Limnology and Oceanography*, 28:801-815.

Dillenberg, H., and M. Dehnel. 1960. Toxic Water Bloom in Saskatchewan 1959. *Canadian Medcial Association Journal*, 83:1151-1154.

Dobberfuhl, D.R. 2003. *Cylindrospermopsis raciborskii* in Three Central Florida Lakes: Population Dynamics, Controls and Management Implications. *Lake and Reservoir Management*, 19:341-348.

Dokulil, M.T., and Mayer, J. 1996. Population Dynamics and Photosynthetic Rates of a *Cylindrospermopsis-Limnothrix* Association in a Highly Eutrophic Urban Lake, Alte Donau, Austria. *Algological Studies* 83:179-195.

Dokulil, M.T., and K. Teubner. 2000. Cyanobacterial Dominance in Lakes. *Hydrobiologia*, 438:1-12.

Downing, J.A., S.B. Watson, and E. McCauley. 2001. Predicting Cyanobacterial Dominance in Lakes. *Canadian Journal of Fisheries Science*, 58:1905-1908.

Drikas, M., C.W.K. Chow, J. House, and M.D. Burch. 2001. Using Coagulation, Flocculation and Settling to Remove Toxic Cyanobacteria. *Journal of American Water Works Association*, 93(2):100-111.

Eaglesham,G.K., R.L. Norris, G.R. Shaw, M.J. Smith, R.K. Chiswell; B.C. Davis, G.R. Neville, A.A. Seawright, and M.R. Moore. 1999. Use of HPLC-MS/MS to Monitor Cylindrospermopsin, a Blue-green Algal Toxin, for Public Health Purposes. *Environmental Toxicology*, 14(1):151-154.

Elliot, J.A., A.E. Irish, and C.S. Reynolds. 2002. Predicting the Spatial Dominance of Phytoplankton in a Light Limited and Incompletely Mixed Eutrophic Water Column using the PROTECH Model. *Freshwater Biology*, 47:433-440.

El Saadi, O., and A. Cameron. 1993. Illness Associated with Blue-Green Algae. *Medical Journal of Australia*, 158:792-793.

Fabbro, L.D., and L.J. Duivenvoorden. 1996. Profile of a Bloom of the Cyanobacterium *Cylindrospermopsis raciborskii* (Woloszynska) Seenaya and Subba Raju in the Fitzroy River in Tropical Central Queensland. *Marine and Freshwater Research*, 47:685-694.

Falconer, I.R. 1991. Tumor Promotion and Liver Injury Caused by Oral Consumption of Cyanobacteria. *Environmental Toxicology and Water Quality*, 6(2):177-184.

Falconer, I.R. 1993. *Algal Toxins in Seafood and Drinking Water*. London: Academic Press.

Falconer, I., A. Choice, and W. Hosja. 1992. Toxicity of the Edible Mussel (*Mytilus edulis)* Growing Naturally in an Estuary During a Water-bloom of the Blue-green Alga *Nodularia spumigena*. *Environmental Toxicology and Water Quality*, 7:119-123.

Fallon, R.D., and T.D. Brock. 1979. Decomposition of Blue-green Algae (Cyanobacteria) Blooms in Lake Mendota, Wisconsin. *Applied and Environmental Microbiology*, 38:499-505.

Farrell, J., and A. Rose. 1967. Temperature Effects on Microorganisms. *Annual Review Microbiology*, 21:101-120.

Fastner, J., R.Heinze, A.R. Humpage, U. Mischke, G.K. Eaglesham, and I. Chorus. 2003. Cylindrospermopsin Occurrence in Two German Lakes and Preliminary Assessment of Toxicity and Toxin Production of *Cylindrospermopsis raciborskii* (Cyanobacteria) Isolates. *Toxicon*, 42:313-321.

Feitz, A.J., T.D. Waite, G.J. Jones, B.H. Boyden, and P.T. Orr. 1999. Photocatalytic Degradation of the Blue Green Algal Toxin Microcystin-LR in a Natural Organic-Aqueous Matrix. *Environmental Science and Technology*, 33:243-249.

Fenchel, T., and B. Jorgensen. 1977. Detritus Food Chains of Aquatic Ecosystems: The Role of Bacteria. *Advanced Microbiology Ecology*, 1:1-58.

Fischer, H.B., E.J. List, R.C.Y. Koh, J. Imberger, and N.H. Brooks. 1979. *Mixing in Inland and Coastal Waters*. New York: Academic Press.

Fournier, J., P. Codaccioni, and G. Soulas. 1981. Soil Adaption to 2,4-D Degradation in Relation to the Application Rates and the Metabolic Behavior of the Degrading Microflora. *Chemosphere*, 10:977-984.

Furusato, E., T. Asaeda, and J. Manatunge. 2004. Tolerance for Prolonged Darkness of Three Phytoplankton Species, *Microcystis aeruginosa* (Cyanophyceae), *Scenedesmus quadricauda* (Chlorophyceae), and *Melosira ambigua* (Bacillariophyceae). *Hydrobiologia*, 527:153-162.

Fushiwaki, Y., and K. Urano. 2001. Adsorption of Pesticides and Their Biodegraded Products on Clay Minerals and Soils. *Journal of Health Science*, 47(4):429-432.

Gates, D.M. 1966. Spectral Distribution of Solar Radiation on the Earth's Surface. *Science*, 151:523-529.

Genty, B., J.M. Briantais, and N.R. Baker. 1989. The Relationship Between the Quantum Yield of Photosynthetic Electron Transport and Quenching of Chlorophyll Fluorescence. *Biochimica et Biophysica Acta*, 990:87-92.

Glibert, P.M., C. Garside, J.A. Fuhrman, and M.R. Roman. 1991. Time-dependent Coupling of Inorganic and Organic Nitrogen Uptake and Regeneration in the Plume of the Chesapeake Bay Estuary and its Regulation by Large Heterotrophs. *Limnology and Oceanography*, 36:895-909.

Goldstein, R., L. Mallory, and M. Alexander. 1985. Reasons for Possible Failure of Inoculation to Enhance Biodegradation. *Applied and Environmental Microbiology*, 50(4):977-983.

Gorham, P., and W. Carmichael. 1988. Hazards of Freshwater Blue-green Algae (Cyanobacteria). In *Algae and Humans Affairs*. Edited by C.A. Lembi and J.R. Warland. Cambridge: Cambridge University Press.

Gorham P.R., J. McLachlan, U.T. Hammer, and W.K. Sivonen. 1964. Isolation and Culture of Toxic Strains of Anabaena Flos-Aquae. *Mitteilungen der Internationalen Vereinigung fir Limnologie*, 15:796-804.

Griffiths, D.J., and M.L. Saker. 2003. The Palm Island Mystery Disease 20 Years On: A Review of Research on the Cyanotoxin Cylindrospermopsin. *Environmental Toxicology*, 18(2):78-93.

Grützmacher, G., G. Böttcher, I. Chorus, and H. Bartel. 2002. Removal of Microcystins by Slow Sand Filtration. *Environmental Toxicology*, 17(4):386-394.

Gude, H. 1985. Influence of Phagotrophic Processes on the Regeneration of Nutrients in two-stage Continuos Culture systems. *Microbial Ecology*, 11:193-204

Habitue, M., and M. Alexander. 1977. Further Evidence for the Regulation of Bacterial Populations in Soil by Protozoa. *Archives of Microbiology*, 113:181-183.

Hallegraeff, G.M. 1993. A Review of Harmful Algal Blooms and Their Apparent Global Increase. *Phycologia*, 32:79-99.

Hamilton, D.H., and G. Schladlow. 1997. Prediction of Water quality in Lakes and Reservoirs. Part 1-Model Description. *Ecological Modelling*, 96:91-110.

Harada, K.-I., S. Imanishi, H. Kato, M. Masayoshi, E. Ito, and K. Tsuji. 2004. Isolation of Adda from Microcystin-LR by Microbial Degradation. *Toxicon*, 44(1):107-109.

Harada, K., I. Ohtani, K. Iwamoto, M. Suzuki, M. Watanabe, M. Watanabe, and K. Terrao. 1994. Isolation of Cylindrospermopsin from a Cyanobacterium *Umezakia natans* and its Screening Method. *Toxicon*, 32(1):73-84.

Harris, G.P. (1980). The Relationship Between Chlorophyll a Fluorescence, Diffuse Attenuation Changes and Photosynthesis in Natural Phytoplankton Populations. *Journal of Plankton Research*, 2:109-127.

Harris, G.P. 1986. *Phytoplankton Ecology*. London: Chapman and Hall.

Harris, G.P., and G. Baxter. 1996. Interannual Variability in Phytoplankton Biomass and Species Composition in a Subtropical Reservoir. *Freshwater Biology*, 35:545-560.

Hašler, P., and A. Poulíčková. 2003. Diurnal changes in Vertical Distribution and Morphology of a Natural Population of *Planktothrix agardhii* (Gom.) Anagnostidis et Komarek (Cyanobacteria). *Hydrobiologia*, 506-509:195-201.

Hawkins, P.R. 1988. The Zooplankton of a Small Tropical Reservoir (Solomon Dam, North Queensland). *Hydrobiologia*, 157: 105-118.

Hawkins, P., N. Chandrasens, G. Jones, A. Humpage, and I. Falconer. 1997. Isolation and Toxicity of *Cylindrospermopsis raciborskii* from an Ornamental Lake. *Toxicon,* 35:341-346.

Hawkins, P., and D.J. Griffiths. 1993. Artificial Destratification of a Small Tropical Reservoir: Effects upon the Phytoplankton. *Hydrobiologia*, 254:169-181.

Hawkins, P.R., and W. Lampert. 1989. The Effect of Body Size on Filtering Rate Inhibition in the Presence of a Filamentous cyanobacterium. *Limnology and Oceanography*, 34:1089-1094.

Hawkins, P., M. Runnegar, A. Jackson, and I. Falconer. 1985. Severe Hepatotoxicity Caused by the Tropical Cyanobacterium (Blue-Green Alga) *Cylindrospermopsis raciborskii* (Woloszynska) Seenaya and Subba Raju Isolated from a Domestic Water Supply Reservoir. *Applied and Environmental Microbiology*, 50(5):1292-1295.

Herzfeld, M., and D. Hamilton. 2000. The CWR Computational Aquatic Ecosystem Dynamics Model CAEDYM. Perth: The Centre for Water Research, University of Western Australia.

Himberg, K., A.-M. Keijola, L. Hiisvirta, H. Pyysalo, and K. Sivonen. 1989. The Effect of Water Treatment Processes on the Removal of Hepatotoxins from *Microcystis* and *Oscillatoria* Cyanobacteria: A Laboratory Study. *Water Research*, 23(8):979-984.

Hoffman, J.R.H. 1976. Removal of *Microcystis* Toxins in Water Purification Processes. *Water SA*, 2(2):58-60.

Ho, L. 2004. *The removal of cyanobacterial metabolites from drinking water using ozone and granular activated carbon*. PhD Dissertation, University of South Australia, Adelaide.

Ho, L, G. Onstad, U. von Gunten, S. Rinck-Pfeiffer, K. Craig, and G. Newcombe. (2006). Chlorination of Four Microcystin Analogues. *Water Research.*

Holst, T., N.O.G. Jørgensen, C. Jørgensen, and A. Johansen. 2003. Degradation of Microcystin in Sediments at Oxic and Anoxic, Denitrifying Conditions. *Water Research*, 37(19):4748-4760.

Horvath, R. 1972. Microbial Co-Metabolism and the Degradation of Organic Compounds in Nature. *Bacteriological Reviews*, 36(2):146-155.

Huber, C. 1972. The Crystal Structure and Absolute Configuration Of 2,9-diacetyl-9-azabicyclo (4,2,1) non-2,3-ene. *Acta Crystallographica*, 28:2577-2582.

Humpage, A. 1999. Cyanobacterial Toxins and Cancer: Assessing the Risks, Water Qaulity News, *The Newsletter of the Cooperative Research Centre for Water Quality and Treatment*, 9.

Humpage, A., J. Rositano, A. Bretag, R. Brown, P. Baker, B. Nicholson, and D. Steffensen. 1994. Paralytic Shellfish Poisons from Australian Cyanobacterial Blooms. *Australian Journal of Marine and Freshwater Research*, 45:761-771.

Humphries, S.E., and J. Imberger. 1982. The Influence of the Internal Structure and Dynamics of Lake Burrinjuck Reservoir on Phytoplankton Blooms. Environmental Dynamics Report ED 82-023. Nedlands: University of Western Australia.

Humphries, S.E., and V.D. Lyne. 1988. Cyanophyte Blooms: The role of Cell Buoyancy. *Limnology and Oceanography*, 33:79-81.

Hunter, P. 1998. Cyanobacterial Toxins and Human Health. *Journal of Applied Microbiology Symposium Supplement*, 84:35-40.

Huszar, V.L.M., L.H.S. Silva, M.M. Marinho, P. Domingos, and C.L. Sant Anna. 2000. Cyanoprokaryote Assemblages in Eight Productive Tropical Brazilian Waters. *Hydrobiologia*, 424:67-77.

Ibelings, B.W. 1996. Changes in Photosynthesis in Response to Combined Irradiance and Temperature Stress in Cyanobacterial Waterblooms. *Journal of Phycology*, 32:549-557.

Imberger, J., J. Patterson, B. Hebbert, and I. Loh. 1978. Dynamics of Reservoir of Medium Size. *Journal of the Hydraulics Division*, 104:725-743.

Imberger, J., and J.C. Patterson. 1981. A Dynamic Reservoir Simulation Model-Dyresm:5. In *Transport Mmodels for Inland and Coastal waters*. Edited by H.B. Fischer. New York: Academic Press.

Imberger, J., and J.C. Patterson. 1990. Physical Limnology. In *Advances in Applied Mechanics*. Edited by T. Wu. Boston: Academic Press.

Ishii, H., M. Nishijima, and T. Abe 2004. Characterization of Degradation Process of Cyanobacterial Hepatotoxins by a Gram-negative Aerobic Bacterium. *Water Research*, 38(11):2667-2676.

Istvanovics, V., H.M. Shafik, M. Presing, and S.V. Juhos. 2000. Growth and Phosphate Uptake Kinetics of the Cyanobacterium, *Cylindrospermopsis raciborskii* (Cyanophyceae) in Throughflow Cultures. *Freshwater Biology*, 43:257-275.

IPCC 1996. Revised 1996 IPCC Guidelines for National Greenhouse Gas Inventories. Inter-Governmental Panel on Climate Exchange, Hadley Centre, UK Meteorological Office, Bracknell, UK 3 vols. http://www.ipcc-nggip.iges.or.jp/public/gl/invs1.htm

Jassby, A.D., and C.R. Goldman. 1974. .Loss Rates from a Phytoplankton Community. *Limnology and Oceanography*, 19:618-627.

Jeffrey, S.W., and N.A. Welshmeyer. 1997. Spectrophotometric and Fluorometric Equations in Common use in Oceanography. In *Phytoplankton Pigments in Oceanography*. Monographs on Oceanographic Methodology No.10. Edited by S.W. Jeffrey, R.F.C. Mantoura, and S.W. Wright. Paris: UNESCO Publishing.

Jochimsen E.M., W.W. Carmichael, J.S. An, D.M. Cardo, S.T. Cookson, C.E.M. Holmes, M.B.D. Antunes, D.A. Demelo, T.M. Lyra, V.S.T. Barreto, S.M.F.O. Azevedo, and W.R. Jarvis. 1998. Liver Failure and Death After Exposure to Microcystins at a Hemodialysis Center in Brazil. *New England. Journal of Medicine*, 338:873–878.

Jones, G. 1990. Biodegradation and Removal of Cyanobacterial Toxins in Natural Waters. Presented at the South East Queensland Water Board Blue-green Algal Seminar, Brisbane, Queensland, 1990.

Jones, G., D. Bourne, R. Blakeley, and H. Doelle. 1994. Degradation of the Cyanobacterial Hepatotoxin Microcystin by Aquatic Bacteria. *Natural Toxins*, 2:228-235.

Jones, G., I. Falconer, and R. Wilkins. 1995. Persistence of Cyclic Peptide Toxins in Dried *Microcystis aeruginosa* Crusts from Lake Mokoan, Australia. *Environmental Toxicology and Water Quality*, 10:19-24.

Jones, G. J., and W. Korth. 1995. In Situ Production of Volatile Odour Compounds by River and Reservoir Phytoplankton Populations in Australia. *Water Science and Technology*, 31(11):145-151.

Jones, G., W. Minatol, K. Craig, and R. Naylor. 1993. Removal of Low Level Cyanobacterial Peptide Toxins from Drinking Water Using Powdered and Granular Activated Carbon and Chlorination – Results of Laboratory and Pilot Plant Studies. In *Proceedings of the 15th Federal AWWA Convention, April 18-23, 1993, Gold Coast, Queensland, Australia*, 579-586.

Jones, G., and A. Negri. 1997. Persistence and Degradation of Cyanobacterial Paralytic Shellfish Poisons (PSPs) in Freshwaters. *Water Research*, 31:525-533.

Jones, G.J., and P.T. Orr. 1994. Release and Degradation of Microcystin Following Algicide Treatment of a *Microcystis aeruginosa* Bloom in a Recreational Lake, as Determined by HPLC and Protein Phosphatase Inhibition Assay. *Water Research*, 28(4):871-876.

Jurczak, T., M. Tarczynska, K. Izydorcsyk, J. Mankiewics, M. Zalewski, and J. Meriluoto. 2005. Elimination of Microcystins by Water Treatment Processes – Examples from Sulejow Reservoir, Poland. *Water Research*, 39:2394-2406.

Kao, C. 1993. Paralytic Shellfish Poisoning. In *Algal Toxins in Seafood and Drinking Water*. Edited by I. Falconer. London: Academic Press.

Keijola, A.-M., K. Himberg, A.L. Esala, K. Sivonen, and, L. Hiisvirta. 1988. Removal of Cyanobacterial Toxins in Water Treatment Processes: Laboratory and Pilot-scale Experiments. *Toxicity Assessment: An International Journal*, 3:643-656.

King, C.R., and R.G. Everson. 1978. North Pine Dam (Lake Samsonvale) Survey (1973-1976) – Final Report. Queensland: Department of Biology and Environmental Science, Queensland Institute of Technology.

Kiviranta, J., K. Sivonen, K. Lahti, R. Luukkainen, and S. Niemela. 1991. Production and Biodegradation of Cyanobacterial Toxins - Laboratory Study. *Archiv fur Hydrobiologie*, 121(3):281-294.

Knoll, L.B., M.J. Vanni, W.H. Renwick. 2003. Phytoplankton Primary Production and Photosynthetic Parameters in Reservoirs Along a Gradient of watershed Land Use. *Limnology and Oceanography*, 48:608-617

Knowles, R. 1990. Acetylene Inhibition Techniques: Development, Advantages, and Potential Problems. In *Denitrification in Soil and Sediment.* Edited by N. P. Revsbech and J. Sørensen. New York: Plenum Press.

Korshin, G.V, C.-W. Li, and M.M. Benjamin. 1997. Monitoring the Properties of Natural Organic Mater Through UV Spectroscopy: A Consistent Theory. *Water Research*, 31(7):1787-1795.

Kromkamp, J., and A.E. Walsby. 1990. A Computer Model of Buoyancy and Vertical Migration in Cyanobacteria. *Journal of Plankton Research*, 12:191-183.

Kuiper-Goodman, T., I. Falconer, and J. Fitzgerald. 1999. Human Health Aspects, In *Toxic Cyanobacteria in Water - A Guide to Their Public Health Consequences, Monitoring and Management*, Edited by I. Chorus and J. Bartram. London: E and F.N. Spon.

Kuiper, J., and A. Hanstveit. 1984. Fate and Effects of 4-Chlorophenol in Marine Plankton Communities in Experimental Enclosures. *Ecotoxicology and Environmental Safety*, 8:15-33.

Lagos, N., H. Onodera, A. Zagatto, D. Andrinolo, S. Azevedo, and Y. Oshima. 1999. The First Evidence of Paralytic Shellfish Toxins in the Freshwater Cyanobacterium *Cylindrospermopsis raciborskii*, Isolated from Brazil. *Toxicon*, 37:1359-1373.

Lahti, K. and L. Hiisvirta. 1989. Removal of Cyanobacterial Toxins in Water Treatment Processes: Review of Studies Conducted in Finland. *Water Supply*, 7:149-154.

Lahti, K., M.R. Niemelä, J. Rapala, and K. Sivonen. 1998. Biodegradation of Cyanobacterial Hepatotoxins – Characterisation of the Toxin Degrading Bacteria. *Harmful Algae, Algas nicovs: Proceedings of the VIII International Conference on Harmful Algae.* Vigo, Spain: International Oceanographic Commission of UNESCO.

Lahti, K., J. Rapala, M. Färdig, M. Niemelä, and K. Sivonen. 1997. Persistence of Cyanobacterial Hepatotoxin, Microcystin-LR in Particulate Material and Dissolved Lake Water. *Water Research*, 31(5):1005-1012.

Lam, A., P. Fedorak, and E. Prepas. 1995. Biotransformation of the Cyanobacterial Hepatotoxin Microcystin-LR as Determined by HPLC and Protein Phosphatase Bioassay. *Environmental Science Technology*, 29:242-246.

Lam, A.K.-Y., E.E. Prepas, D. Spink, S.E. Hrudey. 1995. Chemical Control of Hepatotoxic Phytoplankton Blooms: Implications for Human Health. *Water Research*, 29(8):1845-1854.

Lambert T.W., C.F.B. Holmes, and S.E. Hrudey. 1996. Adsorption of Microcystin-LR by Activated Carbon and Removal in Full Scale Water Treatment. *Water Research*, 30(6):1411-1422.

Lappin, H., M. Greaves, and J. Slater 1985. Degradation of the Herbicide Mecoprop [2-(2-Methyl-4-Chlorophenoxy) Propionic Acid] by a Synergistic Microbial Community. *Applied and Environmental Microbiology*, 49:429-433.

Larson, R.J. 1984. Kinetic and Ecological Approaches for Predicting Biodegradation Rates of Xenobiotic Organic Chemicals in Natural Ecosystems. In *Current Prespective in Microbial Ecology*. Edited by M. Klug and A. Reddy. Washington, DC: American Society for Microbiology.

Lau, R.H., C. Sapienza, and W.F. Doolittle. 1980. Cyanobacterial Plasmids: Their Widespread Occurrence, and the Existence of Regions of Homology Between Plasmids in the Same and Different Species. *Molecular and General Genetics*, 178:203-211.

Lawton, L.A., and P.K.J. Robertson. 1999. Physico-chemical Treatment Methods for the Removal of Microcystins (Cyanobacterial Hepatotoxins) from Potable Waters. *Chemical Society Reviews*, 28:217:224.

Lawton, L.A., P.K.J. Robertson, B.J.P.A. Cornish, and M. Jaspars. 1999. Detoxification of Microcystins (Cyanobacterial Hepatotoxins) Using TiO_2 Photocatalytic Oxidation. *Environmental Science and Technology*, 33:771-775.

Lewis, D. M., J. Antennuci, and J.D. Brookes. 2001. Numerical Simulation of Surface Mixers used for Destratification of Reservoirs. MODSIM 2001,Canberra, Australia, December 2001.

Lewis, D. M., J.D. Brookes, and M.F. Lambert. 2004. Numerical Models for Management of *Anabaena circinalis*. *Journal of Applied Phycology*, 16(6):457-468.

Linden, L.G., D.M. Lewis, M.D. Burch, and J.D. Brookes. 2004. Interannual Variability in Rainfall and its Impact on Nutrient Load and Phytoplankton in Myponga Reservoir, South Australia. *International Journal of River Basin Management*, 2(3):1-11.

Long, B.M., G.J. Jones, and P.T. Orr. 2001. Cellular Microcystin Content in N-limited *Microcystis aeruginosa* can be Predicted from Growth Rate. *Applied and Environmental Microbiology*, 67(1):278-283.

Ludzack, F., and D. Kinkead. 1956. Persistence of Oily Wastes in Polluted Water Under Aerobic Conditions. *Journal of Industrial and Engineering Chemistry*, 48:263-267.

Lyck, S. 2001. Blooms of *Microcystis aeruginosa* in Freshwater – Regulation of Microcystin Production and Natural Degradation of the Toxin. Ph.D. thesis. University of Copenhagen, Denmark.

Mahmood, N., and W. Carmichael. 1986. The Pharmacology of Anatoxin-a(S), A Neurotoxin Produced by the Freshwater Cyanobacteria *Anabaena flos-aquae* NRC 525-17. *Toxicon,* 24:425-434.

Marinho, M. M., and V.L.D.M. Huszar. 2002. Nutrient Availability and Physical Conditions as Controlling Factors of Phytoplankton Composition and Biomass in a Tropical Reservoir (Southeastern Brazil). *Archives Hydrobiology*, 153:443-468.

Matsumura, F. 1974. Microbial Degradation of Pesticides. In *Survival in Toxic Environments.* Edited by M. Khan and J. Bederka. New York: Academic Press.

Mc Auliffe, T.F., and R.F. Rosich. 1990. The Triumphs and Tribulations of Artificial Mixing in Australian Waterbodies. *Water*, Aug:22-23.

McCarthy J.J., W.R. Taylor, and J.L. Taft. 1977. Nitrogenous Nutrition of the Plankton in the Chesapeake Bay. I. Nutrient Availability and Phytoplankton Preferences. *Limnology and Oceanography*, 22:996-1011.

McGregor, G.B., and L.D. Fabbro. 2000. Dominance of *Cylindrospermopsis raciborskii* (Nostocales, Cyanoprokaryota) in Queensland Tropical and Subtropical Reservoirs: Implications for Monitoring and Management. *Lakes and Reservoirs: Research and Management*, 5:195-205.

Meybeck, M. 1982. Carbon, Nitrogen and Phosphorus Transport by World Rivers. *American Journal of Science*, 282:401-450.

Miller, M.J., M.M. Critchley, J. Hutson, and H.J. Fallowfield. 2001. The Adsorption of Cyanobacterial Hepatotoxins from Water onto Soil During Batch Experiments. *Water Research*, 35(6):1461-1468.

Miller, M.J., and H.J. Hallowfield. 2001. Degradation of Cyanobacterial Hepatotoxins in Batch Experiments. *Water Science & Technology*, 43(12):229-232.

Mitchell, A., and D.S. Baldwin. 1998. Effects of Desiccation/Oxidation on the Potential for Bacterially Mediated P Release from Sediments. *Limnology and Oceanography*, 43:481-487.

Montiel, A., and B. Welté. 1998. Preozonation Coupled with Flotation Filtration: Successful Removal of Algae. *Water Science and Technology*, 37(2):65-73.

Moog, D.B., and P.J. Whiting. 2002. Climatic and Agricultural Factors in Nutrient Exports from Two Watersheds in Ohio. *Journal of Environmental Quality*, 31:72-83.

Moore, D., M. O'Donohue, G. Shaw, and C. Critchley. 2003. Potential Triggers for Akinete Differentiation in an Australian Strain of the Cyanobacterium *Cylindrospermopsis raciborskii* (AWT 205/1). *Hydrobiologia*, 506-509:175-180.

Moore, R., I. Ohtani, B. Moore, C. Koning, and W. Yoshida. 1993. Cyanobacterial Toxins. *Gazzetta Chimica Italiana*, 123:329-336.

Mouchet, P., and V. Bonnélye. 1998. Solving Algae Problems: French Expertise and World Wide Applications. *Journal of Water SRT–Aqua*, 47:125-141.

Mukherdee, B., P.N. Pandey, and S.N. Singh. 2002. Mathematical Modelling and System Analysis of Inorganic Carbon in the Aquatic Environment. *Ecological Modelling*, 152:129-143

Negri, A., and G. Jones. 1995. Bioaccumulation of Paralytic Shellfish Poisoning's (PSP) Toxins from the Cyanobacterium *Anabaena circinalis* by the Freshwater Mussel *Alathyria condola*. *Toxicon*, 33:667-678.

Neilan, B.A., M.L. Saker, J. Fastner, A. Törökné, and B.P. Burns. 2003. Phylogeography of the Invasive Cyanobacterium *Cylindrospermopsis raciborskii*. *Molecular Ecology*, 12: 133-140.

Nelson, M., S. Montogomery, E. O'Neill, and P.H. Pritchard. 1986. *Applied and Environmental Microbiology*, 52:383-384.

Nesbitt, H., and J. Watson. 1980. Degradation of the Herbicide 2,4-D in River Water. II. The Role of Suspended Sediment, Nutrients and Water Temperature. *Water Research*, 14:1689-94.

Newcombe, G., D. Cook, S. Brooke, L. Ho, and N. Slyman. 2003. Treatment Options for Microcystin Toxins: Similarities and Differences Between Variants. *Environmental Technology*, 24(3):299-308.

NHMRC and NRMMC (National Health and Medical Research Council and Natural Resource Management Ministerial Council). 2001. *Australian Drinking Water Guidelines*. Canberra: NHMRC and NRMMC.

Nicholson, B.C., J. Rositano, and M.D. Burch. 1994. Destruction of Cyanobacterial Peptide Hepatotoxins by Chlorine and Chloramine. *Water Research*, 28(6):1297-1303.

Ohtani, I., R. Moore, and M. Runnegar. 1992. Cylindrospermopsin, a Potent Hepatotoxin from the Blue-Green Alga *Cylindrospermopsis raciborskii*. *Journal of the American Chemical Society*, 114:7941-7942.

Oliver, R.L., and G.G. Ganf. 2000. Freshwater Blooms. In *The Ecology of Cyanobacteria*. Edited by B.A. Whitton and M. Potts. Netherlands: Kluwer Academic Publishers.

Oliver, R.L., and J. Whittington. 1998. Using Measurement of Variable Chlorophyll-a Fluorescence to Investigate the Influence of Water Movement on the Photochemistry of Phytoplankton. In *Physical Processes in Lakes and Oceans*. Coastal and Estuarine Studies Volume 54: 517-534. Edited by J. Imberger. American Geophysical Union.

Oliver, R.L., J. Whittington, Z. Lorenz, and I.T. Webster. 2003. The Influence of Vertical Mixing on the Photoinhibition of Variable Chlorophyll *a* Fluorescence and its Inclusion in a Model of Phytoplankton Photosynthesis. *Journal of Plankton Research*, 25:1107-1129.

Padisák, J. (1997). *Cylindrospermopsis raciborskii* (Woloszynska) Seenayya and Subba Raju, an Expanding Highly Adaptive Cyanobacterium: Worldwide distribution and Review of its Ecology. *Archiv fur Hydrobiologie. Supplementbund. Monographic Studies*, 107:563-593.

Padisák, J., and C.S. Reynolds. 1998. Selection of Phytoplankton Associations in Lake Balaton, Hungary, in Response to Eutrophication and Restoration Measures, with Special Reference to the Cyanoprokaryotes. *Hydrobiologia*, 384:41-53.

Park H.-D., Y. Sasaki, T. Maruyama, E. Yanagisawa, A. Hiraishi, and K. Kato. 2001 Degradation of the Cyanobacterial Hepatotoxin Microcystin by a New Bacterium Isolated from a Hyereutrophic Lake. *Environmental Toxicology*, 16(4):337-343.

Parkes, R. 1982. Methods for Enriching, Isolating and Analysing Microbial Communities in Laboratory Systems. In *Microbial Interactions and Comminutes.* Edited by J. S. A. Bull. London: Academic Press Inc.

Patterson, J.C., P.F. Hamblin, and J. Imberger. 1984. Classification and Dynamic Simulation of the Vertical Density Structure of Lakes. *Limnology and Oceanography*, 29(4):845-861.

Peterson, H.G., S.E. Hrudey, I.A. Cantin, T.R. Perley, and S.L. Kenefick. 1995. Physiological Toxicity, Cell Membrane Damage and the Release of Dissolved Organic Carbon and Geosmin by *Aphanizomenon Flos-Aquae* After Exposure to Water Treatment Chemicals. *Water Research*, 29(6):1515-1523.

Pfaeder, F., and R. Shimp. 1985. Adaption of Estuarine Ecosystems to the Biodegradation of Nitrilotriacetic Acid: Effects of Pre-Exposure. *Environmental Toxicology and Chemistry*, 4:587-593.

Pike, E., and C. Curds. 1971. The Microbial Ecology of the Activated Sludge Process. In *Microbial Aspects of Pollution*. Edited by G. Sykes and F. Skinner. New York: Academic Press.

Plummer, J.D., and J.K. Edzwald. 2002. Effects of Chlorine and Ozone on Algal Cell Properties and Removal by Coagulation. *Journal of Water Supply – Research and Technology*, 51(6):307-318.

Présing, M., S. Herodek, L. Voros, and I. Kober. 1996. Nitrogen Fixation, Ammonium and Nitrate Uptake During a Bloom of *Cylindrospermopsis raciborskii* in Lake Balaton. *Archives Hydrobiology*, 136:553-562.

Présing, M., S. Herodek, T. Preston, and L. Vörös. 2001. Nitrogen Uptake and the Importance of Nitrogen Loading in Lake Balaton. *Freshwater Biology*, 46:125-139.

Ramberg, L. (1987). Phytoplankton Succession in the Sanyati basin, Lake Kariba. *Hydrobiologia*, 153:193-202.

Rapala, J., K. Lahti, K. Sivonen, and S.I. Niemelä. 1994. Biodegradability and Adsorption on Lake Sediments of Cyanobacterial Hepatotoxins and Anatoxin-a. *Letters in Applied Microbiology*, 19:423-426.

Regel, R.H., J.D. Brookes, and G.G. Ganf. 2004. Vertical Migration, Entrainment and Photosynthesis of the Freshwater Dinoflagellate *Peridinium cinctum* in a Shallow Urban Lake. *Journal of Plankton Research*, 26:143-157.

Regel, R.H., G.G. Ganf, J.D. Brookes, and R. Griffiths. 2004. The Influence of Grid-Generated Turbulence on the Mash01 Unicellular *Microcystis aeruginosa* Strain. *Hydrobiologia*, 517(1):107-120.

Ressom, R., F. Soong, J. Fitzgerald, L. Turczynowicz, O. El Saadi, D. Roder, T. Maynard, and I. Falconer. 1994. Health Effects of Toxic Cyanobacteria (Blue-Green Algae). *National Health and Medical Research Council.* Canberra: Australian Government Publishing Service.

Reynolds, C.S. 1984. *The Ecology of Freshwater Phytoplankton*. Cambridge: Cambridge University Press.

Reynolds, C.S. 1994. The Role of Fluid Motion in the Dynamics of Phytoplankton in Lakes and Rivers. In *Aquatic Ecology: Scales, Patterns and Processes*. Edited by P.S. Giller, A.G. Hildrew, and D. G. Raffaelli. Oxford: Blackwell.

Reynolds, C.S. 1997. *Vegetation Processes in the Pelagic: A model for Ecosystem Theory*. Oldendorf/Luhe Germany: Ecology Institute.

Reynolds, C.S., V.L.M. Huszar, C. Kruk, L. Naselli-Flores, and S. Melo. 2002. Towards a Functional Classification of the Freshwater Phytoplankton. *Journal of Plankton Research* 24:417-428.

Richard, M. 1968. Enzymatic Adaption in Bacteria: Its Biochemical and Genetic Basis. *Essays in Biochemistry*, 4:105-154.

Robson, B.J., and D.P. Hamilton. 2004. Three-dimensional Modelling of a *Microcystis* Bloom Event in the Swan River Estuary, Western Australia. *Ecological Modelling*, 174(1-2):203-222 .

Rositano, J., G. Newcombe, B. Nicholson, and P. Sztajnbok, 2001. Ozonation of NOM and Algal Toxins in Four Treated Waters. *Water Research*, 35(1):23-32.

Rossin, A., J. Lester, and R. Perry. 1982. Removal of Nitrilotriacetic Acid During Primary Sedimentation and its Effect on Metal Removal Raw Sewage. *Environmental Pollution (Series B)*, 4(4):315-330.

Rothlisberg P. C., P. C. Pollard, P. D. Nichols, D. J. W. Moriarty, A.M.G. Forbes, C. J. Jackson, and D. Vaudrey. 1994. Phytoplankton Community Structure and Productivity in Relation to the Hydrological Regime of the Gulf of Carpentaria, Australia, in Summer. *Australian Journal of Marine and Freshwater Research*, 45:265-282

Rubin, H.E., R. Subba-Rao, and M. Alexander. 1982. Rates of Mineralization of Trace Concentrations of Aromatic Compounds in Lake Water and Sewage Samples. *Applied and Environmental Microbiology*, 43(5):1133-1138.

Ryan, E.F., D.P. Hamilton, and G.E. Barnes. 2003. Recent Occurrence of *Cylindrospermopsis raciborskii* in Waikato Lakes of New Zealand. *New Zealand Journal of Marine and Freshwater Research*, 37:829-836.

Runnegar, M., A. Jackson, and I. Falconer. 1988. Toxicity of the Cyanobacterium *Nodularia spumigena* Mertens. *Toxicon*, 26:143-151.

Runnegar, M., S. Kong, Y. Zhong, and S. Lu. 1994. The Role of Glutathione in the Toxicity of a Novel Cyanobacterial Alkaloid Cylindrospermopsin in Cultured Rat Hepatocytes. *Biochemical and Biophysical Research Communications*, 201:235-241.

Saito, T., K. Okana, H.-D. Park, T. Itayama, Y. Inamori, B.A. Neilan, B.P. Burns, and N. Sugiura. 2003. Detection and Sequencing of the Microcystin-LR Degrading Gene, *mlrA*, from New Bacteria Isolated from Japanese Lakes. *FEMS Microbiology Letters*, 229(2):271-276.

Saker, M.L., B.A. Neilan, and D.J. Griffiths. 1999. Two Morphological Forms of *Cylindrospermopsis raciborskii* (Cyanobacteria) Isolated from Solomon Dam, Palm Island, Queensland. *Journal of Phycology*, 35(3):599-606.

Saker, M.L., C.G. Nogueira, V.M. Vasconcelos, B.A. Neilan, G.K. Eaglesham, and P. Pereira. 2003. First Report and Toxicological assessment of the Cyanobacterium *Cylindrospermopsis raciborskii* from Portuguese Freshwaters. *Exotoxicology and Environmental Safety*, 55:243-250.

Sas, H. (1989). *Lake Restoration by Reduction of Nutrient Loading: Expectations, Experiences, Extrapolations*. Sankt Augustin: Academia Verlag Richarz.

Schindler, D. 1971. Carbon, Nitrogen and Phosphorus Eutrophication of Freshwater Lakes. *Journal of Phycology*, 7:321-329.

Schmidt, S.K., and M.J. Gier. 1989. Dynamics of Microbial Populations in Soil Indigenous Microoranisms Degrading 2-4-Dinitrophenol. *Microbial Ecology*, 18(3):285-296.

Schmidt, W., H. Willmitzer, K. Bornmann, and J. Pietsch. 2002. Production of Drinking Water from Raw Water Containing Cyanobacteria – Pilot Plant Studies for Assessing the Risk of Microcystin Breakthrough. *Environmental Toxicology*, 17:375-385.

Seawright, A., Nolan, C., Shaw, G., Chiswell, R., Norris, R., Moore, M., and Smith, M. 1999. The Oral Toxicity for Mice of the Tropical Cyanobacterium *Cylindrospermopsis raciborskii* (Woloszynska). *Environmental Toxicology,* 14:135-42.

Shafik H.M., S. Herodek, M. Présing and L. Vörös. 2001. Factors Affecting Growth and Cell Composition of Cyanoprokaryote *Cylindrospermopsis raciborskii* (Woloszynska) Seenayya et Subba Raju. *Archiv Hydrobiologie*, 103:75-93.

Shaw, G., A. Seawright, M. Moore, and P. Lam. 2000. Cylindrospermopsin, a Cyanobacterial Alkaloid: Evaluation of Its Toxicological Activity. *Therapeutic Drug Monitoring*, 22:89-92.

Shaw, G., A. Sukenik, A. Livne, R. Chiswell, M. Smith, A. Seawright, R. Norris, G. Eaglesham, and M. Moore. 1999. Blooms of the Cylindrospermopsin Containing Cyanobacterium, *Aphanizomenon ovalisporum* (Forti) in Newly Constructed Lakes, Queensland, Australia. *Environmental Toxicology*, 14:167-177.

Sherman, P., I. Tully, and H. Gibson. 1995. Removal of Cyanobacterial Cells and Toxins from Drinking Water with Biologically Active Filters. In *Proceedings of the 16th Federal AWWA Convention, April 2-6, 1995, Sydney, Australia*, 587-592.

Sherman, B.S., J. Whittington, and R.L. Oliver. 2000. The Impact of Destratification on Water Quality in Chaffey Dam. *Archiv fur Hydrobiologie*, 55:15-29.

Shi, L., W.W. Carmichael, and I. Miller. 1995. Immuno-gold Localisation of Hepatotoxins in Cyanobacterial Cells. *Archives Microbiology*, 163:7-15.

Sivonen, K. 1996. Cyanobacterial Toxins and Toxin Production. *Phycologia*, 35(6):12-24.

Sivonen, K., and Jones, G. 1999. Cyanobacterial Toxins. In *Toxic Cyanobacteria in Water - A Guide to their Public Health Consequences, Monitoring and Management*. Edited by I. Chorus and J. Bertram. London: E and F.N. Spon.

Slater, J.H., and Lovatt, D. 1984. Biodegradation and the Significance of Microbial Communities. In *Microbial Degradation of Organic Compounds*. Edited by D.T. Gibson. New York: Marcel Dekker.

Smidt, E., M. Hellwig, and H. Knackmuss. 1983. Degradation of Chlorophenols by a Defined Mixed Microbial Community. *Applied and Environmental Microbiology*, 46:1038-1044.

Smith, M. 2005. Biodegradation of the Cyanotoxin Cylindrospermopsin. Ph.D thesis. Queensland University, Brisbane.

Smith, V.H., G.D. Tilman, and J.C. Nekola. 1999. Eutrophication: Impacts of Excess Nutrient Inputs of Freshwaer, Marine, and Terrestrial Ecosystems. *Environmental Pollution*, 100:179-196.

Sohn, J., G. Amy, J. Cho, Y. Lee, and Y. Yoon. 2004. Disenfectant Decay and Disinfection By-products Formation Model Development: Chlorination and Ozonation By-products. *Water Research*, 38:2461-2478.

Spain, J., P. Prichard, and A. Bourquin. 1980. Effects of Adaption on Biodegradation Rates in Sediment/Water Cores from Estuarine and Freshwater Environments. *Applied and Environmental Microbiology*, 40:726-734.

Spain, J., P. Van Veld, C. Monti, P. Prichard, and C. Cripe. 1984. Comparison of P-Nitrophenol Biodegradation in Field and Laboratory Test Systems. *Applied and Environmental Microbiology*, 48:944-950.

Sprőber, P., H.M. Shafik, M. Présing, A.W. Kovács, and S. Herodek. 2003. Nitrogen Uptake and Fixation in the Cyanobacterium *Cylindrospermopsis raciborskii* Under Different Nitrogen Conditions. *Hydrobiologia*, 506-509:169-174.

Stackebrandt, E., R.G.E. Murray, and H.G. Trueper. 1988. Proteobacteria New Class, a Name for the Phylogenetic Taxon that Includes the Purple Bacteria and Their Relatives. *International Journal of Systematic Bacteriology*, 38(3):321-325.

Stephenson, T., J. Lester, and R. Perry. 1984. Acclimation to Nitrilotriacetic Acid in the Activate Sludge Process. *Chemosphere*, 13:1033-1040.

Stout, J. 1980. The Role of Protozoa in Nutrient Recycling and Energy Flow. *Advances in Microbiology Ecology*, 4:1-50.

Takenaka, S., and M.F. Watanabe. 1997. Microcystin LR Degradation by *Pseudomonas aeruginosa* Protease. *Chemosphere*, 34(4):749-757.

Taylor, W.D., R.F. Losee, M. Torobin, G. Izaguirre, D. Sass, D. Khiari and K. Atasi. 2006. Early warning and management of surface water taste and odor events. Published by Awwa Research Foundation, USA.

Terao, K., S. Ohmori, K. Igarashi, I. Ohtani, M. Watanabe, K. Harada, E. Ito, and M. Watanabe. 1994. Electron Microscopic Studies on Experimental Poisoning in Mice Induced by Cylindrospermopsin Isolated from Blue-Green Alga *Umezakia natans*. *Toxicon*, 32:833-843.

Thomas, R., and M. Meybeck. 1992. The Use of Particulate Material. In *Water Quality Assessments - A Guide to Use of Biota, Sediments and Water in Environmental Monitoring.* Edited by D. Chapman. Cambridge: E and FN Spon.

Thomas, A.D., M.L. Saker, J.H. Norton, and R.D. Olsen. 1998. Cyanobacterium *Cylindrospermopsis raciborskii* as a Probable Cause of Death in Cattle in Northern Queensland. *Australian Veterinary Journal*, 76:592-594.

Tiedje, J. 1977. Influence of Environmental Parameters on EDTA Biodegradation in Soils and Sediments. *Journal of Environmental Quality*, 6(1):21-26.

Torstensson, N., J. Stark, and B. Goransson. 1975. The Effect of Repeated Application of 2,4-D and MCPA on Their Breakdown in Soil. *Weed Research*, 15:159-164.

Toth, L.G., and N.P. Zankai. 1985. Feeding of *Cyclops vicinus* (Uljanin) (Copepoda: Cyclopoida) in Lake Balaton on the Basis of Gut Content Analyses. *Hydrobiologia*, 122:251-260.

Tsuji K., S. Naito, F. Kondo, N. Ishikawa, M.F. Watanabe, M. Suzuki, and K.-I. Harada. 1994. Stability of Microcystins from Cyanobacteria: Effect of Light on Decomposition and Isomerization. *Environmental Science & Technology*, 28(1):173-177.

Tung, S.-C., T.-F. Lin, C.-L. Liu, and S.-D. Lai. 2004. The Effect of Oxidants on 2-MIB Concentration with the Presence of Cyanobacteria. *Water, Science and Technology*, 49(9):281-288.

Turner, P., A. Gammie, K. Hollinrake, and G. Codd. 1990. Pneumonia Associated with Contact with Cyanobacteria. *British Medical* Journal, 300:1440-1441.

UNESCO (United Nations Educational, Scientific and Cultural Organization). 1981: *Tenth Report of the Joint Panel on Oceanographic Tables and Standards*. UNESCO Technical Papers in Marine Science, No. 36. Paris: UNESCO.

Van Buynder, P., T. Oughtred, B. Kirby, S. Phillips, G. Eaglesham, K. Thomas, and M. Burch. 2001. Nodularin Uptake by Seafood During a Cyanobacterial Bloom. *Environmental Toxicology*, 16:468-471.

Varcoe, J.C.R., D.J. Chittleborough, J.W. Cox, and J.A. Van Leeuwen. 2000. Phosphate and NOM Adsorption with Soil Minerals as Influenced by Calcium. In *Proceedings of the* NZSSS/ASSSI Joint Conference. December 3-8, 2000. Rotorua, NZSSS/ASSSI.

Velzevoer, R., M. Drikas, C. Donati, M.D. Burch, and D. Steffensen. 1995. Release of Geosmin by *Anabaena circinalis* Following Treatment with Aluminium Sulphate. *Water Science and Technology*, 31(11):187-194.

Ventullo, R., and R. Larson. 1986. Adaption of Aquatic Microbial Communities to Quaternary Ammonium Compounds. *Applied and Environmental Microbiology*, 51:356-361.

Visser, P.M., B.W. Ibelings, B. Van der Veer, J. Koedoods, and L.R. Mur. 1996. Artificial Mixing Prevents Nuisance Blooms of the Cyanobacterium *Microcystis* in Lake Nieuwe Meer, The Netherlands. *Freshwater Biology*, 36:435-450.

von Gunten, U. 2003. Ozonation of Drinking Water: Part II. Disinfection and By-product Formation in Presence of Bromide, Iodide or Chlorine. *Water Research,* 37:1469-1487.

Walker, R., and A. Newman. 1956. Microbial Decomposition of 2,4-Dichlorophenoxyacetic Acid. *Applied Microbiology*, 4:201-206.

Walsby, A.E., G. Ng, C. Dunn, and P.A. Davis. 2004. Comparison of the Depth Where *Planktothrix rubescens* Stratifies and the Depth Where the Daily Insolation Supports its Neutral Buoyancy. *New Phytologist*, 162:133-145.

Wanninkhof, R. 1992. Relationship Between Wind Speed and Gas Exchange over the Ocean. *Journal of Geophysical Research*, 97:7373-7382.

Watanabe, M.M., K. Kaya, and N. Takanmura. 1992. Fate of the Toxic Cyclic Heptapeptides, the Microcystins from Blooms of *Microcystis* (Cyanobacteria) in a Hypereutrophic Lake. *Journal of Phycology*, 28:761-762.

Watanabe, M., K. Tsuji, Y. Watanabe, K. Harada, and M. Suzuki. 1992. Release of Heptapeptide Toxin (Microcystin) During the Decomposition Process of *Microcystis aeruginosa. Toxicon*, 26:48-53.

Wayne, R., W. Carmichael, S. Brittain, G. Eaglesham, G. Shaw, Y. Liu, and M. Watanabe. 2001. First Report of the Cyanotoxin Cylindrospermopsin and Deoxycylindrospermopsin from *Raphidiosis curvata* (Cyanobacteria). *Journal Phycology*, 37:1121-1126.

Welker, M., C. Steinberg, and G. Jones. 2001. Release and Persistence of Microcystin in Natural Waters. In *Cyanotoxins: Occurrence, Causes, Consequences*. Edited by I. Chorus. Berlin: Springer-Verlag.

Wen, A., M. Fegan, C. Hayward, C. Sukumar, L.I. Sly, A.M. Wen, and S. Chakraborty. 1999. Phylogenetic Relationships Among Members of the Comamonadaceae, and Description of *Delftia acidovorans* (Den Dooren De Jong 1926 and Tamaoka et al. 1987) Gen. Nov., Comb. Nov. *International Journal of Systematic Bacteriology*, 49(2):567-576.

Wetzel, R.G. 2001. *Limnology: Lake and River Ecosystems*. San Diego: Academic Press.

Whittington, J., B. Sherman, D. Green, and R.L. Oliver. 2000. Growth of *Ceratium hirundinella* in a Subtropical Australian Reservoir: the Role of Vertical Migration. *Journal of Plankton Research*, 22:1025-1045.

Wiggins, B., and M. Alexander. 1988. Role of Protozoa in Microbial Acclimation for Mineralization of Organic Chemicals in Sewage. *Canadian Journal of Microbiology*, 34:661-666.

Wiggins, B., S. Jones, and M. Alexander. 1987. Explanations for the Acclimation Period Preceding the Mineralization of Organic Chemicals in Aquatic Environments. *Applied and Environmental Microbiology*, 53(4):791-796.

Zeir, B., G. Harrington, and I. Xagoraraki. 2003. Inactivation of Microcystin-LR by Free Chlorine in Drinking Water Treatment. In *Proceedings. of the AWWA Annual Conference.* Anaheim:AWWA.

ABBREVIATIONS

ADWG	Australian Drinking Water Guidelines
ALF	Alert Levels Framework
AWWA	American Water Works Association
AwwaRF	Awwa Research Foundation
BOM	Bureau of Meteorology
CAEDYM	Computational Aquatic Ecosystem Dynamics Model
CRC	Cooperative Research Centre
CRCWQT	Cooperative Research Centre for Water Quality and Treatment
CYN	Cylindrospermopsin
DONL	Dissolved Organic Nitrogen Parameter
DOC	Dissolved Organic Carbon
DYRESM	Dynamic Reservoir Simulation Model
FABMS	Fast Atom Bombarment Mass Spectral Studies
GAC	Granular Activated Carbon
GALD	Greatest Axial Linear Dimension
GSH	Glutathione
HBRB	Harvey Bay Retardation Basin
HPLC	High Performance Liquid Chromatography
MDS	Multi Dimensional Scaling
MIB	Methylisoborneol
MRM	Multiple Reactant Monitoring
MRW	Myponga Reservoir Water
NOM	Natural Organic Material
NPD	North Pine Dam
PAC	Project Advisory Committee
PAC	Powdered Activated Carbon
PCR	Polymerase Chain Reaction
PHYTOPAM	Pulse Amplitude Modulation Device
PONL	Particulate Nitrogen Concentration
TOC	Total Organic Carbon
WTP	Water Treatment Plant

IWA Publishing's authorised EU representative for General Product Safety Regulations is Diane D'Arras, 15 rue Duret, 75116 Paris, France, e-mail: safety@iwap.co.uk.

Printed and bound by CPI Group (UK) Ltd, Croydon, CR0 4YY
17/09/2026
02218802-0006